EDHF

The Complete Story

EDHF

The Complete Story

Michel Félétou
Paul M. Vanhoutte

CRC Press
Taylor & Francis Group
Boca Raton London New York

CRC Press is an imprint of the
Taylor & Francis Group, an **informa** business

CRC Press
Taylor & Francis Group
6000 Broken Sound Parkway NW, Suite 300
Boca Raton, FL 33487-2742

First issued in paperback 2019

ISBN-13: 978-0-415-33292-7 (hbk)
ISBN-13: 978-0-367-39182-9 (pbk)

Library of Congress Card Number 2005015677

Library of Congress Cataloging-in-Publication Data

EDHF : the complete story / edited by Michel Félétou and Paul M. Vanhoutte.
     p. ; cm.
  Includes bibliographical references and index.
  ISBN-13: 978-0-415-33292-7
  ISBN-10: 0-415-33292-3
  1. Vascular endothelium. 2. Vasodilators.
  [DNLM: 1. Biological Factors—pharmacology. 2 Vasodilator Agents – pharmacology. 3. Biological Factors – physiology. 4. Endothelium, Vascular – drug effects. 5. Endothelium, Vascular – physiopathology. QV 150 E23 2005] I. Félétou, Michel, 1956- II. Vanhoutte, Paul M.

QP110.V34E34 2005
612.1′3—dc22   2005015677

**Visit the Taylor & Francis Web site at
http://www.taylorandfrancis.com**

**and the CRC Press Web site at
http://www.crcpress.com**

Preface

The preface to any book should tell the readers why the authors endeavored to undertake its writing. In our case it came naturally, as we have worked together for twenty years and have tried, over that period of time, to unravel the complexity of endothelium-dependent responses. Why do we think that this is the right time to summarize our experience? When nitric oxide (NO) was proposed and then demonstrated to mediate the endothelium-dependent relaxation, first described by Robert Furchgott, the vascular world seemed convinced that NO was the ultimate and sole explanation for such relaxations. At that time one of us wrote an editorial [Vanhoutte, P.M. (1987) The end of a quest? *Nature* 327: 459–460] expressing the conviction that no, the quest was not over, that NO was not the sole answer. This led to the research in alternative endothelium-derived mediators and in particular to the description and analysis of endothelium-dependent hyperpolarizations, which we attributed to the release of endothelium-derived hyperpolarizing factor (EDHF).

The field has blossomed and matured, but we now believe that the quest is over, and that we understand most, if not all of the ways by which endothelial cells can communicate signals to the underlying smooth muscle cells to tell them to hyperpolarize. Hence we challenged ourselves to write a summary of the knowledge accumulated on that topic during those 20 years.

The advances made in vascular biology in the last 25 years have considerably changed the perception that one could have of the endothelial cells. Once considered as a diffusion barrier preventing the access of the blood cells to the vascular matrix, the endothelium is now recognized as playing a major role in the control of blood fluidity, platelet aggregation, and vascular tone, but also in immunology, inflammation, angiogenesis, and for serving as a metabolizing and an endocrine organ.

In order to control vascular tone, endothelial cells synthesize and release various vasoactive factors. Some relax the underlying smooth muscle cells [e.g., adenosine, prostacyclin (PGI_2), nitric oxide (NO), hydrogen peroxide (H_2O_2), epoxyeicosatrienoic acids (EETs), C-natriuretic peptide (CNP), and the elusive endothelium-derived hyperpolarizing factor (EDHF)], while others contract them [e.g., thromboxane A_2, isoprostanes, 20-hydoxyeicosatetraenoic acid, superoxide anion, H_2O_2, endothelin-1, angiotensin II, and uridine adenosine tetraphosphate]. They can also directly communicate with smooth muscle cells via myoendothelial gap junctions, which allow not only the spread of electronic tone but also the transfer of ions or small molecules such as cyclic nucleotides.

Cardiovascular diseases were the scourge of the twentieth century in the Western world and are most likely to become that of the twenty-first century in both the Western and emerging countries. An endothelial dysfunction linked to an imbalance in the synthesis and/or release of contracting and relaxing factors is often evoked to explain the initiation of the cardiovascular pathology or its development and perpetuation.

The role of NO and prostacyclin in cardiovascular physiology and pathology has been extensively studied since their discovery in the 1970s and identification in the 1980s, in part because inhibitors of their production were available. However, the role of other relaxing factor(s), in particular EDHF, is not as well understood. The purpose of this monograph is to present the recent developments concerning endothelium-dependent hyperpolarizations, which are likely to play a much more important role in cardiovascular physiology and pathology than originally foreseen.

The text is designed for use by graduate students, vascular biologists, and cardiologists interested in the intimate functioning of the blood vessel wall. The monograph is extensively illustrated with original diagrams and schematics that summarize the different steps of endothelium-dependent hyperpolarization. We hope that they will make the complexity of the EDHF world accessible to all.

ACKNOWLEDGMENTS

The authors want to thank Dr. Emmanuel Canet, Institut de Recherches Servier, for providing an educational grant, which made the preparation of the figures possible by our longtime friend and collaborator Robert R. Lorenz. We also would like to thank the staff at the publishers for a most efficient handling of the manuscript.

Contents

1 The Blood Vessel Wall

Blood vessels carry blood throughout the body to deliver oxygen and nutrients, to convey information via the circulating hormones, and also to collect metabolic waste. The blood vessel wall, with the exception of capillaries, contains three layers: the intima, the media, and the adventitia. The capillaries, in order to facilitate exchanges between the circulating blood and the surrounding tissue, are only constituted of the innermost layer, the intima.

The intima is a monolayer of endothelial cells, in direct contact with the blood and the circulating cells, anchored to an 80-nm-thick basal lamina. Endothelial cells and hematopoietic cells are likely to arise from the same precursor cells, the hemangioblast.[1] Endothelial cells are thin, slightly elongated cells that are roughly 70 μm long, 30 μm wide, and 5–10 μm thick. In the blood vessel wall they are orientated along the axis of the vessel, minimizing the shear stress forces exerted by the flowing blood. Although the endothelium consists of a single layer of cells, the total volume of the approximately 10^{13} endothelial cells of the human body is comparable to that of the liver.[2] It occupies a surface measuring approximately 1000 square meters.[3] These morphological data indicate that endothelial cells are a privileged site for exchange and transfer. Important phenotypic variations occur between endothelial cells in the different parts of the arterial tree as well as between arteries and veins.

The media includes mostly smooth muscle cells embedded in a protein matrix composed of collagen, elastin, and other structural proteins. Terminal nerve endings often spread into the media in the vicinity of the smooth muscle cells, particularly in veins. Other cell types include a few mast cells, infiltrated monocytes, and possibly Cajal interstitial-like cells.[4] The capillaries do not contain a media but are surrounded by pericytes, smooth muscle-like cells that control the capillary diameter.

The adventitia is a dense fibro-elastic layer with no or few smooth muscle cells, which contains the terminal nerve endings and, depending on the size of the blood vessels, the vasa vasorum (arterioles, venules, capillaries, and lymphatic vessels), which irrigates the blood vessel itself. The thickness of each of these layers will vary according to the size and the type of a given blood vessel, for instance, arteries vs. veins (Figure 1).[5]

1.1 MULTIPLE FUNCTIONS OF THE ENDOTHELIAL CELLS

Besides modulation of the tone of the underlying vascular smooth muscle cells, which is the main subject of this monograph (Figure 2), the endothelium plays a critical role in a number of other processes.

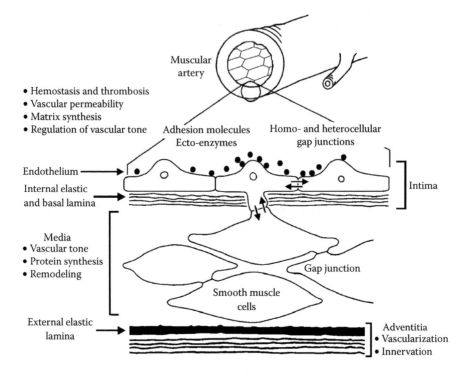

FIGURE 1 Architecture of a muscular artery. The three structural components of the blood vessel, the intima, media, and adventitia, are indicated. The inset shows the privileged role of the endothelial cells at the interface of the flowing blood and the vascular wall.

1.1.1 Synthesis and Degradation of Extracellular Matrix

The basal lamina is an important component of the blood vessel, being the scaffold of every artery, vein, or capillary. The inside of this scaffold is lined with endothelial cells, while the outside is covered with smooth muscle cells or pericytes. The endothelial cells can synthesize virtually all the proteins constituting the basal lamina and produce the relevant enzymes involved in its remodeling. More than 50 different proteins make up the basal lamina, although collagen is the main component. These include type IV, XV, and XVIII collagens, laminin, heparan-sulphate proteoglycans, perlecan, nidogen/entactin, osteopontin, fibronectin, and thrombospondin. Endothelial cells can also produce enzymes such as matrix metalloproteinases that degrade this extracellular matrix, an important action for the plasticity of the blood vessels and for angiogenesis.[6,7]

1.1.2 Fluidity of the Blood

Under normal conditions, in the heart, arteries, capillaries, and veins, the flowing blood is in contact with healthy endothelial cells and remains fluid. The luminal surface of quiescent endothelial cells is anticoagulant and nonthrombogenic, the

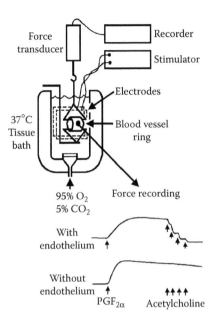

FIGURE 2 Isometric tension recording in an isolated blood vessel. An arterial ring is studied in an organ chamber filled with thermostated and oxygenated physiological salt solution. The inset shows recording traces of an experiment reproducing the seminal observation first made by Furchgott and Zawadzki.[258] Isolated rings of canine coronary artery with and without endothelium are contracted with prostaglandin $F_{2\alpha}$ ($PGF_{2\alpha}$). The addition of acetylcholine produces a concentration-dependent relaxation only in the ring with a preserved endothelial lining.

platelets and leukocytes do not adhere to it, and the coagulation system remains inactivated. By contrast, the macromolecules of the basal lamina, synthesized by the endothelial cells, are strongly thrombogenic and activated endothelial cells promote thrombus activation. The endothelial cells, therefore, regulate the equilibrium between thrombosis, hemostasis, and thrombo-resistance.

Endothelial cells maintain blood fluidity via different anticoagulant mechanisms, the most important being the protein C/protein S pathway. The interaction of thrombin with endothelial thrombomodulin facilitates the activation of protein C, providing that the latter has formed a complex with protein S. The activated protein C inactivates two essential components of the blood coagulation cascade, factor VIIIa and factor Va. The endothelial cell surface is rich in heparin-like sulphated glycosaminoglycans molecules that bind antithrombin and are a major sink for the inactivation of activated thrombin. The endothelium contributes to fibrinolysis by releasing tissue-type and urokinase-type plasminogen activator (t-PA, u-PA) and by provoking the activation of plaminogen into plasmin. Endothelial cells also prevent adhesion, aggregation, and activation of platelets and promote platelet de-aggregation by expressing 13-hydroxyoctadecadienoic acid (13-HODE) on their cell surface, by releasing prostacyclin and NO, metabolizing ATP and ADP by membrane ectonucleotidases, and preventing the action of thrombin.

After endothelial cell activation following vessel injury (e.g., exposure to throm-bin or activation by lipopolysaccharides and/or cytokines), the balance established by the endothelium can be tipped in favor of adhesion and aggregation of platelets and thus clot formation. Endothelial cells synthesize and express tissue factor, receptors for factor IX/IXa, factor Xa, and also release the natural inhibitor of t-PA, plasminogen activator inhibitor type I (PAI-1). Furthermore, thrombin causes the release of von Willebrand factor and the expression of P-selectin. Finally, endothelial cells produce pro-aggregating substances such as thromboxane A_2, 15-hydroxy-tetraenoic acid (15-HETE), and platelet-activating factor (PAF).[8]

1.1.3 VASCULAR PERMEABILITY

The endothelium forms a semipermeable barrier between the blood and the inter-stitium. At the level of the capillaries, the control of solute and macromolecule transfer across the blood vessel wall is another major function of the endothelium. Three types of microvascular endothelium can be distinguished according to the degree of continuity of the endothelial cells. In most vascular beds, the endothelium is continuous, while in glands, mucosae, and renal cortex, perforations, called fenes-trae (diameter up to 70 μm), occur. In the liver and renal glomeruli the endothelium is discontinuous.

Aberrations of endothelial barrier function lead to an abnormal extravasation of fluid and macromolecules, resulting in edema and tissular dysfunction. An increase in microvascular permeability takes place in inflammation, ischemia-reperfusion, atherosclerosis, sepsis, diabetes, thermal injury, angiogenesis, or tumor metastasis. Inflammatory mediators released under these pathological conditions [e.g., hista-mine, bradykinin, thrombin, platelet-activating factor, cytokines and growth factors such as vascular endothelial growth factor (VEGF), reactive oxygen species] can increase microvascular permeability.[9–12] In the course of inflammation, this is the first recognized step occurring at the level of postcapillary venules.

Plasma extravasation becomes complicated with the simultaneous action of several vasoactive agents and by the subsequent interactions of leukocytes with the endothelium. This involves three different steps, first the rolling of leukocytes on the surface of the endothelium, followed by their activation, their adhesion to the endothelial surface, and finally the infiltration in the surrounding tissue. These various steps require the induced or constitutive expression of different adhesion proteins such as E-selectin, P-selectin, ICAM, and VCAM, which are involved in endothelial junctions, leukocyte adhesion, and transendothelial migration.[13,14]

1.1.4 ANGIOGENESIS

Angiogenesis, or the formation of new capillaries and blood vessels derived from pre-existing blood vessels, is fundamental in reproduction, development, and repair. Angiogenesis can also be pathological. Both physiological and pathological angio-genesis primarily involve the endothelial cells.[15] These cells are normally quiescent and are among the most genetically stable cells of the body as their turnover time is over hundreds of days. However, during angiogenesis, endothelial cells can proliferate rapidly with a turnover time of less than five days.[16] This increase in proliferation is associated with degradation of the basal lamina by several matrix-degrading enzymes such as matrix metalloproteinases. This leads to sprouting of pre-existing microvessels,[17] a complex phenomenon tightly regulated in a spatial and temporal manner. Endothelial cells make numerous interactions with their environment, not only with the matrix but also with circulating cells, other endothelial cells, pericytes, smooth muscle cells, or fibroblasts. Endothelial cells then migrate, proliferate, and establish tubular structures, the basis of the new blood vessel. Numerous factors with angiogenic or anti-angiogenic properties have been identified. Growth factors such as PDGF, FGF, TGFβ, VEGF, angiopoietin, ephrins, and their respective receptors play a crucial role at each step of the angiogenic process. Depending on the origin of the neovascularization (physiological, pathological) or the vascular bed involved, the mechanisms leading to the angiogenic process can differ markedly. Some of the growth factors and cytokines involved can be synthesized and released by the endothelium itself.[18]

1.1.5 METABOLISM

Endothelial cells can metabolize or conversely activate numerous circulating factors with or without vasoactive properties, including polypeptide hormones, amines, nucleotides, lipoproteins, and metabolites of arachidonic acid. For instance, endothelial cells express monoamine oxidase (MAO) and -o-methyltransferase (COMT) (inactivating epinephrine and serotonin), angiotensin converting enzyme (ACE) (activating angiotensin-II and degrading bradykinin, substance P or enkephalins), neutral endopeptidase (NEP) (inactivating natriuretic peptides, substance P, tachykinins or bradykinin), endothelin converting enzymes (ECE) (producing endothelins), and dipeptidyl-peptidase IV (DPP-IV) [cleaving glucagon-like peptide-1(GLP-1)].[19–23]

1.2 VASCULAR SMOOTH MUSCLE

Smooth muscle cells are phenotypically and functionally heterogeneous. Developmental studies suggest that the vascular system can be considered as a mosaic structure made up of independently arising smooth muscle cell populations. During development, most of the smooth muscle cells of the systemic vascular tree are supplied from the neural crest and the nodose placode. Smooth muscle cells can also arise from endothelial cells and from multipotent stem cells in the bone marrow. The coronary artery smooth muscle cells derive from a unique lineage, progenitors located in a transient organ called the proepicardial organ, situated outside the heart itself.[24] Thus, some of the structural and functional diversity of blood vessels (vascular heterogeneity) results, at least partially, from the multiple origins of vascular smooth muscle cells during development.

The principal function of vascular smooth muscle cells is to maintain vascular tone. The degree of vascular constriction helps to determine the resistance to flow, the capacitance of the circulation, and the rigidity of the arterial wall. The other functions of these cells are related to environmental stresses and the repair of the blood vessel wall after injury. Vascular cells can adopt either a contractile or a synthetic phenotype both *in vivo* and *in vitro*. Spindle-shaped smooth muscle cells, associated with the contractile phenotype, are typical of differentiated arteries and veins. Epitheloid or rhomboid smooth muscle cells, associated with the synthetic phenotype, appear in developing blood vessels or pathological arteries (atherosclerosis, restenosis after angioplasty, and coronary artery bypass grafting). However, both spindle-shaped and epitheloid phenotypes coexist in the media of healthy arteries, the proportion of which varies according to the size of the blood vessel and the the age of the animal, as well as the circulating and micro-environmental factors.[25,26]

Elastin is the principal protein synthesized by the vascular smooth muscle cells and can contribute up to 50% of the dry weight of a given blood vessel. Each elastic lamina alternates with a ring of smooth muscle cells, forming the functional and structural unit of the arterial wall. The elastic lamina absorbs the pulse pressure during the cardiac systole and releases this energy during the diastole to maintain blood pressure (windkessel function). Furthermore, by interacting with a 67-kDa binding protein (splice variant of beta-galactosidase), elastin also plays an important role in the proliferation, migration, and differentiation of vascular smooth muscle.[27,28] Disruption of the elastin fibers by mechanical injury or following activation of matrix metalloproteinases by endothelial injury and inflammation is a common event in various proliferative vascular diseases (atherosclerosis, restenosis).

Resistance arteries with an internal diameter less than 400 μm constitute the major site of generation of resistance to flow. Both the arteriolar density and the artery diameter set the level of peripheral vascular resistance. According to Poisseuille's law, the resistance to flow varies inversely with the fourth power of the radius. Thus, minor changes in the lumen diameter of small arteries markedly influence resistance. These changes in lumen size can be structural, mechanical (stiffness), and functional. Smooth muscle cells play a major role in these three processes. The structural changes, as seen, for instance, in hypertension, with the remodeling of the vascular wall, involve a combination of smooth muscle growth and apoptotic processes in eutrophic remodeling or an increase in smooth muscle size and density (hyperplasia and hypertrophy) in hypertrophic remodeling.[29–32] Both structural changes and mechanical changes are associated with alterations in the protein composition of the extracellular matrix of the media, mostly secreted by the vascular smooth muscle cells.

The fine tuning, the moment-to-moment regulation of vascular diameter, involves the contractile properties of the smooth muscle cells. This functional activity of the cells is heavily regulated by the sympathetic system and circulating hormones, as well as by autocrine and paracrine factors.[33]

1.3 MEMBRANE POTENTIAL AND CALCIUM HOMEOSTASIS IN VASCULAR SMOOTH MUSCLE AND ENDOTHELIAL CELLS

The contraction of the vascular smooth muscle cells can be elicited by vasoconstrictor agonists, depolarization of the cell membrane, or mechanical stimulation. It is initiated, and to a lesser extent maintained, by an increase in the intracellular free calcium concentration ($[Ca^{2+}]_i$). Contractions of vascular smooth muscle activated by neurohumoral mediators usually involve a combination of calcium entry and calcium release, although the contribution of each pathway differs markedly depending on the stimulating agonist or the vascular bed studied.[34,35] The two major sources that can provide the calcium ions are the extracellular space and the internal stores, particularly the sarcoplasmic reticulum. This increase in $[Ca^{2+}]_i$ activates myosin light chain kinase, which in turn phosphorylates the regulatory light chains of myosin II to generate contraction. This calcium-dependent phosphorylation of the light chains of myosin II is modulated in a calcium-independent manner by the constitutively active myosin light chain phosphatase. This enzyme is inhibited by monomeric GTPase Rho and the Rho-associated kinase, as well as protein kinase C, and is activated by cyclic-GMP. Furthermore, the myosin light chain kinase activity is also controlled by various kinases (e.g., protein kinase A, calmodulin-dependent protein kinase II, and p21-activated kinase), indicating that even if an increase in $[Ca^{2+}]_i$ plays a dominant role in the contraction of smooth muscle, vascular tone is also extensively regulated in both a calcium-dependent and -independent manner, by a complex network of activating and inactivating kinase cascades (Figure 3).[36-38]

Similarly, the activity of the endothelial cells, including their ability to synthesize and release vasoactive factors, depends heavily on changes in $[Ca^{2+}]_i$. In both cell types, the level of membrane potential and the activity of various ionic pumps and channels control calcium homeostasis.

1.3.1 CELL MEMBRANE POTENTIAL

The cell membrane not only separates physically the intracellular components of the cell (e.g., cytoplasm, organelles, proteins, and nucleus) from the outside world, but also creates a difference in potential between the internal and the external media (Figure 4). The sarcoplasmic membrane is an electric condenser that is semipermeable to some ions. Potassium, sodium, and to a lesser extent, chloride and calcium are the preponderant ionic species involved in the establishment of the cell membrane potential. The concentration of potassium is higher in the intracellular milieu than in the extracellular space, while the intracellular concentrations of sodium, calcium, and chloride ions are lower in the intracellular fluid (in vascular smooth muscle cells: K^+: 150 to 200 mM vs. 4 to 7 mM; Na^+: 15 to 25 mM vs. 150 mM; Ca^{2+}: less than 100 nM vs. 1 to 2.5 mM; and Cl^-: 40 to 60 mM vs. 130 mM, for the intracellular vs. extracellular concentrations, respectively).[39] The asymmetrical ionic concentrations between the intracellular and extracellular spaces generate the flow of charges, i.e., the electric current responsible for the cell membrane potential. This potential can be computed following Goldman's equation, and is a function of the respective ionic concentrations on each side of the membrane and the intrinsic permeability for each considered ion.[40,41]

$$Em = (RT/F) \, Log \, \{[P_K^+(K^+)_o + P_{Na}^+(Na^+)_o + P_{Ca}^{2+}(Ca^{2+})_o - P_{Cl}^-(Cl^-)_i$$
$$+ \cdots]/[P_K^+(K^+)_i + P_{Na}^+(Na^+)_i + P_{Ca}^{2+}(Ca^{2+})_i - P_{Cl}^-(Cl^-)_o + \cdots]\}$$

Em = membrane potential
R = perfect gas constant

T = temperature in Kelvin degrees

F = Faraday's constant

P_K^+, P_{Na}^+, P_{Ca}^{2+}, P_{Cl}^-, = membrane permeability of the given ionic species

$(K^+)_o$, $(Na^+)_o$, $(Ca^{2+})_o$, $(Cl^-)_o$ = extracellular concentration of the given ion

$(K^+)_i$, $(Na^+)_i$, $(Ca^{2+})_i$, $(Cl^-)_i$ = intracellular concentration of the given ion

If the membrane is selectively permeable to a single ionic species, the ion will diffuse following its concentration gradient, modifying accordingly the cell membrane

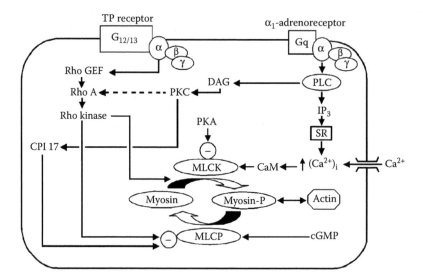

FIGURE 3 Calcium-dependent and -independent contractions in vascular smooth muscle cells. The stimulation of G-protein-coupled cell membrane receptors evokes a contractile response that, depending on the G-protein involved in the coupling, involves different mechanisms. Gq-coupled receptors (e.g., α_1-adrenergic receptor) produce calcium-dependent and -independent contractions. The activation of phospholipase C (PLC) and the subsequent production of inositol trisphosphate (IP_3) release calcium from intracellular stores (sarcoplasmic reticulum, SR). The increase in the concentration of intracellular calcium ($[Ca^{2+}]_i$) activates calmodulin (CaM), which allows the phosphorylation of myosin (myosin-P) by the myosin light chain kinase (MLCK) and the establishment of the bond with actin. The production of diacylglycerol (DAG) activates the calcium-independent protein kinase C (PKC) pathway. PKC activates the PKC-potentiated inhibitory protein of 17 kDa (CPI-17), which inhibits the myosin light chain phosphatase (MLCP), preventing the dephosphorylation of myosin. Additionally, PKC may activate RhoA and Rho kinase, which, in a calcium-independent manner, activates MLCK and inhibits MLCP. The stimulation of the $G_{12/13}$-coupled receptors (e.g., receptor for thromboxane A_2, TP receptor) produces a virtually calcium-independent contraction via the activation of Rho guanine nucleotide exchange factor (Rho-GEF) and the Rho-kinase pathway. The depolarization, following the activation of Gq-coupled receptors and, to a lesser extent that of $G_{12/13}$-coupled receptors, opens voltage-gated calcium channels, which contributes to the increase in $[Ca^{2+}]_i$. The various proteins involved in the contraction of the vascular smooth muscle cells are heavily regulated by, for instance, the cyclic-AMP-dependent protein kinase A (PKA) or the cyclic-GMP (cGMP)-dependent protein kinase G.

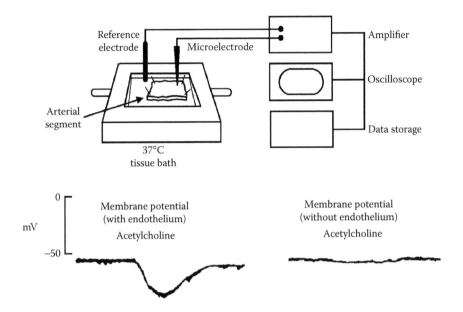

FIGURE 4 Membrane potential recording in an isolated blood vessel with an intracellular microelectrode. An isolated strip or ring of a blood vessel is studied in an organ chamber filled with thermostated and oxygenated physiological solution. A sharp glass microelectrode is inserted inside a cell of the blood vessel wall and the difference in potential between the extracellular and the intracellular compartments is recorded. The traces shown below the schematic are recordings obtained in vascular smooth muscle cells of isolated canine coronary artery. They show that the presence of acetylcholine produces a hyperpolarization of the smooth muscle cells only in the presence of endothelial cells.

potential until the value of the latter does not allow any further diffusion of the ion. This membrane potential value at which a given ion no longer passively diffuses is called its equilibrium potential and can be computed using the Nernst's equation.

$$E_X = (RT/zF) \text{ Log } [P_X^+(X^+)_o]/[P_X^+(X^+)_i]$$
E_X = equilibrium potential for a given ion
z = valency of the ion

In smooth muscle cells, under physiological conditions of ionic concentrations and temperature, the equilibrium potentials for potassium, sodium, calcium, and chloride ions are approximately -100, $+60$, $+100$, and -20 to -30 mV, respectively.[42,43] The value of the resting membrane potential of vascular smooth muscle and endothelial cells is generally between -40 and -70 mV, indicating that the membrane at rest is mainly, but not exclusively, permeable to potassium ions. Increasing the permeability to potassium ions (for instance by opening a potassium channel) will drive the membrane potential toward the equilibrium potential for potassium ions, and will thus hyperpolarize the cell. In contrast, increasing the permeability to sodium, calcium, or chloride ions will depolarize the cell.

1.3.1.1   Na$^+$/K$^+$ ATPases

The surface membrane of virtually every animal cell expresses hundreds or even millions of copies of Na$^+$/K$^+$ ATPases. This "sodium pump," as it is also frequently named, tirelessly rejects the sodium toward the extracellular space and accumulates potassium in the intracellular medium. This vital process maintains the sodium and potassium ionic gradients and is achieved at the expense of a substantial fraction of the ATP produced by the cell. These ionic gradients drive numerous co- and counter-transporters allowing glucose and amino acids intake, regulation of cell volume, pH, and calcium homeostasis, and underlie the electrical activity of all excitable cells.[44] Na$^+$/K$^+$ ATPases are electrogenic as three sodium ions are extruded toward the extracellular medium, while two potassium ions are transported. The activity of the pump contributes to the regulation of the cell membrane potential.[45]

Na$^+$/K$^+$ ATPase is composed of a noncovalently linked α subunit and a glycosilated β subunit. Four different isoforms of the α subunit (α1 to α4) and three isoforms of the β subunit (β1 to β3) have been identified in mammalian cells. The enzymatic function has been totally assigned to the multispanning membrane α subunit, which also contains the binding sites for ATP and for the cardiac glycoside ouabain. The β subunit serves as a chaperone molecule to facilitate appropriate insertion of the α subunit in the plasmalemnal membrane. In addition, the β subunit modulates the cation affinity of Na$^+$/K$^+$ ATPase. Each combination of α and β subunit produces a functionally active enzyme that possesses distinct affinities for Na$^+$ and K$^+$ and different ouabain sensitivities.[46] A third protein, termed γ subunit, with a single transmembrane domain and an invariant motif, phenylalanine-X-tyrosine-aspartate (FXYD) has also been identified. While this γ subunit is not required for the Na$^+$/K$^+$ ATPase activity, it regulates the affinity of the enzyme for Na$^+$, K$^+$, and ATP, influences ouabain-sensitivity, and modifies the voltage-dependence of K$^+$ activation.[47–49] This family of proteins contains seven members FXYD1 to FXDYD7 (the first six are also known as phospholemnan, Na$^+$/K$^+$-ATPase γ subunit, mammary tumor protein of 8 K-Da, corticosteroid hormone-induced factor, subunit related to ion channel, and phosphohippolin, respectively).[46]

In mammalian arteries both vascular smooth muscle and endothelial cells express the housekeeping form of the Na$^+$/K$^+$-ATPase, which comprises the α1 subunit. This isoform is nearly fully activated at the physiological concentration of extracellular potassium (5 mM) and in the rat is very poorly sensitive to ouabain (IC$_{50}$ > 10^{-5} M). However, depending on the species and/or vascular bed studied, both the endothelial and vascular smooth muscle cells can express the α2 and/or α3 isoforms. These isoforms are activated by extracellular concentrations of potassium in a window compatible with a physiological role for small increases in potassium concentrations (between 3 and 15 mM) and are markedly more sensitive to the inhibitory action of ouabain (in the rat, IC$_{50}$ ranging from 20 to 500 nM).[46,50,51]

1.3.1.2   Potassium Channels in the Vascular Wall

Ion channels are membrane protein complexes that span lipid bilayers to form pores in order to allow the passive diffusion of selected ionic specie(s). Potassium channels, which selectively pass potassium ions, are the largest and most diverse subgroup of ion channels. Up to 75 different genes related to potassium channels have been identified in the human genome. The function of all potassium channels is to allow the formation of a transmembrane "leak" extremely specific for potassium ions. Since cells maintain a much higher intracellular concentration of potassium than that present in the extracellular medium, the opening of potassium channels implies a change in membrane potential toward more negative values. Potassium channels set the resting membrane potential and regulate cell volume. They play a key role in the cellular signaling processes such as regulation of neurotransmitter release, heart rate, hormone secretion, smooth muscle tone, or epithelial electrolyte transport.[52]

K$^+$ channels are classified in four subgroups according to their membrane topology (NC-IUPHAR subcommittee on potassium channels). The largest group is the voltage-gated potassium channel subtype family with six transmembrane-helix domains and one pore domain (38 identified genes in the human genome, K$_V$). The second group is the calcium-activated potassium channel subtype family, which contains six or seven transmembrane-helix domains and one pore domain (5 identified genes in the human genome, K$_{Ca}$). The third group is the inward rectifier subtype family with two transmembrane-helix domains and one pore domain (15 identified genes in the human genome, K$_{ir}$). The fourth group is the two-pore-domain potassium channel subtype family, which contains four transmembrane-helix domains and two pore domains (14 identified genes in the human genome, K$_{2P}$) (Figure 5).[53–55]

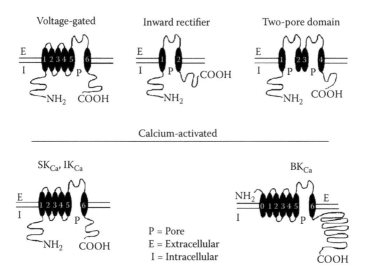

FIGURE 5 Potassium channel family. The potassium channel family has been classified in four subgroups according to their membrane topology (NC-IUPHAR subcommittee on potassium channels). The first and largest group is the voltage-gated potassium channel subtype family with six transmembrane domains and one pore domain. The second group is the calcium-activated potassium channel subtype family, which contains six or seven transmembrane domains and one pore domain. The third group is the inward rectifier subtype family with two transmembrane domains and one pore domain. The fourth group is the two-pore-domain potassium channel subtype family, which contains four transmembrane domains and, as its name indicates, two pore domains.

1.3.1.2.1   Voltage-Gated Potassium Channel Family

The voltage-gated potassium channel family is a homotetramer or heterotetramer family composed of four α subunits, each containing six transmembrane segments and a conducting pore. This large family of potassium channels is subdivided into three groups. The first contains the K_V1-K_V6 and K_V8-K_V9 families (25 different genes homologous to those first cloned in *Drosophila*: *Shaker, Shab, Shaw, Shal*), the second is the K_V7 family (5 different genes: KCNQ1-5), and the third contains the K_V10-K_V12 families (ether-a-go-go-like, homologous to the *Drosophila ether-a-go-go* or *eag*; 8 different genes). Some of these channels include an auxiliary β subunit, a cytoplasmic protein associated with the intracellular N-terminal region of the α subunit.[52,54]

1.3.1.2.1.1   Kv Channels

Vascular smooth muscle cells can express Kv1.1, Kv1.2, Kv1.3, Kv1.5, Kv1.6 (*Shaker* family), Kv2.1, Kv2.2 (*Shab* family), and the electrically silent Kv9.3 (*Shal* family). The expression pattern depends on the vascular bed and the size of the artery. In smooth muscle cells, the voltage-activated potassium current is an important component of the outward potassium conductance and can be divided into two different types depending on its inactivation properties: the delayed rectifier current (K_{DR}), which displays no or slow time-dependent inactivation and the A-type or rapidly inactivating, transient outward current (K_{TO}), which exhibits fast inactivation.[56,57] Vascular smooth muscle cells do not necessarily show A-type potassium current. For instance, the smooth muscle cells of the guinea pig carotid artery do not express this current, while those of the rabbit carotid artery and pericytes do (Figures 6, 7, and 8).[58,59]

The identification of the various potassium channels involved in the global potassium current has been generally achieved pharmacologically and by determining the biophysical properties of the channels. These techniques are not precise enough to allow differentiation among the Kv channel subtypes, for instance, between Kv1.1 and Kv1.5.[60] Furthermore, comparing the characteristics of K_{DR} observed in isolated smooth muscle cells of the rabbit portal vein to those of Kv1.5 (isolated from those cells and then re-expressed in the human embryonic kidney cell line, HEK 293) suggests that the native Kv channels expressed in vascular smooth muscle cells are likely to be heteromultimers.[61] A high degree of complexity can be achieved since modulatory β subunits[62] are also expressed in vascular smooth muscle cells. Indeed, α subunits and/or β subunits of different Kv families can form heteromultimers.[63,64]

In vascular smooth muscle, the K_{DR} current, as observed in the whole-cell configuration of the patch-clamp technique, has the following characteristics: voltage-dependent activation for potential above the physiological membrane potential (–50 mV), slow and incomplete inactivation, slowly deactivating tail currents, and inhibition by 4-aminopyridine. The single-channel conductance depends on the K_V expressed in the vascular tissue studied but is consistent with that determined for cloned and re-expressed channels (5–20 pS).[56] Some Kv channels are sensitive to charybdotoxin, a toxin present in the venom of the Middle East scorpion *Leuirus quinquestriatus* var. *hebraeus*. The A-type potassium current is blocked by this toxin. However, in vascular smooth muscle of most arteries, including that of the guinea

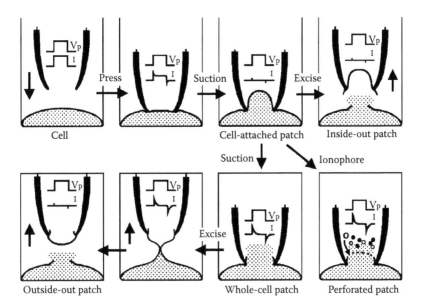

FIGURE 6 The various configurations of the patch-clamp technique. The patch-clamp technique is an electrophysiological technique that allows the study of ionic channels expressed in the cell membrane or in a lipid bilayer. A polished glass pipette is brought into contact with a given cell. When the tip of the pipette touches the cell membrane, a tight seal is formed (giga-seal). Then, by applying various maneuvers, suction, excision, and permeation of the membrane with toxins, the global electrophysiological properties of the ionic channel expressed by the cell and/or the properties of individual channels can be studied.

pig carotid artery, the K_{DR} current is usually unaffected by charybdotoxin.[65] The activity of Kv channels underlying the K_{DR} current is voltage-dependent but is also heavily regulated by kinases such as protein kinase A (PKA), PKG, and PKC.[56] The activity of this channel plays a predominant role in the control of the cell membrane potential and thus the tone of vascular smooth muscle. Its activation upon depolarization, which can be caused by physical (intraluminar pressure-induced myogenic tone) or neurohumoral mediators (e.g., norepinephrine, endothelin, angiotensin II), is a useful protective mechanism in restoring membrane potential and preventing excessive contraction of the smooth muscle and thus vasospasm. This protective mechanism is referred to as the voltage-dependent brake.[56]

There are very few reports showing the expression of K_V in freshly isolated endothelial cells. $K_V1.5$ can be expressed in the endothelial cells of the rat aorta and a decreased expression of this channel has been observed in the genetically hypertensive and stroke-prone SHR-SP rat.[66]

1.3.1.2.1.2   KCNQ and Ether-A-Go-Go-Like Voltage-Dependent Potassium Channels

KCNQ and ether-a-go-go-related genes (ERG) are expressed mainly in the brain and in the heart. These two families of potassium channels regulate the electrical

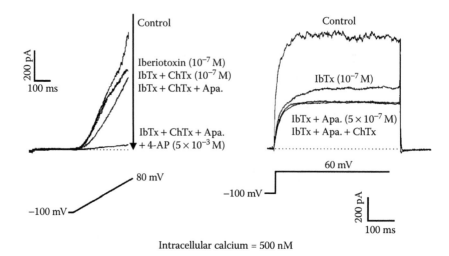

FIGURE 7 Potassium channels in smooth muscle cells of the guinea pig carotid artery. Effect of the combination of different inhibitors of potassium channels in freshly isolated smooth muscle cells of the guinea pig carotid artery (whole-cell configuration of the patch-clamp technique). Left panel: Original traces of the large global currents observed in the presence of intracellular calcium (5×10^{-7} M) for a ramp depolarization from -100 to $+80$ mV. This current is partially inhibited by iberiotoxin (IbTx), indicating the presence of BK_{Ca}. The addition of charybdotoxin (ChTx) does not produce any further inhibition, suggesting that the A-type or rapidly inactivating, transient outward current (K_{TO}) is not expressed in those cells. The addition of apamin (Apa) produces a further inhibition, indicating the presence of an "SK_{Ca}-like" current. The subsequent addition of 4-aminopyridine (4-AP) produces the virtual abolition of the remaining global outward current, demonstrating the contribution of K_V. Right panel: Original traces of the large outward current observed under similar conditions for a step depolarization from -100 to $+60$ mV, confirming the contribution of BK_{Ca}, "SK_{Ca}-like," and K_V channels in the recorded current. The effect of 4-aminopyridine is not shown for the sake of clarity. (By permission from the Nature Publishing Group; modified from Quignard et al.[65])

excitability of neurones and of the cardiac myocytes.[67] Both families are associated with familial long QT syndrome (HERG for *human eag-r*elated gene).[68] In the spontaneously active smooth muscle cells of the murine portal vein, ERG1 (the predominant isoform), ERG2, and ERG3, as well as KCNQ1, are expressed. ERG1 contributes to the resting membrane conductance while KCNQ1 contributes to the delayed rectifier current, regulating the time course of the action potential.[69,70] Whether or not some of these channels are also expressed in smooth muscle cells of the tonic-type arteries and/or in endothelial cells and playing a role in the control of vascular tone is uncertain.

1.3.1.2.2 Calcium-Activated Potassium Channels

This family of potassium channels, once included in the K_V potassium channel family, is divided into two subfamilies, the $K_{Ca}2.1$, $K_{Ca}2.2$, $K_{Ca}2.3$, $K_{Ca}3.1$ family

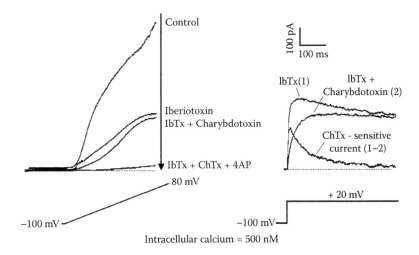

FIGURE 8 Potassium channels in smooth muscle cells of the rabbit carotid artery. Effect of the combination of different inhibitors of potassium channel in freshly isolated smooth muscle cells of the rabbit carotid artery (whole-cell configuration of the patch-clamp technique). Left panel: Original traces of the large global currents observed in the presence of intracellular calcium (0.5 μM) for a ramp depolarization from −100 to +80 mV. This current is partially inhibited by iberiotoxin (IbTx: 10^{-7} M), indicating the presence of BK_{Ca}. The addition of charybdotoxin (ChTx: 10^{-7} M) produces a further inhibition, indicating the presence of an A-type or rapidly inactivating, transient outward current (K_{TO}). The subsequent addition of 4-aminopyridine (4-AP: 5×10^{-3} M) produces the virtual abolition of the remaining global outward current demonstrating the contribution of K_V. Apamin did not produce any inhibition in those cells, indicating that the unidentified channel responsible for "SK_{Ca}-like" current is not expressed (data not shown). Right panel: Original traces of the large outward current observed in the presence of iberiotoxin for a step depolarization from −100 to +60 mV. The inhibitory effect of charybdotoxin is confirmed and the arithmetic subtraction of the trace recorded in the presence of the combination of charybdotoxin plus iberiotoxin (trace 2) from that in the presence of iberiotoxin alone (trace 1) allows the visualization of the kinetic of the A-type current (trace 1–2).

(small and intermediate conductance calcium-activated potassium channels also known as SK1, SK2, SK3, and IK1) and the $K_{Ca}1.1$, $K_{Ca}4.1$, $K_{Ca}4.2$, $K_{Ca}5.1$ family (large conductance calcium-activated potassium channels also known as BK_{Ca}).

1.3.1.2.2.1   Small and Intermediate Conductance Calcium-Activated Potassium Channels

SK_{Ca} channels were first identified on the basis of their pharmacology. They are blocked by apamin, a toxin present in the venom of the European bee, *Apis mellifera*, but are insensitive to charybdotoxin. Three α subunits of the SK_{Ca} channel family, $K_{Ca}2.1$, $K_{Ca}2.2$, and $K_{Ca}2.3$ (or SK1, SK2, and SK3, respectively) share a highly conserved transmembrane region (80–90% identity), but vary significantly in the amino and carboxy terminal domains.[71] Alternative splicing occurs in human SK1 and SK3.[72,73] The calcium sensitivity of the SK channel is ascribed to the association of the calcium binding protein calmodulin with the α subunit.[74.] The single channel conductance of the SK_{Ca} channels is less than 20 pS. They are sensitive to an increase in intracellular calcium concentration (0.6–0.7 μM), and are virtually voltage-independent.[75–77]

Besides apamin, scyllatoxin (also named leiurotoxin I), a toxin isolated from the venom of the scorpion, *Leiurus quinsquestriatus*, the plant alkaloid tubocurarine, and the synthetic compound UCL-1684[78] are potent and reasonably specific blockers of SK_{Ca} channels.[79–82] The SK1 subtype is not or is poorly sensitive to apamin.[83] However, the human SK1 channel, when cloned and re-expressed in various cell types, is sensitive to apamin in the nanomolar range.[81,84] In contrast, the human SK3 isoform, hSK3_ex4, is insensitive to either apamin or scyllatoxin.[73] Tamapin, a venom peptide from another scorpion species, *Mesobuthus tamulus*, appears to be selective toward SK2 vs. SK1 and to a lesser extent vs. SK3.[85] 1-EBIO and chlor-zoxazone-related compounds can activate recombinant SK_{Ca},[86] but there is no specific activator of these channels available (Table 1, Figure 9).

Recombinant SK_{Ca} channel α subunits can form heteromultimers that are also sensitive to apamin.[87] Whether native SK_{Ca} channels form heteromultimers and whether their expression involves the association of the α subunits with regulatory β subunits is not clear.[79,88]

Another channel with a significant homology (approximately 50%) with the α subunits of the SK_{Ca}, $K_{Ca}3.1$ (also known as SK4 or IK1), has been cloned. This protein is also a calcium-activated potassium channel, associated with calmodulin and highly sensitive to changes in intracellular calcium concentrations (0.1 to 0.3 μM). Its conductance (20–60 pS), intermediate between SK_{Ca} and BK_{Ca}, is at the origin of its name, intermediate conductance calcium-activated potassium channel.[89–92] This IK_{Ca} is also known as the Gardos channel involved in the regulation of the cell volume of erythrocytes.[93] This channel is insensitive to apamin or iberiotoxin but is blocked by charybdotoxin and clotrimazole, a cytochrome P450 mono-oxygenase inhibitor. TRAM 34 and TRAM 39, nonpeptidic analogues of clotrimazole, devoid of cytochrome P450 mono-oxygenase inhibitory properties, are specific blockers of IK_{Ca}.[94] 1-EBIO is a rather

---

TABLE 1
Calcium-Activated Potassium Channels and Pharmacological Tools

BK_{Ca}	IK_{Ca} (or SK4)	SK_{Ca} (SK1, SK2, SK3)
Large conductance	Intermediate conductance	Small conductance
> 100 pS	15–60 pS	3–15 pS
	BLOCKERS	
TEA, TBA, Charybdotoxin	TEA, TBA, Charybdotoxin, Clotrimazole	TEA, TBA, Tubocurarine
Iberiotoxin	TRAM 34, TRAM 39	Apamin, Scyllatoxin,
UCL 1684		Tamapin
	OPENERS	
NS 004, NS 1619	1-EBIO	1-EBIO, Riluzole

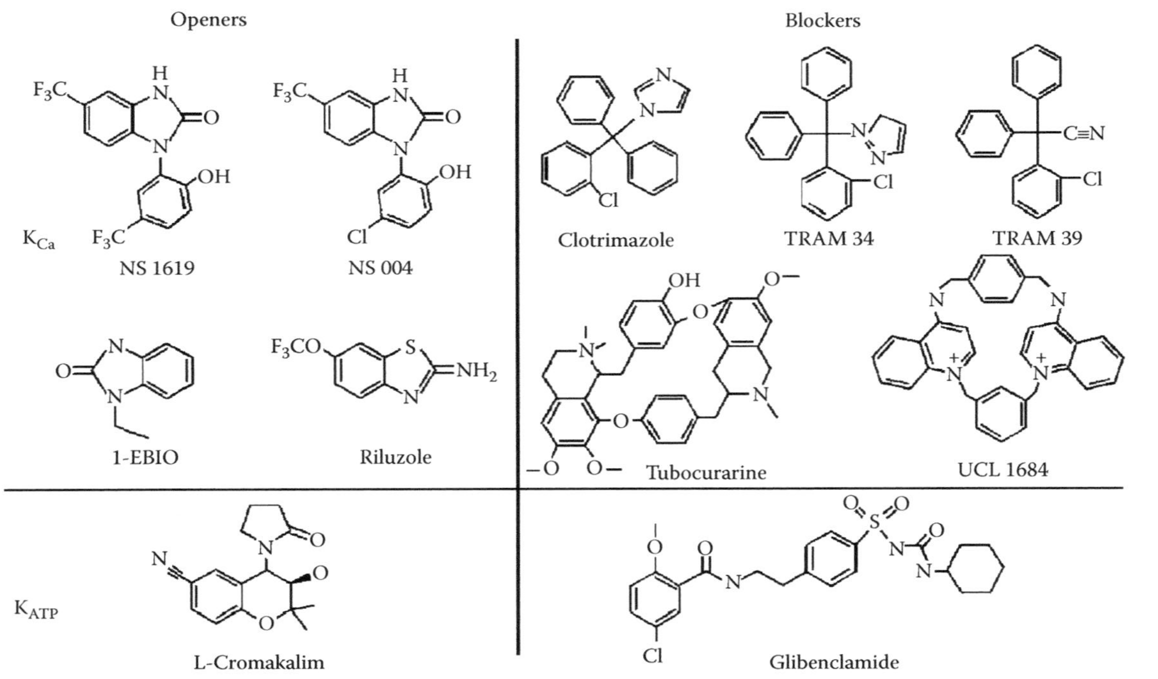

FIGURE 9 Molecular structure of some nonpeptidic openers and blockers of potassium channels. NS 1619 and NS 004 are specific openers of BK_{Ca}, TRAM 34 and TRAM 39 are specific blockers of IK_{Ca}, and UCL 1684 is a specific blocker of SK_{Ca}.

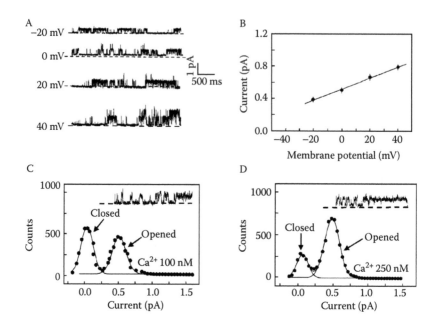

FIGURE 10 SK_{Ca} in freshly dissociated porcine coronary endothelial cells (outside-out configuration of the patch-clamp technique). (A) A small-conductance channel, as observed in single-channel recordings (250 nM free Ca^{2+} pipette solution, in the presence of 10^{-7} M charybdotoxin, physiological K^+ gradient). (B) The unitary conductance was determined by variance analysis of activity recorded at −20, 0, 20, and 40 mV (with the assumption of a single channel). Data were fitted to give slope conductance of a 6.8 pS channel. The Ca^{2+}-dependency of the SK_{Ca} channel is shown in C and D. (By permission from the Nature Publishing Group; modified from Burnham et al.[99])

weak and poorly specific activator of the IK_{Ca} channel. However, as in certain vascular tissues it does not activate SK_{Ca} or BK_{Ca}, it can be regarded as a reasonably selective tool to activate this channel.[95] Whether or not native IK_{Ca} channels are associated with regulatory β subunits is not clear. This population of potassium channels can theoretically also form heteromultimers with the α subunit SK1, SK2, or SK3.[79] Knock-out mice for the $K_{Ca}3.1$ gene appear normal and are fertile, but the volume regulation of T-lymphocytes and erythrocytes is severely impaired in these animals.[96] The impact of the gene deletion on cardiovascular function and the endothelial function is unknown.

In the vascular wall, SK_{Ca} channels are expressed in endothelial cells (Figure 10 and Figure 11).[97–102] In the porcine coronary artery,[99] in the rat carotid artery,[101] and in the murine mesenteric artery and aorta,[102] the expression of the SK3 α subunit is linked to the functional SK_{Ca} channel (Figure 11). The SK_{Ca} channels are also expressed in smooth muscle cells of the gastrointestinal tract[103] and the vasculature.[100] However, there is little evidence of a functional role of this channel in vascular smooth muscle cells. An apamin-sensitive conductance has been identified in renal arterioles[104] and in the smooth muscle cells of the guinea pig carotid artery (Figure 7).[65]

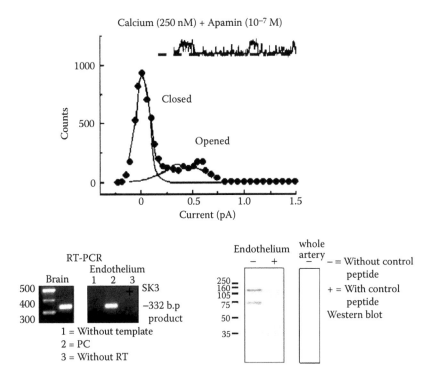

FIGURE 11 SK$_{Ca}$ in porcine coronary endothelial cells. Top left: The inhibitory effect of apamin (10^{-7} M) on the 6.8 pS channel activity (outside-out configuration of the patch-clamp technique, under a physiological K$^+$ gradient, with 250 nM Ca^{2+} pipette solution and in the presence of charybdotoxin). Bottom left: The detection of SK$_{Ca}$ subunit mRNA in porcine coronary arterial endothelium (RT-PCR analysis of SK3 expression in samples of endothelium as positive control, pig brain). The polymerase chain reaction (PC) was performed with (lane 2) or without reverse transcription (RT). Size markers in base pairs (b.p.) are indicated. Bottom right: Western blot analysis showing the SK3 protein expression in the endothelial cells. Primary antibodies were used with (+) and without (–) pre-incubation with the control peptide. Molecular weight markers are indicated (kDa). (By permission from the Nature Publishing Group; modified from Burnham et al.[99])

However, in both cases this current is voltage-dependent and therefore unlikely to involve any of the known SK channels isoforms. The molecular identification of the potassium channel, responsible for the apamin-sensitive potassium current observed in vascular smooth muscle cells, is not available.

The IK$_{Ca}$ channel is constitutively expressed in endothelial cells (Figure 12 and Figure 13)[97,101,105,106] and in nonvascular smooth muscle cells.[107] However, in freshly isolated vascular smooth muscle cells this potassium channel is not or is very poorly expressed (Figure 7),[65,108] while in proliferating cells, as seen in culture or after vascular injury, its expression increases.[108,109] By contrast, in native endothelial cells this channel is expressed, but becomes less preponderant in culture.[97,105,106,110–112]

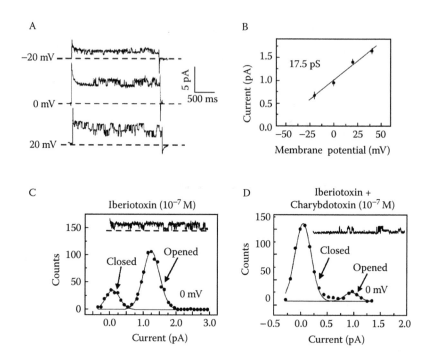

FIGURE 12 IK_{Ca} in freshly dissociated porcine coronary endothelial cells (outside-out configuration of the patch-clamp technique). (A) Single-channel activity was recorded at the indicated holding potentials under asymmetrical K^+ gradients with Ca^{2+} fixed at 250 nM in the pipette solution. Representative traces are shown. (B) Mean unitary currents were plotted against holding potential and fitted with a linear function. The slope conductance computed was 17.1 ± 0.4 pS. (C) The effects of iberiotoxin and (D) the subsequent addition of charybdotoxin were recorded. Distributions of unitary conductance amplitude (recorded at 0 mV and using 250 nM free Ca^{2+} in the pipette solution) are shown, together with representative traces. A Gaussian fit was performed for every single channel experiment, and the amplitude of single channels was determined with the best fit. Then the mean of the obtained amplitudes was calculated. These data show that the current is insensitive to iberiotoxin (IbTX, 10^{-7} M) but blocked by charybdotoxin (ChTX, 10^{-7} M). (By permission from the Nature Publishing Group; modified from Bychkov et al.[106])

1.3.1.2.2.2 Large-Conductance Calcium-Activated Potassium Channels

BK_{Ca} are both voltage- and calcium-regulated potassium channels indicating that they play an important role in limiting the entry of calcium and the cell excitability. They are characterized by a high unitary conductance, usually defined as being in excess of 100 pS (in vascular smooth muscle cells, between 250 to 300 pS when recorded with equimolar concentration of potassium on either side of the membrane, i.e., symmetrical potassium concentration), and are widely expressed in tissues of many species. The BK_{Ca} activity is increased when depolarizing voltages are applied and when the intracellular calcium concentration is raised. However, they do not

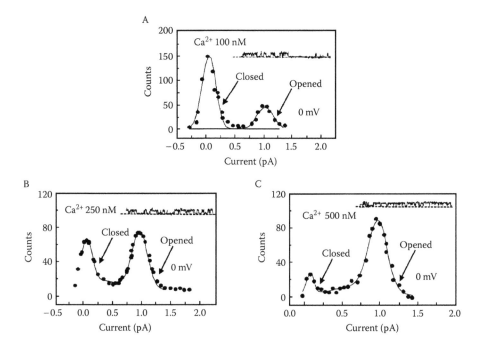

FIGURE 13 Effect of Ca^{2+} on the open probability of IK_{Ca}. The 17.1 pS channel activity was recorded from outside-out patches taken from freshly dissociated porcine coronary artery endothelial cells. Distributions of unitary conductance amplitude recorded at 0 mV are shown, together with representative traces (closed state is indicated by C). A Gaussian fit was performed for every single channel experiment, and the amplitude of single channels was determined with the best fit. Then the mean of the obtained amplitudes was calculated. The effect of Ca^{2+} was determined using pipette solutions with free Ca^{2+} fixed at 100, 250, and 500 nM. (By permission from the Nature Publishing Group; modified from Bychkov et al.[106])

have an absolute calcium requirement since they can be activated by depolarization at virtually zero calcium, and since their open-state probability is independent of calcium in the range of 0 to 100 nM. Their gating is facilitated by intracellular calcium concentrations in excess of 100 nM. Unlike SK_{Ca} and IK_{Ca}, the calcium sensitivity is not linked to an association with calmodulin, but presumably lies in the region of negatively charged aspartate residues situated in the unusually long intracellular C-terminal portion, the so-called "Ca^{2+} bowl." BK_{Ca} are selectively blocked by the scorpion toxin iberiotoxin, but also by charybdotoxin and low concentrations of tetraethylammonium. They are activated by synthetic compounds such as the benzimidazolone derivatives, NS-1619 and NS-004 (Table 1, Figure 9).[113,114]

The first cloned structural component encoding a calcium-activated potassium channel was characterized in *Drosophila* with a flight muscle mutation called *Slowpoke*. The identification of its cDNA sequence, known as *dSlo*, has allowed the subsequent cloning and expression of mammalian homologs $K_{Ca}1.1$ (*Slo1*).[114] BK_{Ca} channels appear similar to the other members of the K_V channel family as they are formed by a tetrameric association of pore-forming α subunits. However, in contrast to the other K_V channels, which possess six transmembrane helix domains, BK_{Ca}, in its N-terminal region, has a seventh membrane-spanning domain (S0), resulting in an extracellular N-terminal portion.[115] Also in contrast to the other K_V channels, apart from the *Slo1* α subunit, only three additional genes that belong to the BK_{Ca} channel family, $K_{Ca}4.1$ (*Slack* or *Slo2*), $K_{Ca}5.1$ (*Slo3*), and $K_{Ca}4.2$, have been identified. Although structurally homologous to the *Slo* α subunit, $K_{Ca}4.1$ and $K_{Ca}5.1$ encode potassium channels with markedly different properties in term of conductance, calcium sensitivity, and pharmacology, as well as tissular distribution.[114] When *Slo2* and *Slo1* are expressed together, they appear to form calcium-activated heteromultimeric potassium channels of intermediate conductance.[116]

The *Slo1* α subunits from vertebrates and *Drosophila* display numerous isoforms generated by alternative splicing. These splice variants show significant changes in calcium and voltage sensitivity as well as in channel kinetics and tissular distribution. This suggests an adaptive expression in order to fulfill the physiological requirements of specific cells and tissues.[113] In addition to alternative splicing of the *Slo1* α subunit, the expression of accessory β subunits can lead to channel diversity. Four β subunits (β1 to β4) have been cloned and they apparently all modulate the pharmacology and kinetics of the *Slo1* α subunit. The β subunits possess two transmembrane domains and probably interact with the S0 domain of the α subunit. The combination of *Slo1* α subunit splice variants with one or more possible splice variants of up to four β subunits potentially can produce a large number of BK_{Ca} channels with distinctive properties.[52,114] The β1 and β2 are the two subunits that have been the most extensively characterized. The β1 subunit is abundantly expressed in vascular and nonvascular smooth muscle cells and, like the β2 subunit, increases the apparent voltage and calcium sensitivity of *Slo1*. However, the β1, when coexpressed with *Slo1*, yields a non-inactivating current, while the β2 subunit causes inactivating currents.[113]

BK_{Ca} channels are expressed in virtually all vascular smooth muscle cells and pericytes (Figure 14). However, at membrane potentials close to the resting state, the open-state probability of BK_{Ca} is very low, suggesting that their contribution in determining the resting membrane potential is minimal.[65,117] The role of BK_{Ca} should be seen rather as a physiological brake, a feedback inhibitor of contraction, and/or increase in intracellular calcium concentration in response to humoral (e.g., norepinephrine, angiotensin II) or physical stimuli such as an increase in intravascular pressure.

Spontaneous transient outward currents (STOC) are observed in coronary and cerebral arteries, as well as in small myogenically active arteries. They are caused by the activation of 20–100 clustered BK_{Ca} channels in response to localized, elemental calcium release events from internal calcium stores, the calcium sparks.[118–120] Thanks to the BK_{Ca} channel activation, these calcium sparks paradoxically lead to a decreased overall intracellular calcium concentration and thus to relaxation of arterial smooth muscle. In mice with a disrupted gene for the auxiliary β1 subunit, the generation of calcium sparks in vascular smooth muscle cells are of normal amplitude and frequency, but the frequency of the spontaneous transient outward currents is reduced. The transgenic mice, when compared to the wild-type controls, have a higher systemic arterial blood pressure. The contractile responses of isolated

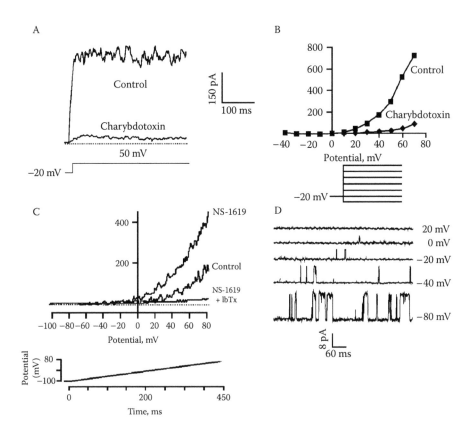

FIGURE 14 BK_{Ca} in cultured bovine retinal pericytes. (A) Original traces showing the effects of the nonspecific BK_{Ca} blocker, charybdotoxin (10^{-7} M), on a step depolarization (whole-cell configuration of the patch-clamp technique, holding potential –20 mV, step depolarizations, high intracellular calcium concentration: 170 nM). Under those experimental conditions K_V channels were inactivated and do not contribute to the global outward current. (B) Current-voltage relationship of the outward potassium current recorded before (control) and after application of charybdotoxin. (C) Original traces showing the effects of BK_{Ca} opener, NS1619 (20 µM), and its inhibition by iberiotoxin (10^{-7} M). Whole-cell configuration of the patch-clamp technique, holding potential –20 mV, ramp depolarization, high intracellular calcium concentration. (D) Unitary large conductance calcium-activated potassium current, recorded in the cell-attached patch-clamp configuration, with different holding potential applied externally by the pipette (20, 0, –20, –40, –80 mV). (By permission from Lippincott Williams and Wilkins; modified from Quignard et al.[59])

aortic rings to agonists and KCl in the former are increased when compared to aortas from the latter.[121] These results suggest that spontaneous transient outward currents, elicited in response to calcium sparks, contribute to the general regulation of vascular tone and, thus, that the β1 subunit plays an essential role in this process.

In cultured endothelial cells,[122] in the rabbit aorta,[123] and in the porcine renal artery[124] a functional BK_{Ca} channel is expressed. In most endothelial cells, when freshly isolated, BK_{Ca} channels are poorly expressed and iberiotoxin-sensitive currents are observed only at very positive potentials, while no single channel activity can be detected.[97,106,125] This can be attributed possibly to the absence in these cells of regulatory BK_{Ca} β subunits that enhance Ca^{2+} sensitivity.[123,125,126]

1.3.1.2.3   Inward Rectifiers

The inward rectifier potassium channel (Kir) gene family is divided into seven subfamilies (Kir1.0 to Kir7.0) with fifteen different genes identified. This family is organized as tetramers, each subunit containing two transmembrane segments and one pore domain. When expressed in heterologous expression systems, most of these genes form potassium channels with various degrees of inward rectification.[54] This means that the channel conducts potassium current more readily into than out of the cell over a wide range of potentials. When the membrane potential is negative compared to the equilibrium potential for potassium ions (E_K), the driving force for the flux of potassium ions is in the inward direction and potassium ions readily flow through K_{IR}. However, for positive membrane potentials (compared to E_K) the outward flow of potassium ions through K_{IR} is smaller. Under physiological conditions, the membrane potential of vascular cells is always positive compared to E_K, so it is the relatively small efflux of potassium ions that plays a physiologically relevant role. The inward rectification is at least partially conferred by channel block from intracellular positively charged ions, such as magnesium, and by polyamines.[127,128]

The molecular composition of K_{IR} channels in endothelial and smooth muscle is far from being clarified. However, beyond any doubt, two families play a major role in the vascular cells: the Kir2 and Kir6 families. Additionally, the Kir3 family could possibly be involved in the regulation of vascular tone.

1.3.1.2.3.1   Kir2, the strong or "classical" inwardly rectifying
potassium channel

This subfamily includes four identified genes (Kir2.1 to Kir2.4). In both endothelial and smooth muscle cells, Kir2.1 appears to be the most relevant channel. Transcripts for Kir2.2 and Kir 2.3 have been detected in some but not all studies. Whether or not these channels play a role in the endothelial and/or smooth muscle remains to be determined.[129,130]

In smooth muscle cells, the expression of the K_{IR} channel increases as the diameter of the artery decreases. In peripheral vascular beds,[131–133] but also in autoregulatory vascular beds such as the coronary and cerebral circulations,[134–135] K_{IR} channels contribute significantly to the resting membrane potential of the smooth muscle cells. K_{IR} channels are blocked by micromolar concentrations of barium. This inorganic cation is the most specific, albeit imperfect, blocker available (Figure 15).

A unique feature of K_{IR} channels is the action of extracellular potassium on its gating. A moderate increase in potassium concentration, in the range of 1 to 20 mM, enhances potassium efflux through K_{IR} at physiologically relevant potentials.[59,136] This leads to hyperpolarization and relaxation of the arterial smooth muscle cells. The K_{IR} channel can be regarded as a metabolic sensor producing vasodilatation and increases in blood flow when potassium accumulates in the circulation during neuronal activity or exercise, for instance. Potassium-induced dilatations are absent in

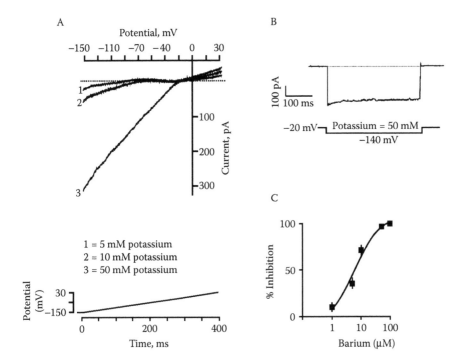

FIGURE 15 K_{IR} in cultured bovine retinal pericytes. (A) Original traces showing the effects of different concentrations of extracellular potassium (5, 10, 50 mM) on the inward potassium current. Whole-cell configuration of the patch-clamp technique, holding potential –20 mV, step depolarizations, high intracellular calcium concentration 170 nM, ramp depolarization from –150 to +30 mV. (B) Original trace showing the inward potassium current produced by a hyperpolarizing step (from –20 to –140 mV). (C) Concentration-dependent inhibitory effect of barium on the inward potassium current (at –140 mV, 50 mM of extracellular potassium; IC_{50} for barium chloride was approximately 6 µM). (By permission from Lippincott Williams and Wilkins; modified from Quignard et al.[59])

cerebral arteries from knockout mice in which the Kir2.1 has been deleted, confirming the role of this channel subtype in vascular smooth muscle cells.[137]

K_{IR} channels are observed in virtually all endothelial cells and are the most prominent channels in these cells where they contribute substantially to the resting membrane potential. The single-channel conductance of endothelial K_{IR} ranges from 23 to 30 pS (symmetrical potassium concentration) and is similar to that reported in other cells.[138,139] The Kir2.1 is the isoform also expressed in endothelial cells.[140,141] Endothelial K_{IR} channels are activated not only by potassium ions but also by shear stress.[142]

1.3.1.2.3.2   Kir3, the G-Protein-Gated Inwardly Rectifying Potassium Channel

This subfamily includes four different genes (Kir3.1 to Kir3.4, or GIRK1-4). These channels are controlled by G-protein-coupled receptors and are expressed predominantly in the heart, the central and peripheral nervous system, and in endocrine tissues. Upon stimulation of G-protein-coupled receptors, the βγ subunit of the associated G-protein dissociates from the α subunit and both subunits act as downstream effectors. The βγ subunit directly binds and activates the G-protein-gated inwardly rectifying potassium channels (GIRK).[143] This family of potassium channels mediates the vagal-induced slowing of heart rate by muscarinic acetylcholine M2 receptor stimulation (IKACh).[144]

In vascular smooth muscle and endothelial cells the functionality of this population of potassium channels remains hypothetical. The mRNA of Kir3.1 has been detected in rat aortic smooth muscle[145] and it may be involved in some of the relaxing effects of natriuretic peptides.[146]

1.3.1.2.3.3   Kir6, the ATP-Sensitive Potassium Channel

ATP-sensitive potassium channels (K_{ATP}) are weakly rectifying, high-conductance, potassium-selective channels. Their level of activation is inversely related to the absolute value of the intracellular ATP/ADP ratio and therefore K_{ATP} channels set the membrane potential according to the metabolic state of the cells.[147] They are expressed in numerous cell types including pancreatic β cells, neurons, and cardiac, skeletal, and smooth muscle cells. Besides the sarcolemnal K_{ATP} channel, a putative channel in the inner mitochondrial membrane has been characterized pharmacologically, but not cloned. This K_{ATP} mitochondrial channel could play an important role in ischemic preconditioning.[148]

K_{ATP} channels are composed of the pore-forming α subunit Kir6 and the regulating sulfonylurea receptor (SUR) subunit. Two genes encoding for the Kir6 α subunit (Kir6.1 and Kir6.2) and for the SUR-associated regulatory subunits (SUR1 and SUR2) have been identified. Two splice variants of each SUR subunit exist. SUR subunits are complex proteins, with 17 transmembrane domains clustered in three major domains, and two long intracellular loops that contain the nucletotide binding domains. The SUR proteins are not only the site of regulation for intracellular nucleotides, but also the targets of most of the K_{ATP} blockers (the antidiabetic agents sulfonylureas and glinides) and K_{ATP} openers (diazoxide, pinacidil, nicorandil, and cromakalim derivatives).[149]

Four subunits of the Kir6 associate with four SUR subunits to form a functional K_{ATP} channel octamer. The pancreatic β cell K_{ATP} channel is formed by the association of Kir6.2 and SUR1 while the cardiac K_{ATP} channel is formed by the association of Kir6.2 and SUR2A. In vascular smooth muscle cells, the K_{ATP} channels are formed by the association of either Kir6.2 or Kir6.1 with SUR2B.[147] The predominant K_{ATP} channel subtype expressed in vascular smooth muscle cells has a low unitary conductance (20 to 50 pS in symmetrical potassium concentration) and, based on experiments using recombinant channel expression, is composed of Kir6.1 and SUR2B. Neurohumoral substances regulate its activity. However, in vascular smooth muscle cells, a wide range of single-channel conductance coexists, suggesting that other K_{ATP} channels are expressed, not only Kir6.2 and SUR2B, but possibly also heteromeric Kir6.1 and Kir6.2 channels.[149–151]

Freshly dissociated endothelial cells of the rabbit aorta and guinea pig cardiac capillaries express functional K_{ATP} channels. In the latter, like in vascular smooth muscle cells, the channels are formed of SUR2B with Kir6.1 and/or Kir6.2 subunits.[152,153] However, in several vascular preparations, the hyperpolarization of the endothelial cells in response to cromakalim derivatives does not result from direct activation of endothelial K_{ATP} channel, but rather is indirect and linked to the transmission of the hyperpolarization from the smooth muscle to the endothelium by means of gap junctions.[154–156]

1.3.1.2.4 Two-Pore-Domain Potassium Channel Family

The last group of potassium-selective pore forming α subunits are formed by proteins with four transmembrane segments and two pore domains, an unusual feature that is the origin of their name, two-pore-domain potassium channels or tandem-pore-domain potassium channels (K_{2P}). The first member of this potassium channel family was originally cloned in yeast[157] before being identified in mammals.[158] Fourteen human sequences have so far been identified, **T**andem of P domain in **W**eak **I**nward rectifier potassium (**K**+) channels (K_{2P}1.1, K_{2P}6.1, and K_{2P}7.1 or TWIK-1, TWIK-2, and KCNK7), **TWIK RE**lated potassium (**K**+) channels (K_{2P}2.1 and K_{2P}10.1 or TREK-1 and TREK-2), **TWIK R**elated **A**rachidonic **A**cid-stimulated potassium (**K**+) channels (K_{2P}4.1 or TRAAK), **TWIK** related **A**cid-**S**ensitive potassium (**K**+) channels (K_{2P}3.1, K_{2P}5.1, K_{2P}9.1, K_{2P}17.1, and K_{2P}15.1 or TASK-1 to TASK-5), **TWIK** related **AL**kaline-activated potassium (**K**+) channels (K_{2P}16.1 and K_{2P}17.1 or TALK-1 and TALK-2), and **TWIK** related **H**alothane **I**nhibited potassium (**K**+) channels (K_{2P}13.1 and K_{2P}12.1 or THIK-1 and THIK-2). A functional expression has been demonstrated for only seven of these genes (TWIK-2, TASK-1 to 3, TREK-1 to 2, and TRAAK). The other proteins could be silent subunits that may need to be associated with a yet-to-be-identified protein in order to form functional channels. Some of these proteins/channels are likely targets for volatile anesthetics.[53,159]

Mature functional K_{2P} channels are homodimers since four potassium domains are necessary to form a potassium-selective pore. However, in contrast to the other potassium channels, which are formed by a noncovalent tetrameric association, K_{2P}, with the exception of TASK-1, are formed by cystein-dependent disulfide-bridged homodimers.[160] Some K_{2P} subunits may co-assemble into heterodimers with specific characteristics,[161] again illustrating the large diversity of the potassium channel family. The unitary conductance of K_{2P} channels varies widely from 10 to 100 pS.

TWIK and TASK-1 channels produce constitutive, non-inactivating potassium currents of weak amplitude at all potentials. The expression of these potassium channels is widely distributed and TASK-1 has been identified in vascular smooth muscle cells.[162] Their functional characteristics suggest that they carry background potassium currents and that they play, along with the inward rectifier potassium channel family, an important role in the setting of the cell membrane potential and in the regulation of cell excitability. There is no specific inhibitor of these potassium channels. They are relatively insensitive to classic potassium channel blockers such as tetraethylammonium and 4-aminopyridine.[53,158,162] The most notable characteristics of TASK-1 are their extreme sensitivity to external pH at pH values within the physiological range and their activation by volatile anesthetics, such as halothane and isoflurane, at concentrations used in human general anesthesia.[163–165]

TREK-1, TREK-2, and TRAAK have a low basal activity, but they can be activated strongly by arachidonic acid and other unsaturated fatty acids such as oleic, linoleic, eicosapentaenoic, and docosahexaenoic acid and by physical stimuli such as stretch, cell swelling, and negative pressure. TREK-1 is activated by NO and cyclic-GMP-dependent protein kinase and internal acidification. Both TREK-1 and TREK-2 are activated by volatile anesthetics. TREK-1 and TRAAK channels are highly flickering and have a unitary conductance of 100 and 45 pS, respectively.[166–170] The expression of both TREK-1 and TREK-2 has been detected in vascular smooth muscle cells.[171] These channels are blocked by barium and micromolar concentrations of gadolinium, but are resistant to classic potassium channels blockers. TREK-1 is blocked by elevated concentrations of quinidine.[53,159,171]

The expression of these channels in endothelial cells has not been determined.

1.3.1.3 Chloride Channels

Chloride ions, unlike calcium, do not play a role as intracellular messengers. However, they are involved in the regulation of intracellular pH, cell volume, transepithelial transport, electrical excitability, and the control of resting membrane potential. Chloride channels are channels that allow the passive diffusion of negatively charged anions along their electrochemical gradient. Some of these channels may conduct other anions (I^-, NO_3^-, Br^-, SCN^-) better than chloride itself, but are nevertheless referred to as chloride channels since Cl^- is the most abundant anion in the organism.[172] There is a large variety of chloride channels, expressed on the plasma membrane or/and intracellular organelles, which have been identified according to their biophysical characteristics. However, the molecular structure of the chloride channels is only known for a few of them, suggesting that entire gene families of chloride channels remain to be discovered.

Three molecularly distinct chloride channel families are well established, the CLC gene family (9 members in mammals), the cystic fibrosis transmembrane conductance regulator (CFTR), and the ligand-gated γ-aminobutyric acid- and glycine-receptor chloride channels (19 and 4 members in mammals, respectively). In addition, a family of putative intracellular chloride channels ($CLIC_1$ to $CLIC_5$) has been proposed. However, the genes encoding the swelling-activated chloride current ($I_{Cl,swell}$) and the Ca^{2+}-activated chloride channels (CLCA) remain to be identified.[172]

The intracellular distribution of Na^+, K^+, and Ca^{2+} in various cell types is fairly homogeneous. However, the distribution of $[Cl^-]_i$ is markedly different depending on the cell type. For instance, in neurons, $[Cl^-]_i$ is lower than that predicted from passive distribution and the equilibrium potential for Cl^- ions is close to or more negative than the resting membrane potential. Therefore, activation of ligand-gated γ-aminobutyric acid- and glycine-receptor chloride channels either stabilizes or hyperpolarizes the cell membrane.[42] Similarly, in smooth muscle cells, Cl^- ions are not passively distributed across the cell membrane. However, in contrast to neurons, in smooth muscle cells the $[Cl^-]_i$ is higher than that predicted from passive distribution and averages 30 to 60 mM. The equilibrium potential for chloride anions is, therefore, considered to be between −20 and −30 mV, which is positive when compared to the resting membrane potential. Consequently, the opening of chloride channels produces an efflux of chloride anions and depolarization.[42] The accumulation of chloride ions beyond their equilibrium potential is associated with a low chloride permeability and is maintained by different transport systems: the chloride/bicarbonate exchanger, the Na^+, K^+, Cl^- co-transport, and possibly a third unidentified mechanism called "pump III." A substantial amount, possibly up to 40%, of the energy spent for the activity of the Na^+/K^+-ATPase is required to drive the Na^+, K^+, Cl^- co-transport. This indicates that the energetically expensive Cl^- accumulation must have a physiological purpose for the smooth muscle cells.[173] Two major Cl^- currents are recorded in vascular smooth muscle cells: CLCA and $I_{Cl,swell}$. The CLC-3 gene may encode the latter channel although no consensus has been reached.[172,174] $I_{Cl,swell}$ is activated by low osmotic pressure and by mechanical stress produced, for instance, by vascular distension during a rise in blood pressure. NO tonically downregulates the activity of $I_{Cl,swell}$ and so could regulate the contractions of the smooth muscle cells in various vascular beds including the coronary arteries.[175–177] Whether this Cl^- channel (or a swelling-activated cation channel) regulates myogenic tone in cerebral arteries is uncertain.[178,179] CLCA are likely to play a major role in the contractions of vascular smooth muscle cells. Following receptor activation, the release of calcium from intracellular stores opens these channels leading to depolarization, the opening of voltage-gated Ca^{2+} channels, and the subsequent increase in $[Ca^{2+}]_i$. Niflumic acid is the best-known inhibitor of CLCA.[180,181]

In the endothelial cells, the $[Cl^-]_i$ is also elevated (>30 mM) and the activation of chloride channels produces cell depolarization.[139,174] $I_{Cl,swell}$ or VRAC (volume-regulated anion channel) is a multifunctional channel that is expressed constitutively in endothelial cells. Its molecular identity is unknown. It contributes to the establishment of the resting membrane potential and its blockade provokes hyperpolarization of the endothelial cells.[182] $I_{Cl,swell}$ is activated by hypoosmolarity, but also by mechanical stimuli, including shear stress, and it contributes to intracellular pH and $[Ca^{2+}]_i$ homeostasis.[139] This channel can therefore be involved in the paracrine and endocrine activity of the endothelial cells, as well as with the proliferation and the differentiation of these cells.

CLCA is also expressed in endothelial cells. The activation of this channel by the increase in $[Ca^{2+}]_i$ depolarizes the cell and counteracts the hyperpolarization evoked by the activation of K_{Ca}. The role of CLCA is markedly different in various endothelial cells, minor in some but predominant in others such as in the bovine pulmonary artery where the activation of CLCA virtually abolishes the hyperpolarizing effects of opening K_{Ca}.[183,184] The molecular identity of CLCA in endothelial cells has not yet been determined.

Additionally, high-conductance Cl^- channels have also been observed in endothelial cells (single-channel conductance: 100 to 400 pS). Although the physiological role of these channels remains elusive, their activity is inhibited by β-estradiol, suggesting that this property could be involved in the beneficial cardiovascular effects of the hormone.[139,185]

1.3.2 CALCIUM HOMEOSTASIS IN VASCULAR SMOOTH MUSCLE AND ENDOTHELIAL CELLS

Calcium ions are ubiquitous intracellular second messengers involved in the regulation of numerous cellular processes including contractile and secretory activities. The regulation of the intracellular calcium concentration both in the endothelial and smooth muscle cells is therefore of utmost importance for the control of vascular tone. At rest, in both cell types, the intracellular calcium concentration is very low (less than 100 nM). Calcium ions are sequestrated in specific storage sites (mostly the sarcoplasmic reticulum and the mitochondria) or are extruded by specific transport systems to the extracellular space. Intracellular calcium can be raised via calcium entry from the extracellular space or calcium release from the intracellular organelles.

1.3.2.1 Contraction of Vascular Smooth Muscle

1.3.2.1.1 Calcium Entry

Calcium entry can occur through voltage-dependent calcium channels or several types of calcium-permeable channels that are non-voltage gated.

1.3.2.1.1.1 Voltage-Dependent Calcium Channels

The voltage-dependent calcium channel (Ca_V) is a superfamily of proteins that is encoded by at least 10 different genes organized in three main subfamilies (Ca_V1 to Ca_V3). The nomenclature is based on the similarities in the amino-acid sequences of the α subunits, following that used to describe the potassium channel family.[186] In vascular smooth muscle, the Ca_V predominantly expressed is a splice variant of the $Ca_V1.2$ also expressed in cardiac muscle ($Ca_V1.2a$ and $Ca_V1.2b$, for cardiac and smooth muscle, respectively). This channel is the classic dihydropyridine-sensitive calcium channel, formerly named L-type voltage-dependent calcium channel, and is widely expressed in the smooth muscle cells of virtually all vascular beds, although some other calcium channels that are dihydropyridine-insensitive, such as $Ca_V3.1$ (formerly T-type) or calcium channels with an unknown molecular identity, can be expressed in some vascular smooth muscle cells.[187,188]

The molecular structure of Ca_V is similar to that of sodium channels and is constituted by the pore-forming and voltage-sensitive α_1 subunit. This α_1 subunit contains four repeats, each with six transmembrane domains. The calcium channel is a heteromeric complex composed of the α_1 subunit and associated regulatory subunits (β and α_2/δ). In mammalian cells, four genes encoding the intracellular β subunits, each able to generate two splice variants, have been identified. Vascular smooth muscle cells can express β_1, β_2, and β_3. Additionally, an α_2/δ complex consisting of an extracellular α_2 subunit linked by a disulfide bond to a membrane spanning δ subunit is also associated to the α_1 subunit to form the voltage-dependent calcium channel of the smooth muscle cells.[189]

Ca_V are activated by depolarization. In some vascular smooth muscles, such as those of the longitudinal layers of the portal vein, if a threshold is reached, an action potential can be elicited.[190,191] However, in most of the arteries, the open probability of Ca_V is too low to observe the generation of action potentials but sufficient to allow a sustained calcium influx during depolarization. The presence of physiological brakes, such as K_V and BK_{Ca} channels, prevents the membrane potential from reaching the threshold values required to trigger the action potential.[192] The inhibition of these potassium channels with poorly specific blockers (e.g., tetraethylammonium) unmasks the Ca_V-dependent action potentials in normally quiescent smooth muscle cells.[193,194]

1.3.2.1.1.2 Voltage-Independent Calcium Entry (TRP Channels)

Several types of calcium-permeable channels have been described in vascular smooth muscle cells. They include receptor-operated calcium channels (ROC), activated by agonists acting on their receptors, and store-operated calcium channels (SOC) that are activated following the calcium depletion of the internal stores (sarcoplasmic reticulum). ROC and SOC may be closely related and belong to the transient receptor potential channel or TRP channel family.[195]

TRP channels were first identified in the *Drosophila* where a mutation led to impaired vision caused by the lack of a specific calcium influx pathway into photoreceptors. Homologues of this protein were identified in other species, including more than 20 in humans. Mammalian TRP-related proteins are classified into three subfamilies TRPC (TRPC1 to TRPC7), TRPV (TRPV1 to TRPV6), and TRPM (TRPM1 to TRPM8). C stands for canonical as this subfamily of protein has the highest homology with the *Drosophila* TRP channels, V for vanilloid as this subfamily is closely related to the vanilloid recptor TRPV1, and M for melastatin since this subfamily has the highest homology with the tumor suppressor melastatin (TRPM1).[196] TRP channels have six transmembrane domains and the pore region is included between the fifth and the sixth transmembrane domains. They form homo- and possibly heterotetramers.[197] TRP channels are much more than SOC or ROC as they are primary sensors for physical (heat, cold, mechanical stresses) or chemical (e.g., pH, pheromones, capsaicin, bitter and sweet taste) stimuli.[196]

All the members of the TRP channel family are membrane proteins but whether they all are plasmalemnal proteins and whether they all form functional channels is not known. Depletion of intracellular calcium stores initiates or modulates various TRP channels. However, re-expression of TRP channels in diverse expression systems may mimic some but not all the properties of SOC or ROC channels in native cells.[196]

All the members of the TRPC and possibly some members of the TRPV families can be expressed in vascular smooth muscle cells. TRPC4 is the most abundant, but the pattern of expression varies depending on the species and the vascular bed studied. So far, most of the evidence for a role as an ROC is linked to TRPC6 and possibly TRPC1.[198,199] Although it is generally accepted that members of the TRP protein family form the channels corresponding to SOCs and ROCs, to date there is no firm and conclusive evidence showing exactly which TRP channel underlies the molecular identity of either in a given vascular smooth muscle cell. One possible explanation could be the occurrence in native cells of alternative splicing of TRP channel genes and the formation of heteromeric TRP tetramers.[195,200,201] For instance, a channel made up of two TRPC1 plus one TRPC3 and one TRPC6 subunits may form a functional SOC activated by store depletion, while a channel consisting of two TRPC3 and two TRPC6 proteins may form a functional ROC opening predominantly by receptor-activated G protein.[202] The proper identification and characterization of the TRP channels expressed in vascular smooth muscle is a significant challenge for the improved comprehension of the regulation of calcium homeostasis in these cells.

1.3.2.1.2 *Calcium Release*

The sarcoplasmic reticulum is the main organelle capable of taking up, storing, and releasing calcium ions. The mitochondria also play an important role in calcium homeostasis, especially in situations where the intracellular calcium concentration is elevated.[203] In vascular smooth muscle cells, the sarcoplasmic reticulum is often closely associated with the plasmalemna forming a superficial buffer barrier that allows spatial differences in the intracellular calcium concentration.[204] Furthermore, some sarcoplasmic reticulum compartments lie just beneath the specialized domain of the plasma membrane containing the Na^+/K^+-ATPase with high ouabain affinity ($\alpha2/\alpha3$ subunits), the $Na^+/Ca2^+$ exchanger, and store-operated calcium channels (TRP channels). These microdomains associated with caveolae form functional units (plasmerosome) specialized in calcium-regulation that have a marked influence on the signaling role in smooth muscle cells.[205–207]

Specialized calcium pumps (SarcoEndoplasmic Reticulum Ca^{2+}-ATPase, SERCA) generate and maintain the calcium gradient between the inside of the sarcoplasmic reticulum and the surrounding cytoplasm. Three different genes encode the SERCA pumps and the smooth muscle cells generally express SERCA2 and SERCA3.[208] Thapsigargin and cyclopiazonic acid are specific inhibitors of these calcium pumps.[192,209] The calcium is stored inside the sarcoplasmic reticulum by binding to specialized proteins such as calsequestrin.

Stimulation of G-protein-coupled receptors by neurohumoral substances or exogenous agonists activates phospholipase C. This leads to the subsequent formation of diacylglycerol and inositol trisphosphate (IP_3). IP_3 activates specific receptors/channels situated on the sarcoplasmic reticulum, named IP_3 receptors. Three genes encode the elementary subunits composing the IP_3 receptors. Functional IP_3 receptors are composed by the assembly of four subunits forming homo- or heterotetramers.[210] The activation of IP_3 receptors is regulated by the $[Ca^{2+}]_i$. Calcium by itself can activate a sarcoplasmic reticulum receptor/channel to induce calcium release.

Calcium-induced calcium release is due to the activation of ryanodine receptors (RyR). These channels are tetrameric entities and three isoforms have been cloned (Ry1-3). Smooth muscle cells express preferentially RyR-2 and RyR-3.[211] Finally, in vascular smooth muscle cells, plasmalemnal Ca_v, in addition to its well-known role as voltage-dependent selective calcium channel, acts as a voltage sensor that, in the absence of calcium influx, triggers fast G-protein-dependent calcium release from the sarcoplasmic reticulum and subsequently elicits the contraction of the myocytes.[212]

The $[Ca^{2+}]_i$ is restored by various mechanisms. The calcium can be taken up again in the sarcoplasmic reticulum and/or pumped out of the cells. Calcium extrusion can be achieved by a specific Plasma Membrane Ca^{2+}-ATPase (PMCA) and possibly exchangers such as the Na^+/Ca^{2+} exchanger. The PMCA are encoded by at least four genes, PMCA1 and 4 being the most widely expressed. This pump is not electrogenic since each calcium extruded is exchanged for two protons.[213]

The close spatial relationship of the sarcoplasmic reticulum with the plasma membrane, the direct and indirect activating role of calcium on its own release, and the activation of Ca_V and SOC show that calcium entry and calcium release are interdependent phenomena that concur to achieve calcium homeostasis in vascular smooth muscle. For instance, contractile and relaxing agents modulate the incidence of calcium sparks elicited by clustered RyR, activated by calcium entry through Ca_V. Calcium sparks can act as a positive feedback to augment the contractility of the smooth muscle directly by increasing $[Ca^{2+}]_i$ and indirectly by activating calcium-sensitive chloride channels. Conversely, they also have a relaxing effect by activating BK_{Ca}, which produces spontaneous transient outward currents and hyperpolarization of the smooth muscle cells (Figure 16).[192,214,215]

1.3.2.1.3 Membrane Potential and Tone of Vascular Smooth Muscle

The membrane potential of the vascular smooth muscle cells is determined by potassium channels, two pore domains, K_V, BK_{Ca}, and K_{IR} potassium channels. The contribution of each of those conductances varies widely depending on the species and the blood vessel studied. Additionally, Na^+/K^+-ATPase and possibly $I_{Cl,swell}$ can contribute to the setting of the resting membrane potential. Depolarization of the vascular smooth muscle cells can be achieved by inhibiting (K_V, BK_{Ca}, K_{IR}, Na^+/K^+-ATPase) and/or activating ($I_{Cl,swell}$) these channels and pumps. Furthermore, the activation of calcium-activated chloride channels, nonselective cation channels, and Ca_V contribute to the depolarization (Figure 16).[174,216–218]

A fast-activated tetrodotoxin-sensitive Na^+ channel is present in various vascular smooth muscle cells, including in those of the human coronary artery.[219–221] However, the physiological and/or pathophysiological role of this channel is far from being established. Under physiological conditions, most vascular smooth muscle cells do not appear to express this Na^+ channel, indicating that this conductance is not required for depolarization and/or contraction.

The hyperpolarization of the smooth muscle cells, whatever mechanism involved, is a powerful means to produce relaxation. It decreases not only Ca^{2+} influx by reducing the open probability of Ca_V, but also the release of Ca^{2+} from intracellular stores by decreasing the turnover of intracellular phosphatidylinositol and the Ca_V-dependent activation of the sarcoplasmic reticulum.[212,222,223]

1.3.2.2 Calcium Homeostasis in Endothelial Cells

Endothelial cells in general do not express fast-activated tetrodotoxin-sensitive Na^+ channels or Ca_V and are considered to be "nonexcitable cells." Nevertheless, cytoplasmic $[Ca^{2+}]_i$ is a key regulator of endothelial function, including the synthesis of NO and prostacyclin and EDHF-mediated responses.[224] For instance, acetylcholine increases $[Ca^{2+}]_i$ by activating both an IP_3-dependent calcium release from intracellular stores and a calcium influx from the extracellular space.[225,226] This increase in

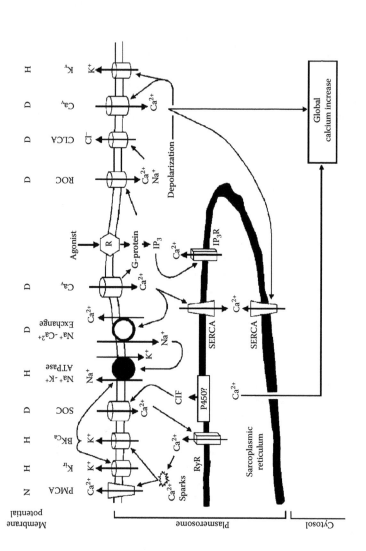

FIGURE 16 Regulation of the intracellular calcium concentration ([Ca²⁺]ᵢ) in vascular smooth muscle cells. The regulation of [Ca²⁺]ᵢ is differently regulated in the cytosol as a whole and in specialized domains constituted by the portion of the cytosol included between the plasma membrane and closely situated sarcoplasmic reticulum (plasmerosomes). The contraction of the smooth cells is induced by the activation of a G-protein-coupled receptor, for instance a Gq-coupled receptor. A global increase in [Ca²⁺]ᵢ is achieved due to the complex summation of calcium entry and calcium release, both phenomena being intensively regulated. Inositol trisphosphate (IP₃), produced by the activation of phospholipase C, stimulates the IP₃ receptor (IP₃R), situated on the membrane of the sarcoplasmic reticulum, and produces calcium release.

$[Ca^{2+}]_i$ is associated with the hyperpolarization of endothelial cells due to the activation of calcium-activated potassium channels.[225,227,228] Agonist-induced hyperpolarization constitutes a positive feedback mechanism for the entry of calcium through receptor-operated channels (ROC) since the electrical driving force for calcium is enhanced. Depletion of the sarcoplasmic reticulum calcium stores with thapsigargin (or cyclopiazonic acid), a specific inhibitor of the calcium pump (SERCA), promotes an increase in $[Ca^{2+}]_i$ via the activation of nonselective cation currents and Ca^{2+}-selective entry pathways, SOC.[229–231] The molecular identity of ROC and SOC in the endothelial cells is still uncertain.

Endothelial cells express six of the seven known TRPC channels. TRPC7 has not been detected so far in these cells.[230,232] The most compelling evidence for the involvement of a TRP channel in calcium influx comes from experiments involving TRPC4(–/–) knock-out mice. In the aortic endothelial cells of these genetically modified animals, the agonist-induced calcium entry is reduced and this is associated with an impairment of agonist-induced endothelium-dependent relaxations.[233] In pulmonary endothelial cells, the changes in $[Ca^{2+}]_i$ elicited by agonists are also decreased in this strain of mice and this is associated with the abolition of actin stress fiber formation. The TRPC4(–/–) knock-out mice show an alteration in lung microvascular permeability demonstrating that TRPC4s are also involved in the regulation of endothelial barrier function.[234] In pulmonary endothelial cells, antisense

FIGURE 16 (Continued) The emptying of the calcium stores elicits the refilling of these stores by the opening of store-operated channels (SOC, most likely constituted of TRP channels) under the control of a calcium influx factor (CIF), possibly 5,6 epoxyeicosatrienoic acids (5,6-EET) produced by a ctytochrome P450 monooxygenase (P450). Calcium entry elicits calcium release (calcium-induced calcium release) by activating the sarcoplasmic reticulum ryanodine receptor (RyR). Additionally, the G-protein activates receptor-operated channels (ROC most likely constituted of TRP channels), nonspecific cation channels, which promote calcium entry and cell depolarization. Calcium activates chloride channels (CLCA), the opening of which also depolarizes the cell. Depolarization activates voltage-dependent calcium channel (Ca_V) leading to further calcium entry and depolarization. Ca_V acts also as voltage sensors capable, in the absence of calcium influx, of triggering G-protein-dependent calcium release from the sarcoplasmic reticulum. However, voltage-activated potassium channels (K_V) and large-conductance calcium-activated potassium channels (BK_{Ca}) are stimulated simultaneously, acting as a brake mechanism. Potassium extrusion activates both the inward rectifying potassium channel (K_{IR}) and the Na^+-K^+-ATPase, which would reinforce the hyperpolarizing effect. Calcium is either pumped out of the cell by a specific plasma membrane Ca^{2+}-ATPase (PMCA) or back into the sarcoplasmic reticulum by another specific pump (SERCA). Additionally, calcium ions can be extruded by the Na^+/Ca^{2+} calcium exchanger, and the resulting increase in intracellular sodium activates the Na^+/K^+-ATPase. Even in the absence of receptor stimulation, "spontaneous" calcium release and/or entry may generate rapid changes in intracellular calcium concentrations in the plasmerosomes (calcium sparks). The subsequent activation of a spontaneous outward current (STOC) linked to the activation of BK_{Ca} or of spontaneous inward current linked to the activation of CLCA will regulate membrane potential and local $[Ca^{2+}]_i$, the $[Ca^{2+}]_i$ concentration in the bulk cytosol remaining unaffected. On the top of the graph the electrogenicity of each membrane protein is indicated: N = neutral; H = hyperpolarization; D = depolarization.

oligonucleotides directed against TRPC1 also produce a significant inhibition of the SOC-dependent current elicited by thapsigargin, supporting the idea that endogenous TRPC1 contributes directly to this current.[235] TRPC1 can also be activated by the phospholipase C pathway, which provides a store-independent pathway. It is uncertain whether TRPC1 and TRPTC4 form homotetramers or are subunits included in the formation of heteromultimers.[230]

TRPV4 is highly expressed in the endothelial cells.[236] It was originally described as a volume-sensitive cation channel but is also ligand-activated, in particular by epoxyeicosatrienoic acids.[237] TRPV4 is a potential constituent of a mechano-transducer.[238] However, whether or not TRPV4 plays such a role in endothelial cells remains unclear.

TRPM channels are also expressed in endothelial cells. This is particularly the case for TRPM4, which is activated by an increase in calcium, is calcium impermeable, and evokes depolarization.[239] A nonselective cation current, which possesses identical features and is modulated by NO, has been characterized electrophysiologically in endothelial cells.[240] The depolarization may act as a negative feedback to reduce the driving force for calcium entry.

The physiological importance of calcium signaling is regulated differently in conduit and microvascular endothelium possibly because, as shown in the pulmonary circulation, they arise from distinct embryological origins.[241] The calcium response elicited by thapsigargin is smaller in pulmonary microvascular endothelial cells than in the endothelium of large pulmonary arteries, a phenomenon attributed to a significant decrease in the SOC-dependent current. However, this is not associated with a decreased expression of TRPC1 or TRPC4, but rather with structural changes in the microdomains that enclose the sarcoplasmic reticulum and the caveolin-containing plasma membrane. These specialized microdomains are absent in microvascular endothelium.[230]

These different possibilities in the regulation of endothelial calcium signaling are likely to have a major impact on the responses of endothelial cells to agonists and physical stimuli. This may explain the differential contribution of the various pathways in endothelium-dependent relaxations. For instance, the contribution of EDHF-mediated responses as a mechanism for endothelium-dependent relaxation increases as the vessel size decreases,[242,243] apart from the coronary and renal vascular beds in which EDHF plays a major role even in conduit arteries.[224,244]

1.3.3 INTEGRATION AND COORDINATION OF RESPONSES IN THE VASCULAR WALL: GAP JUNCTIONAL COMMUNICATION

Integration and coordination of responses among the various cells composing a tissue is essential for the proper function of any given organ, including the blood vessel wall. Cells can communicate by different means, on the one hand by the release of various hormones, mediators, or other substances and on the other by direct electrical and chemical intercellular communications by means of gap junction channels. These channels are the only class of channels that span the closely apposed membranes of two different cells.[245,246]

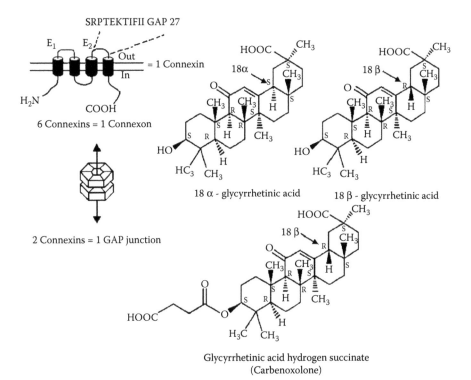

FIGURE 17 Gap junctions and gap junction inhibitors. Representation of connexins assembling to form gap junctions. Gap peptides such as Gap27, are connexin mimetics that possess conserved sequence homology with the second extracellular loop domains of the connexins thought to be involved in gap junction communication in the vascular wall. The glycyrrhetinic acid derivatives are nonpeptidic inhibitors of gap junction communication. Although poorly specific, they are useful tools.

Gap junctions are composed of subunit proteins called connexins. Six connexins in one cell are assembled to form a hemichannel or connexon. Two connexons (12 connexins), one on each cell membrane, are linked to form the functional gap junction (Figure 17). In most tissues, these gap junctions are organized as plaques that are aggregates consisting of a few to over thousands of individual channels. Connexin proteins belong to a highly conserved multigene family with at least 20 identified members in humans and mice.[247] They can be classified according to their molecular mass in kDa.[248] Depending on the vascular bed, smooth muscle cells express connexin (Cx) 37, Cx40, Cx43, and/or Cx 45, while endothelial cells express Cx37, Cx40, and/or Cx43.[245,249–252] In endothelial cells, these various connexins are present in the very same gap-junctional plaque.[253] There are three possible assemblies of these connexins to form a gap junction channel. The homotypic type is the assembly of two identical connexons expressed at the membrane surface of each cell type and is therefore a dodecameric structure of identical connexin subunit proteins. A second type is the heterotypic gap junctional channel consisting of two distinct connexins, six in one hemichannel and six of another type in the other hemichannel. The third type is a heteromeric gap junction where at least one of the two connexons contains more than one connexin. The number of different gap junction channels that theoretically can be expressed is virtually limitless. Some of these heteromeric channels are expressed in native cells and they are likely to have specific characterictics and functions, the extent of which remains to be determined.[254,255]

The docking of two connexons leads to the creation of an aqueous central pore, the gap junction channel, which permits the transfer of ions and polar molecules and provides an electrical continuity allowing a uniform membrane potential among coupled cells. Gap junctions are therefore permeable not only to ions, such as calcium, but also to second messengers such as cyclic-AMP, IP3 and nucleotides (ADP, ATP), and small peptides up to 10 amino acids in length ($M_r < 2$ kDa).[256] Thus, both electrotonic current spread and diffusion of signal molecules via gap junctions are likely to coordinate a functional syncytium such as vascular smooth muscle.[255] There is growing evidence that gap junctions coordinate not only the contraction of vascular smooth muscle but also the relationship between endothelial and smooth muscle cells. From a few cells activated by the synaptic release of a transmitter, the passage of a neurohumoral substance in the flowing blood, or an iontophoretically applied agonist in the vicinity of an endothelial or smooth muscle cell, the diffusion of a message by means of gap junctions allows the synchronized contraction or relaxation of the entire vascular wall.[245,257]

2 Endothelium-Derived Mediators

The term "endothelium-dependent responses" was coined after the seminal work of Robert Furchgott in 1980,[1] although endothelium-derived vasoactive substances had been identified previously (e.g., prostacyclin, adenosine). However, it was only after this princeps publication that the fundamental role of the endothelial cells in controlling the tone of the underlying vascular smooth muscle was perceived fully. Endothelial cells exert this physiological mission by various means that involve different pathways including the metabolism of arachidonic acid, NO-synthase activity, and the so-called EDHF pathway.

2.1 METABOLISM OF ARACHIDONIC ACID

The fatty acid arachidonic acid, the most common precursor of prostaglandins, is released from the cell membrane phospholipids. Two major phospholipases are implicated in prostanoid formation, phospholipase A2 (PLA_2) acting on phosphatidyl-ethanolamine, phosphatidyl-choline, or plasmalogens, as well as phospholipase C, which together with the diacylglycerol lipase acts sequentially on phosphatidyl-inositols derivatives.[2] There are, so far, at least 19 enzymes identified with PLA_2 activity subdivided into four main groups: secreted PLA_2 (ten enzymes identified), cytosolic PLA_2 (three enzymes, $cPLA_2\alpha$, β and γ), calcium-independent PLA_2, and the platelet-activating factor acetylhydrolase (four enzymes).

The calcium-independent PLA_2 is a crucial molecular determinant in the activation of SOC channels and capacitive Ca^{2+} influx in various cell types including vascular smooth muscle cells.[3] $cPLA_2\alpha$ is the only PLA_2 enzyme that shows significant selectivity toward phospholipids containing arachidonic acid. This intracellular enzyme is activated by submicromolar calcium concentrations and is expressed ubiquitously and constitutively in most cells and tissues. $cPLA_2\alpha$ is an essential component of the initiation of the metabolism of arachidonic acid.[4]

Following its release, arachidonic acid can be metabolized by several enzymatic systems including cyclooxygenases, lipoxygenases, and cytochrome P450 monooxygenases.

2.1.1 CYCLOOXYGENASES

PGH-synthase, the first enzyme involved in the biosynthetic pathway of prostanoid formation, possesses both a cyclooxygenase catalytic activity leading to the formation of prostaglandin G_2 (PGG_2) and a peroxidase activity catalyzing the reduction of PGG_2 to prostaglandin H_2 (endoperoxide, PGH_2). Although this single protein molecule is associated with both cyclooxygenase and peroxidase activities, PGH-synthase is usually termed cyclooxygenase.[5]

Cyclooxygenase is an integral membrane protein found in microsomal membranes. It was first purified in 1976, cloned by three separate groups in 1988, and is known as cyclooxygenase-1 or COX-1.[6-9] However, in 1991, several laboratories identified the product of a second gene with cyclooxygenase and peroxidase activities, which was named cyclooxygenase-2 or PGH-synthase-2 (COX-2).[10,11] COX-2 can be induced by several stimuli associated with cell activation and inflammation.[5] In endothelial cells, COX-1 is expressed constitutively, but can also be induced, for instance, by shear stress.[12] In these cells COX-2 expression is induced by various cytokines and an increase in cyclic-AMP.[13] Both endothelial and vascular smooth muscle cells contain cyclooxygenase; however, endothelial cells contain twenty times more of the enzyme than smooth muscle cells.[14]

Various biologically active eicosanoids are formed from the ephemeral, but also biologically active, PGH_2 through the action of a set of synthases, namely PGD, PGE, PGF, PGI, and thromboxane synthases (Figure 18). Due to their chemical and metabolic instability, prostanoids are believed to work as autocrine or paracrine mediators. Furthermore, activation of cyclooxygenases is a source of superoxide anions because of their ability to co-oxidize substances such as NADPH.[15]

Isoprostanes are prostaglandin isomers that are produced from the oxidative modification of polyunsaturated fatty acids via a free radical-catalyzed mechanism.[16] In addition to the nonenzymatic generation of isoprostanes, a cyclooxygenase-dependent production of 8-isoprostanes (8-*iso*-prostaglandin F2α) can occur in endothelial cells.[17] 8-Isoprostane can be a direct product of cyclooxygenase or an indirect consequence of superoxide anion production by cyclooxygenase-mediated metabolism and the subsequent oxidation of membrane lipids. Isoprostanes are not only markers of oxidative stress,[18] but also bioactive molecules that can produce either vasoconstriction or relaxation (Figure 19).[19,20]

2.1.1.1   PGD-Synthases

Two distinct types of PGD-synthases have been identified, the lipophilic ligand-carrier protein- (lipocalin) type enzyme and the hematopoietic enzyme. PGD_2 is a major prostaglandin in the central nervous system and in immune cells. It can also be synthesized in the vascular wall, as both types of PGD-synthases are expressed in endothelial cells where they are up-regulated in response to an increase in fluid shear stress.[21] However, PGD_2 is mainly involved in the regulation of sleep and in allergic responses. PGD_2 is dehydrated to produce the J series of prostaglandin (PGJ_2 analogues). PGJ_2 acts as a ligand for the PPARγ nuclear receptor, which is involved in the differentiation of adipocytes.[22] PPARγ is also expressed in endothelial and smooth muscle cells.[23,24] The thiazolidinediones, anti-diabetic drugs acting as PPARγ agonists, prevent proliferation and migration of smooth muscle and inhibit the interaction between leukocytes and endothelial cells.[25]

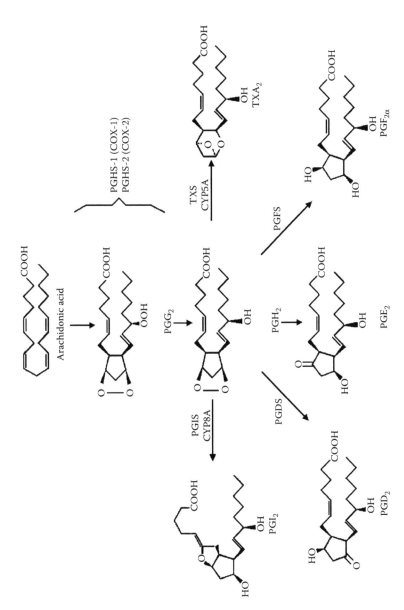

FIGURE 18 Arachidonic acid metabolism: the cyclooxygenase pathways. COX: cyclooxygenase; PG: prostaglandin; PGDS: prostaglandin D synthase; PGFS: prostaglandin F synthase; PGHS: prostaglandin synthase; PGIS: prostacyclin synthase; CYP: cytochrome P450 monooxygenase.

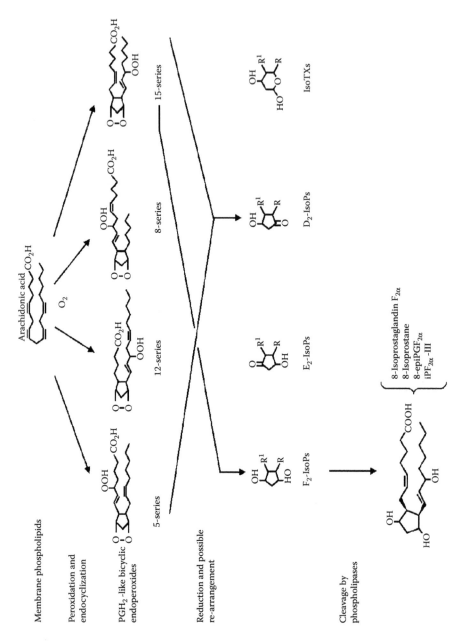

FIGURE 19 Nonenzymatic formation of isoprostane from membrane bound arachidonic acid. The molecular structure and origin of 8-isoprostane, a vasoactive isoprostane, is shown as an example. The synonyms of 8-isoprostane derived from various nomenclatures are also mentioned. Isop: isoprostaglandin; IsoTX: isothromboxane.

2.1.1.2 PGE-Synthases

Cytosolic and membrane bound forms of PGE-synthases have been identified and both require glutathione as a co-factor. The cytosolic form is a constitutive enzyme expressed ubiquitously and in abundance in many tissues and cell types. This cytosolic form is strictly associated with COX-1 suggesting that this isoform contributes physiologically to the production of PGE_2.[26] The membrane-bound form of PGE-synthase shows significant homology with the superfamily of membrane-associated proteins involved in eicosanoid and glutathione metabolism (MAPEG), which includes 5-lipoxygenase-activating protein (FLAP) and leukotriene C4 synthase.[27] This isoform is up-regulated in response to stimuli that induce COX-2 expression and is preferentially associated with COX-2 in order to produce PGE_2. In the vascular wall, both the endothelium and the vascular smooth muscle can express this isoform. [26]

The two PGE_2 synthetic pathways, the COX-1-cytosolic PGE synthase and the COX-2-membrane-bound PGE synthase are linked to various physiological and pathological events. PGE_2, depending on the concentration, the vascular bed, and the experimental conditions, can produce both contraction and relaxation of vascular smooth muscle.[28–30]

2.1.1.3 PGF-Synthases

Prostaglandin F_2 isoforms are synthesized from PGH_2 by the membrane-associated 9,11-endoperoxide reductase (a glutathione-dependent enzyme), but also from PGD_2 and PGE_2 by cytosolic PGD_2 11-ketoreductase and PGE_2 9-ketoreductase (or 20α-hydroxysteroid deshydrogenase), respectively. Prostaglandin $F_{2\alpha}$ is produced in the vascular wall and is a potent vasoconstrictor.[31]

2.1.1.4 PGI-Synthase

Prostacyclin synthase belongs to the cytochrome P450 superfamily (in the human CYP8A1). In most blood vessels, prostacyclin is the principal metabolite of arachidonic acid, with the endothelial cells being the predominant site of its synthesis.[6,32] Prostacyclin is generally described as an endothelium-derived vasodilator that involves the stimulation of specific cell surface receptors (IP receptors) and activation of adenylyl cyclase leading to an elevation of intracellular cyclic-AMP.[33,34] Prostacyclin is also a potent inhibitor of platelet adhesion to the endothelial cell surface, and of platelet aggregation.[6,35–37]

2.1.1.5 Thromboxane Synthase

Thromboxane synthase, first described in platelets, also belongs to the cytochrome P450 superfamily (in the human, CYP5). It catalyzes two distinct reactions with PGH_2 as a substrate, on the one hand the formation of thromboxane A2, and on the other the formation of 12-hydroxyheptadecatrienoic acid plus malondialdehyde. The human thromboxane synthase is a single-copy gene and is widely expressed in mammalian tissues, including vascular smooth muscle.[38–40] The presence of low levels of thromboxane synthase activity has also been demonstrated in purified endothelial cells,[38] but in most cases the production of thromboxane A2 by the endothelial cells themselves is likely to be negligible. However, in the presence of platelets, arachidonic acid released from stimulated endothelial cells can be metabolized into thromboxane A2 by the platelet thromboxane A2 synthase.[41] Thromboxane A_2 is a potent vasoconstrictor and a powerful inducer of platelet aggregation.

2.1.1.6   Prostaglandins and Thromboxane Receptors

Prostaglandins interact with seven specific transmembrane, G-protein coupled receptors that are classified in five subtypes, DP, EP, FP, IP, and TP, as a function of their sensitivity to the five primary prostanoids, prostaglandins D_2, E_2, $F_{2\alpha}$, I_2, and thromboxane A_2, respectively. There are four subtypes of the EP receptor (EP_1 to EP_4) with identified splice variants for the EP1 and EP3 receptor subtypes. No subtypes have been identified for the other prostanoid receptors, but two splice variants have been described for the FP and TP receptors.[42] The two splice isoforms of the TP receptors (TPα and TPβ) can produce different functional responses. Both isoforms are coupled to phospholipase C activation but TPα activates adenylyl cyclase, whereas TPβ inhibits it.[43] Activation of IP receptors stimulates adenylyl cyclase but can also produce inositol-trisphospate and, thus, increases $[Ca^{2+}]_i$.[44]

Depending on the vascular bed, the gender, and the species, each of these receptors can be expressed in the vasculature. Schematically, the stimulation of the DP, EP1, EP2, EP4, and IP receptors produces vasodilatation, while the activation of the EP3, FP, and TP receptors induces vasoconstriction. The eight prostanoid receptors have been individually knocked out in mice and the observation of the various phenotypes will allow a better understanding of the contribution of each of these receptors to the regulation of vascular tone.[45]

The prostanoid receptor family may encompass more than these eight characterized receptors. PGD_2 has recently been identified as the cognate ligand for an orphan seven transmembrane receptor, CRTH2, which does not belong to the prostanoid but to the chemokine receptor family.[46]

2.1.2   Lipoxygenases

Lipoxygenases are a class of nonheme iron dioxygenases that catalyze the hydroperoxidation of polyunsaturated fatty acids. Mammalian lipoxygenases insert oxygen into the 5-, 8- (mice and rat only), 12-, and 15-positions of arachidonic acid with various stereo configurations (S or R). In humans, six functional lipoxygenase genes and three pseudogenes have been identified: 5-lipoxygenase, platelet type 12-lipoxygenase, 12(R)-lipoxygenase, 15-lipoxygenase type I, 15-lipoxygenase type II, and epidermis-type lipoxygenase-3 (Figure 20).[47]

12R-Lipoxygenase forms 12R-hydroperoxy-eicosatetraenoic acid (12R-HPETE) from arachidonic acid with a high specificity. In the human, the pattern of expression of this enzyme seems very restricted and appears limited to the skin and the tonsils. The catalytic activity of epidermis-type lipoxygenase-3 is not properly defined at present.[48]

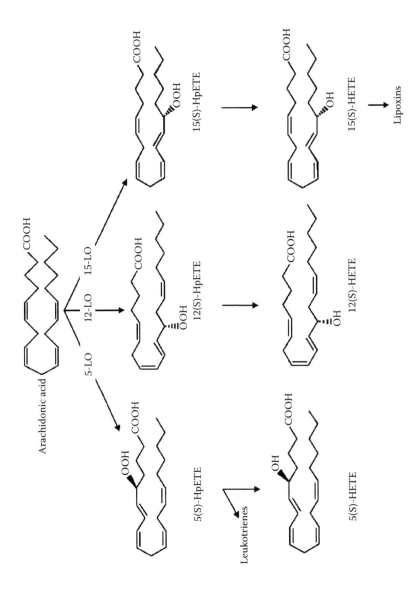

FIGURE 20 Arachidonic acid metabolism: the lipoxygenase pathways. LO: lipoxygenase; HpETE: hydroperoxy-eicosatetraenoic acid; HETE: hydroxyeicosatetraenoic acid.

5-Lipoxygenase converts arachidonic acid to 5(S)-hydroperoxy-6-*trans*-8,11,14-*cis*-eicosatetraenoic acid (5-HPETE) and then 5-HPETE to leukotriene A4 (LTA$_4$). 5-Lipoxygenase is a calcium-sensitive enzyme that needs to be translocated to the nuclear membrane and, to a lesser extent, to other organelles, in order to show leukotriene synthesis activity. A 5-lipoxygenase-activating protein (FLAP), once thought to be required for the anchoring of the 5-lipoxygenase to the cell membrane, is essential for the activation of the enzyme.[49] Cysteinyl leukotrienes have a plethora of vascular (changes in permeability for instance) and nonvascular effects. 5-Lipoxygenase knock-out mice and rats treated with a FLAP inhibitor develop less right ventricular hypertrophy and less pulmonary hypertension under chronic hypoxic conditions than wild-type animals. This suggests that in rodents the enzyme contributes to the generation of chronic pulmonary hypertension.[47,50]

12S-lipoxygenase introduces molecular oxygen into arachidonic acid to form 12S-hydroperoxy-5,8,10,14-eicosatetraenoic acid (12S-HPETE). There are three isoforms of 12S-lipoxygenases named after the cells in which they were identified first: platelet, leukocyte, and epidermis.[51] In the human, only the platelet type 12S-lipoxygenase is expressed. The production of 12S-HPETE and the expression of platelet-type 12-lipoxygenase are increased in hypertension.[52] There is a marked up-regulation of 12-lipoxygenase mRNA in the wall of arteries injured by a balloon catheter, which affects mainly smooth muscle and infiltrating inflammatory cells.[53] Leukocyte-type 12S-lipoxygenase (rat, mouse, pig) shows higher identity with the human and rabbit 15-lipoxygenase-type I than with the other 12-lipoxygenase enzymes, and therefore they are called 12/15-lipoxygenases.[51]

15-Lipoxygenases are lipid-peroxidizing enzymes that catalyze the stereo-selective introduction of molecular oxygen at carbon 15 of arachidonic acid to form 15S hydroperoxy-5,8,11,13-eicosatetraenoic acid (15S-HPETE). Subsequently, 15S-HPETE becomes the substrate for the synthesis or more complex bioactive lipids such as lipoxins. Two types of 15-lipoxygenases have been cloned in the human. Type I, or reticulocyte-type, 15-lipoxygenase, as mentioned above, shares a highly phylogenic relatedness with the rat, murine, and porcine leukocyte-type 12-lipoxygenase. Type II, or epidermis-type 15-lipoxygenase, is related to the murine 8-lipoxygenase, which only shares a low degree of homology with type I.[54] Type I is expressed ubiquitously, while the pattern of expression of the type II appears to be more restricted (skin, prostate, lung, and cornea). Type I has a broad substrate specificity, which includes linoleic acid with the introduction of oxygen at carbon 13 giving rise to 13-HODE (13-hydroperoxyoctadecadienoic acid), an inhibitor of platelet adhesion to the endothelial cell surface.[55] Liberation of arachidonic acid from the phospholipid stores is not an absolute prerequisite for the enzymatic activity of type I 15-lipoxygenase. This isoform may oxygenate polyenoic fatty acid esterified in the membrane and subsequently hydro(pero)xy derivatives can be released by the phospholipases.[56,57] NO may act as a reversible inhibitor of 15-lipoxygenase and can be effectively consumed during 15-lipoxygenase-catalyzed oxidation of polyenoic fatty acids, suggesting that this enzyme can be considered as a regulator of cellular NO metabolism.[58,59] The biological and pathophysiological roles of 15-lipoxygenases are multiple and include cell differentiation and maturation, inflammation, hyperreactivity of smooth muscle (e.g., asthma), carcinogenesis, apoptosis, and atherosclerosis.[54] Mouse knock-out for the leukocyte-type 12-lipoxygenase, the enzyme that is most closely related to human platelet type 15-lipoxygenase, when crossed with other strains susceptible to develop atherosclerosis (apolipoprotein E deficient, low-density lipoprotein deficient) shows a 50% reduction in aortic atherosclerotic lesions.[47]

2.1.3 CYTOCHROME P450 MONOOXYGENASES

Cytochrome P450 monooxygenases are expressed ubiquitously in bacteria, plants, and animals. More than 2500 cytochrome P450 genes have been identified so far and they are subdivided into 78 families. In the human, 57 cytochrome P450 genes and 58 pseudogenes have been sequenced. They are classified into 18 families that are subdivided into 43 subfamilies. Cytochrome P450s are heme-containing membrane-bound enzymes that catalyze NADPH-dependent oxidation. Most of these enzymes are expressed in the liver, but both liver and extrahepatic enzymes are involved in the biosynthesis and/or metabolism of endogenous substrates such as cholesterol, steroids, bile acids, vitamins, and fatty acids, including arachidonic acid. They catalyze the breakdown of xenobiotics (drugs, toxic chemicals, carcinogens). Some are constitutively expressed in an organ-, gender-, age-, and species-dependent manner, while others are regulated by hormones, cytokines, food intake, disease (e.g., diabetes, hypertension), and intake of xenobiotics.[60,61]

Mammalian cytochrome P450 monooxygenases do not metabolize membrane-bound arachidonic acid and are therefore under the control of phospholipases.[62] They oxidize arachidonic acid by the following reactions: (1) bis-allylic oxidation (lipoxygenase-like reaction) to generate any of the six regio-isomeric hydroxyeicosatetraenoic acids (5-, 8-, 9-, 11-, 12-, and 15-HETE); (2) hydroxylation at or near the terminal sp^3 carbon (AA ω/ω-1 hydroxylase) yielding 16-, 17-, 18-, 19-, or 20-HETE; and (3) olefin epoxidation (AA epoxygenase) producing four epoxyeicosatrienoic acids (5,6-, 8,9-, 11,12-, 14,15-EET).[60,61] Cytochrome P450 monooxygenases can catalyze either one or several type of reactions. For instance, CYP 4A1 catalyzes only the ω/ω-1 hydroxylation, while CYP 4A2 and CYP4A3 catalyze both ω/ω-1 hydroxylation and epoxygenation (Figure 21).[61]

2.1.3.1 Bis-Allylic Oxidation (Lipoxygenase-Like Reaction)

The products of bis-allylic oxidation are structurally similar to those of lipoxygenases, with no evidence that hydroperoxide intermediates are formed. These reactions are catalyzed by cytochrome P450 of the 1A, 3A, 2C, 4F families.[61] 12R-HETE is the predominant enantiomer generated by cytochrome P450 catalyzed reactions and is a powerful and enantioselective inhibitor of Na^+/K^+-ATPase.[60,63]

2.1.3.2 Arachidonic Acid ω/ω-1 Hydroxylase

The ω/ω-1 hydroxylation of arachidonic acid is catalyzed mainly by the cytochrome P450s of the 4A, 4B, and 4F families. Cytochrome 4A is expressed in blood vessels mostly in the smooth muscle cells and is responsible for the synthesis of 20-HETE.[61,64] 20-HETE blocks large conductance calcium-activated potassium channels

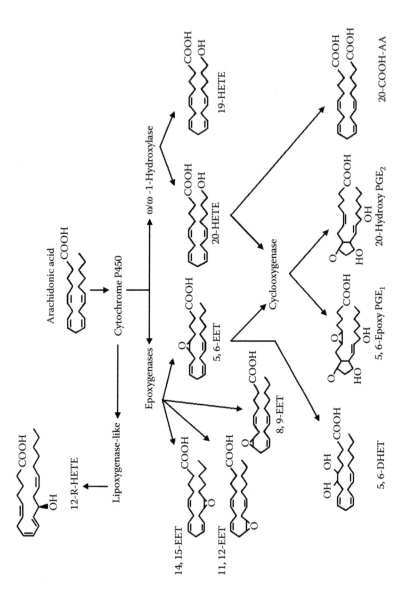

FIGURE 21 Arachidonic acid metabolism: the cytochrome P450 pathways. AA: arachidonic acid; HETE: hydroxyeicosatetraenoic acid; EET: epoxyeicosatrienoic acid; DHET: dihydroxyeicosatetraenoic acid. EET and HETE are themselves substrates for epoxide hydrolases (DHET) and or cyclooxygenases.

(BK$_{Ca}$) and is a potent vasoconstrictor of small arteries.[65,66] The endogenous levels of 20-HETE in some vascular beds such as in the cerebral and renal circulations are elevated and are responsible for an elevated intrinsic vascular tone. In small porcine coronary arteries, the cytochrome P450 4A-dependent production of 20-HETE functionally antagonizes EDHF-mediated relaxations by a protein kinase

Cα-dependent mechanism involving the inhibition of the Na$^+$/K$^+$-ATPase.[67] In these vascular beds, the inhibition of cytochrome P450 activity produces vasodilatation.[64,65,68] 20-HETE plays a significant role in the myogenic response of various arterioles in response to elevated transmural pressure and in the autoregulation of renal and cerebral blood flow. Additionally, 20-HETE may act as an oxygen sensor in the microcirculation.[61] The structural analogues of 20-HETE, 5-, 15-, and 19-HETE, antagonize its vasoconstrictor action and may serve as endogenous inhibitors.[69]

2.1.3.3 Olefin Epoxidation (Arachidonic Acid Epoxygenase)

The formation of EETs by the epoxygenase reaction is catalyzed by numerous cytochrome P450s, namely the 1A, 1B, 2B, 2C, 2D, 2E, 2J, 3A, and 4A families. Human endothelial cells express the 2C8, 2C9, 2J2, 3A, and 2B1 families that all produce EETs. Only a limited number of observations suggest that vascular smooth muscle cells can produce EETs.[61] However, red blood cells store and release EETs, suggesting a vasoregulatory role for these circulating cells.[70] EETs are potent vasodilators. They increase the open probability of BK$_{Ca}$ in vascular smooth muscle cells and produce hyperpolarization and a subsequent relaxation.[71–73] The mechanism by which EETs activate BK$_{Ca}$ in vascular smooth muscle cells is not clearly understood. They do not directly activate these channels as has been previously shown with other fatty acids.[74] EETs may interact with an intracellular "receptor" to activate ADP-ribosyltransferase. This enzyme catalyzes the ADP-ribosylation of G$_s\alpha$, which then promotes the phosphorylation and the activation of BK$_{Ca}$.[75,76] Furthermore, EETs increase [Ca^{2+}]$_i$ in endothelial cells,[77] which could evoke or facilitate the release of NO and/or prostacyclin. EETs are converted by epoxide hydrolases into dihydroxyeicosatetraenoic acids (DiHETE). DiHETEs are in general less potent vasodilators than the corresponding EETs[78] with the exception of the coronary circulation where EETS and DiHETE are equipotent.[79]

Additionally, the metabolism of arachidonic acid by cytochrome P450 is a significant source of oxygen-derived reactive species and therefore can contribute to oxidative stress. For instance, in endothelial cells of the porcine coronary artery, CYP2C9, which produces vasodilator EETs, paradoxically also produces superoxide anions.[80]

2.2 NITRIC OXIDE SYNTHASES

2.2.1 Historical Notes

Furchgott and Zawadzki[1] first showed that the presence of the endothelial lining is required to observe the relaxation of isolated strips of rabbit aorta in response to acetylcholine. These authors attributed the relaxation to the calcium-dependent release of an endothelial factor of unknown origin and termed it endothelium-derived relaxing factor abbreviated as EDRF (Figure 2). The main target of EDRF in the smooth muscle cells is the soluble guanylyl cyclase.[81,82] The mechanism of EDRF-induced relaxation is similar to that of nitrovasodilators[83] and its half-life in physiological solution is short, in the order of a few seconds.[84,85] EDRF is inactivated by superoxide anion[86] and scavenged by oxyhemoglobin.[87] These observations prompted Furchgott and Ignarro to propose independently, at the meeting "Mechanism of Vasodilatation" held in Rochester (MN) in July 1986, seven years after Furchgott's seminal observation, that EDRF and NO are one and the same, or very similar molecules.[88,89] Subsequently, Moncada and colleagues demonstrated that the amount of NO released by the endothelial cells could account for the biological activity of EDRF.[90] Finally, the same group identified the enzymatic origin of NO production in the endothelial cells: the L-arginine-NO-synthase pathway.[91] Diverse analogues of L-arginine have been identified as fairly specific inhibitors of NO-synthase.[92] Some of these guanido-substituted L-arginine analogues, NG monomethyl-L-arginine, asymmetric dimethylarginine, and symmetric dimethylarginine, are naturally occurring and have the potential to affect L-arginine handling and/or NO synthesis in various biological systems (Figure 22).[93]

2.2.2 CHARACTERIZATION OF THE THREE NITRIC OXIDE SYNTHASES

NO derived from L-arginine has been involved in the responses of macrophages to inflammatory stimuli such as bacterial lipopolysaccharides and interferon-γ, and in their ability to kill tumor cells or bacteria.[94–98] A NO-forming enzyme is present in the brain and the L-arginine-NO-cyclic-GMP pathway is activated by neurotransmitters in various neurons.[99,100] These different observations prompted the conclusion that distinct types of NO production were associated with three different subtypes of NO biosynthetic enzymes. These were cloned, characterized, and termed according to the chronology of their characterization: NO-synthase I (or neuronal, nNOS, NOS-1),[101,102] NO-synthase II (or inducible, iNOS, NOS-2),[103,104] and NO-synthase III (or endothelial, eNOS, NOS-3).[105,106] In fact, the expression of the various NOS isoforms is much more diverse than their name would suggest. For instance, in the vascular wall the three NOS can be present. NOS-1 is expressed not only in perivascular nerves but can also be detected in endothelial and smooth muscle cells.[107–109] The expression of NOS-2 has been documented in all nucleated cells in the cardiovascular system. In the blood vessel wall this includes endothelial and smooth muscle cells as well as fibroblasts, leukocytes, and mast cells.[110] Actually, NOS-3 is expressed not only in endothelial cells but also in cardiac myocytes and platelets.[111,112]

It is generally assumed that the two constitutive isoforms (NOS-1 and NOS-3) are stimulated by an increase in intracellular calcium concentration, while NOS-2 is calcium-independent.[113,114] However, in response to stimuli such as shear stress, pulsatile stretch, and isometric contraction, the NO production by NOS-3 is virtually calcium-independent. The increase in enzyme activity evoked by these mechanical stimuli involves protein kinase Akt/PKB-dependent phosphorylation of serine and threonin residues located in the reductase and calmodulin-binding domain of NOS-3, respectively. The phosphorylation of NOS-3 alters the sensitivity of the enzyme to calcium, rendering its activity maximal at subphysiological concentrations of the ion.[115–118] Endothelial nitric-oxide synthase (NOS-3) is regulated by signalling pathways involving multiple sites of phosphorylation.[119] Various endogenous mediators,

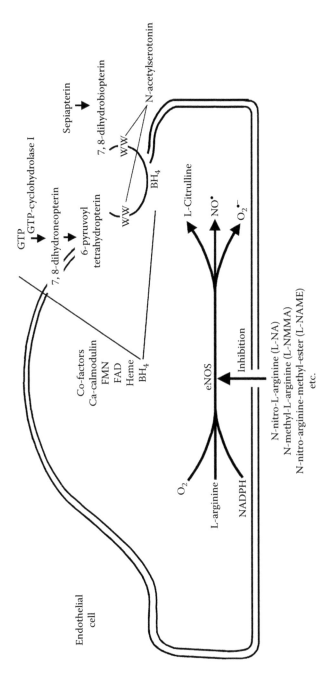

FIGURE 22 The L-arginine-NOS-3 pathway. The synthetic pathway of tetrahydrobiopterin (BH$_4$), an essential cofactor, is also shown. WW: inhibitory effect of N-acetyl serotonin

including cytokines such as vascular endothelial growth factor (VEGF), and exogenous substances, such as flavonoids and polyphenols, elicit NO release by this mechanism.[120–122]

The three NOS enzymes are homodimers containing a reductase and an oxygenase domain. In the presence of molecular oxygen, they catalyze a five-electronoxidation of one of the guanido nitrogens of L-arginine to form NO with the stoichiometric formation of L-citrulline. They all require the cofactors NADPH, FAD, FMN, heme, and tetrahydrobiopterin (BH_4) in order to catalyze this reaction.[123] The conversion of L-arginine to NO by NOS-3 involves two independent steps. The first one is the NADPH-dependent hydroxylation of L-arginine to N^G-hydroxy-arginine, a reaction requiring calcium-calmodulin, which is accelerated by the presence of BH_4. This hydroxylation step resembles cytochrome P450 monoxygenation. The second step is also calcium-calmodulin-dependent, requires O_2 and NADPH, and is also accelerated by BH_4.[124] An essential feature of NOS, despite the ability of the reductase or the oxygenase domain to function independently, is that the NO synthase activity requires dimerization. The heme plays an essential role in the formation of the homodimer and BH_4 stabilizes the dimer once formed.[125] Thus, this pteridine cofactor critically controls NOS-3 activity and decreased endothelial levels of BH_4 are responsible for a reduction in NO production. Activation of NOS-3, in the absence of BH_4 but in the presence of calcium/calmodulin, stimulates production of superoxide anions.[126,127] When L-arginine is deficient, NOS-3 can generate both superoxide anions and nitric oxide leading to the detrimental production of peroxynitrite.[128]

2.2.3 NITRIC OXIDE SYNTHASES AND CARDIOVASCULAR FUNCTION

Although NO has long been known to be synthesized by bacteria, it was completely unexpected that this potentially noxious and reactive free radical would play such a vital role in mammals. Indeed, endothelium-derived NO is not only a potent vasodilator but also a powerful inhibitor of platelet adhesion and aggregation. In fact, NO and prostacyclin show a remarkable synergistic interaction in inhibiting platelet aggregation.[35–37,129]

The evaluation of the role of each of the NOS isoforms in pathophysiology had to await the availability of genetically engineered nitric oxide synthase knock-out mice [(NOS(-/-)], since conventional pharmacological inhibitors are poorly isoform-specific.[130] NOS-3 is clearly the preponderant isoform in the control of vascular tone, since NOS-3(-/-) mice have elevated systemic and pulmonary arterial pressures and since large arteries (but not the more peripheral blood vessels) of these animals no longer show endothelium-dependent relaxations in response to acetylcholine.[131–133] By contrast, mice overexpressing NOS-3 are hypotensive.[134] Blood vessels are under the control of excitatory and inhibitory nerves, and NO derived from NOS-1, expressed in nonadrenergic noncholinergic nerves, is an inhibitory neurotransmitter.[135–137] NO from both NOS-3 and NOS-1 provides parallel pathways for vasodilatation and, under some circumstances, one pathway can compensate for the dysfunction of the other. NO derived from NOS-2 also plays a role in vasodilatation, especially in pathological states. For instance, NOS-2 is responsible for the severe systemic hypotension observed in toxic shock as NOS-2(-/-) mice are resistant to sepsis-induced hypotension.[138]

In diseases such as hypertension, hypercholesterolemia, atherosclerosis, heart failure, and diabetes, endothelium-dependent relaxations are reduced. In most of these cases, this is due to a reduced bioavailability of NO. This does not necessarily mean a decrease in NO production, but a loss of its availability because of oxidative inactivation by excessive superoxide anion generation in the vascular wall. Different enzymes can produce superoxide anions including cytochrome P450 monooxygenase,[80] xanthine oxidase,[139] NOS-3,[140] and NADPH oxidase.[141] The latter is likely to be the most significant contributor to oxidative stress in the blood vessel wall. However, each enzyme can play an important role in the generation of oxidative stress depending on the circumstances. For instance, supplementation with the substrate of NOS-3, L-arginine,[142,143] or with BH_4 itself (or its precursors such as sepiapterin[144]) can restore endothelium-dependent relaxations in various models of cardiovascular diseases, although the mechanisms by which the supplementation, with either L-arginine or pteridins derivatives, exert their beneficial effects are unclear.

2.3 THE OTHER ENDOTHELIUM-DERIVED VASOACTIVE FACTORS

Besides prostacyclin and NO, endothelial cells release other vasoactive factors. An additional relaxing pathway(s), associated with the hyperpolarization of the underlying smooth muscle, is observed in the presence of inhibitors of cyclooxygenase and NO synthases. This was originally attributed to the release of an endothelium-derived hyperpolarizing factor (EDHF), which was supposed to diffuse to and activate K^+ channels on the vascular smooth muscle. In fact, the activation of cell membrane receptor in endothelial cells by certain neurohumoral substances opens endothelial cell K^+ channels. Several mechanisms have been proposed to link this pivotal step to the subsequent hyperpolarization of the vascular smooth muscle.[145] The main concepts are considered in detail in the next parts of this monograph.

In addition, endothelial cells also can synthesize and release contracting factors (EDCF) of diverse nature: peptides, metabolites of arachidonic acid, and reactive oxygen species.[146,147]

3 Endothelium-Dependent Hyperpolarizations

A lot of confusion has arisen in the field of endothelium-dependent hyperpolarizations due to the use of the acronym "EDHF," which implies that a single diffusible substance underlies this novel mechanism of relaxation of the vascular wall. In fact, numerous endothelium-derived factors can produce such hyperpolarization of the underlying smooth muscle cells.

3.1 THE THIRD PATHWAY: HISTORICAL NOTES

Very early in the saga of endothelium-dependent relaxations, pharmacological studies using various inhibitors of the metabolism of arachidonic acid suggested that at least three different pathways were involved.[1] At that time only one of these was clearly identified, the cyclooxygenase-dependent production of prostacyclin,[2] but the others were unknown.

De Mey et al.[1] first proposed the existence of a third pathway because, in canine femoral arteries, the endothelium-dependent relaxations to acetylcholine obtained in the presence of indomethacin (excluding the involvement of prostacyclin) were abolished by mepacrine, whereas the responses to thrombin and ATP were unaffected (Figure 23). Similarly, Rubanyi et al.[3] and Rubanyi and Vanhoutte[4] showed under bioassay conditions that the biphasic endothelium-dependent relaxation to acetylcholine exhibited different susceptibility to antioxidants or to inhibitors of the arachidonic acid cascade, suggesting the involvement of different relaxing factors.

The Na^+/K^+-ATPase blocker, ouabain, became a precious tool for discriminating between the different non-prostacyclin-mediated relaxations. High concentrations of ouabain inhibit the acetylcholine-induced relaxation of canine femoral arteries[5] and rat aortas.[6] In canine coronary arteries, the endothelium-dependent, indomethacin-insensitive relaxation to arachidonic acid was also blocked by ouabain, while that to acetylcholine, ADP, or thrombin was not.[7] In a superfusion bioassay system,[3] rings of canine coronary artery (without endothelium, bioassay ring) were superfused with a perfusion solution flowing out of an isolated segment of canine coronary artery with endothelium (donor artery). Relaxations caused by the basal release of EDRF, or to that elicited by acetylcholine, were reduced by incubation of the bioassay ring with ouabain. However, the relaxations elicited by bradykinin were not affected by the cardiac glycoside.[8] In a modified version of this bioassay system, where the donor coronary artery was replaced by cultured endothelial cells grown on microcarrier beads, the incubation of the endothelial cells with ouabain did not affect the bioassay ring under basal conditions or upon stimulation with ADP, but prevented the relaxation induced by bradykinin or the calcium ionophore A23187.

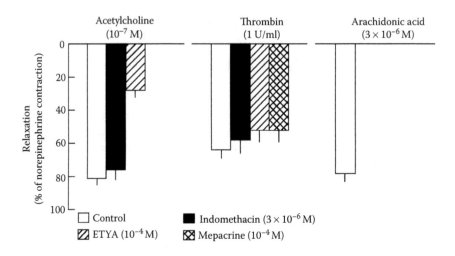

FIGURE 23 Endothelium-dependent relaxations in canine femoral arteries. Acetylcholine (10^{-7} M)-induced relaxations are unaffected by indomethacin (3×10^{-6} M), partially inhibited by 5,8,11,14-eicosatetrienoic acid (ETYA: 10^{-4} M), and abolished by mepacrine, while that to thrombin (1U/ml) is unaffected by the three inhibitors, and that to arachidonic acid (3×10^{-6} M) is blocked by each inhibitor. These results indicate that, in the canine femoral artery, the endothelium can release at least three different relaxing substances.

By contrast, treatment of the bioassay ring with ouabain reduced the relaxation to the basal release of EDRF and that stimulated with ADP without affecting that elicited by bradykinin or the calcium ionophore.[9]

Taken in conjunction, these results indicate that endothelial cells, after inhibition of cyclooxygenase, release more than one relaxing factor. Ouabain and the inhibitors of the arachidonic acid metabolic cascade, at the concentrations used in these various studies, were poorly selective tools and the effects observed were not necessarily related to Na^+/K^+-ATPase blockade or inhibition of the metabolism of arachidonic acid. Nevertheless, they were useful, albeit imperfect, tools, and ouabain is still used to explore EDHF-mediated responses in the human.[10,11]

Independently, in the late 1970s, Kuriyama and Suzuki[12,13] and Kitamura and Kuriyama[14] had observed that acetylcholine can produce a contraction in guinea pig and rabbit coronary and mesenteric arteries with, paradoxically, a simultaneous hyperpolarization of the vascular smooth muscle cells. Later, in the guinea pig mesenteric artery, Bolton et al.[15] demonstrated that acetylcholine not only causes direct contraction of the smooth muscle cells but also a concomitant endothelium-dependent relaxation that was accompanied by hyperpolarization of the cell membrane of vascular smooth muscle. This initial observation has been confirmed in various blood vessels from different species (Figure 24)[16–20] including humans.[21,22] When L-arginine analogues became available as specific inhibitors of the production of NO,[23,24] it became obvious that endothelium-dependent relaxations and endothelium-dependent hyperpolarizations were more or less resistant to inhibitors of cyclooxygenase and NO synthase (Figure 25).[25–34] Furthermore, endothelium-dependent

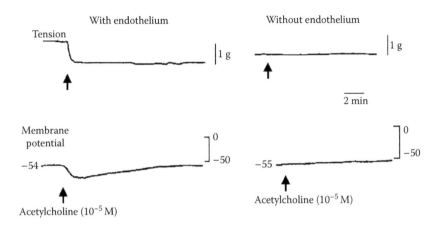

FIGURE 24 Changes in cell membrane potential and tension in isolated strips of canine coronary artery. Acetylcholine produces a relaxation and a hyperpolarization of the vascular smooth muscle only when the endothelial cells are present. (By permission from the Nature Publishing Group; modified from Félétou and Vanhoutte.[17])

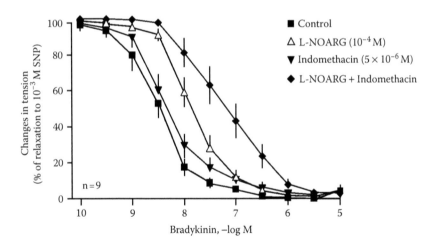

FIGURE 25 Relaxations to bradykinin in isolated canine coronary arteries with endothelium. Bradykinin produces a concentration-dependent and endothelium-dependent relaxation in canine coronary arterial rings (expressed in percentage of the maximal relaxation evoked by sodium nitroprusside, SNP; data in rings without endothelium not shown). The cyclooxyge- nase inhibitor, indomethacin, does not affect the relaxation to bradykinin, while the NO- synthase inhibitor L-nitro-arginine (L-NOARG) produces a significant shift to the right of the concentration-response curves, indicating a contribution of NO in the overall relaxation mechanism. The combination of the two inhibitors produces a further shift. However, even in the presence of the two inhibitors, bradykinin still evokes the complete relaxation of the isolated arteries. This relaxation resistant to the presence of the two inhibitors is associated with the endothelium-dependent hyperpolarization of the vascular smooth muscle cells. (By permission from Lippincott Williams and Wilkins; modified from Corriu et al.[771])

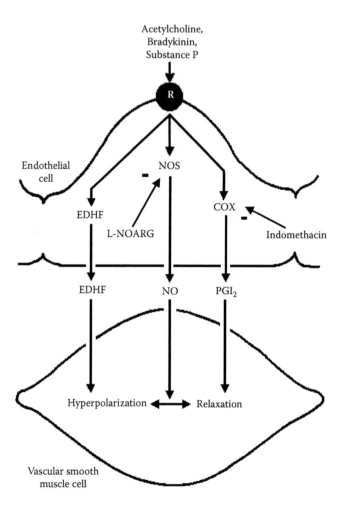

FIGURE 26 Schematic showing the endothelial pathways responsible for endothelium-dependent relaxations. Two pathways are properly identified, the NO synthase and the cyclooxygenase pathways. The third one, involving the hyperpolarization of the underlying vascular smooth muscle cells, will be discussed in detail in the following chapters of this monograph. EDHF: endothelium-derived hyperpolarizing factor; PGI₂: prostacyclin.

responses, which are resistant to inhibitors of NO synthase and cyclooxygenase, are observed without an increase in intracellular levels of cyclic nucleotides (cyclic GMP and cyclic AMP) in the smooth muscle cells.[19,27,35–38] Therefore the existence of an additional pathway that involved the hyperpolarization of the smooth muscle was suggested and attributed to a noncharacterized endothelial factor termed "EDHF" for endothelium-derived hyperpolarizing factor (Figure 26).[17,39–44]

However, under certain circumstances, both prostacyclin and NO can be considered to be endothelium-derived hyperpolarizing factors. If available cyclooxygenase inhibitors such as indomethacin are potent compounds that abolish the production of prostacyclin in vascular tissues, NO synthase inhibitors do not necessarily fully inhibit the endothelial production of NO. Therefore, in order to be confident that a third pathway truly exists, it is essential to first understand the mechanisms of the hyperpolarization produced by either prostacyclin or NO and then to eliminate beyond any reasonable doubt that the production and/or action of these two known vasodilators contributes to an observed endothelium-dependent hyperpolarization.

3.2 PROSTACYCLIN

Prostacyclin is the principal metabolite of arachidonic acid produced by cyclooxygenase in the endothelial cells.[45,46] Other endothelium-derived prostaglandins such as prostaglandin E_2 can show vasodilatator properties, albeit to a lesser extent. Endogenously synthesized and released prostacyclin, as well as exogenously added prostacyclin and its stable analogues (iloprost, beraprost, or cicaprost), activate IP receptors on, and induce hyperpolarization of, the vascular smooth muscle cells in arteries from various species including rat, hamster, guinea pig, rabbit, sheep, and dog (Figure 27). In most of these blood vessels, the hyperpolarizations involve the opening of K-ATP channels and are blocked by sulfonylureas such as glibenclamide (Figure 28).[47-55]

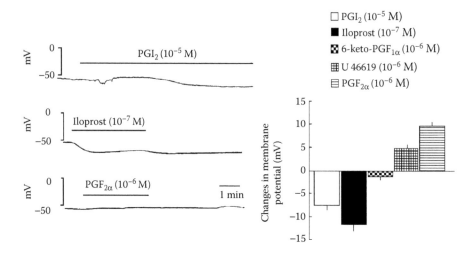

FIGURE 27 Effects of various prostanoids on the cell membrane potential of the vascular smooth muscle of isolated carotid arteries of the guinea pig (in the presence of inhibitors of NO synthases and cyclooxygenases). Left: Original traces showing the hyperpolarizations produced by prostacyclin (PGI$_2$) and its stable analogue and agonist at the IP receptors, iloprost, as well as the small but significant depolarization produced by prostaglandin $F_{2\alpha}$ (PGF$_{2\alpha}$). PG: prostaglandin. Right: Summary bar graphs showing the hyperpolarizing effects of prostacyclin and iloprost. The metabolite of prostacyclin, 6-Keto-PGF$_{1\alpha}$ is inactive. Besides prostaglandin $F_{2\alpha}$ and the synthetic analogue of thromboxane A_2 and agonist at TP receptors, U 46619 causes depolarization. (By permission from Elsevier Publishing Group; modified from Corriu et al.[53])

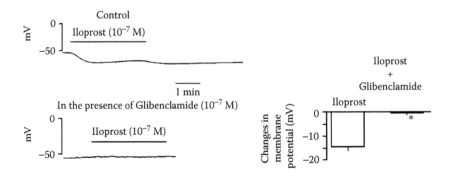

FIGURE 28 Effects of iloprost on the cell membrane potential of vascular smooth muscle isolated carotid artery of the guinea pig (in the presence of inhibitors of NO synthases and cyclooxygenases). The hyperpolarizing effect of iloprost is fully inhibited by the blocker of K_{ATP}, glibenclamide. Left: Original traces. Right: Summary bar graph. (By permission from Elsevier Publishing Group; modified from Corriu et al.[53])

It is not yet completely clear whether or not the opening of the K-ATP is dependent on the intracellular cyclic-AMP accumulation or on the activation of a G-protein.[56] Other activators of adenylyl cyclase, such as β-adrenergic agonists, hyperpolarize vascular smooth muscle cells by opening K-ATP,[57] and cell-permeable analogues of cyclic-AMP also produce hyperpolarization,[49] suggesting that cyclic nucleotide-dependent protein kinases are involved in the hyperpolarization produced by prostacyclin.

In some preparations, such as the tail artery of the rat, prostacyclin and iloprost activate not only K-ATP but also BK_{Ca} by a mechanism involving the protein kinase A-dependent phosphorylation of the potassium channels.[58,59] In the guinea pig aorta, the relaxations produced by iloprost, cicaprost, and beraprost involve BK_{Ca}, but not K-ATP, via a cyclic AMP-dependent, as well as a Gs-dependent and cyclic AMP-independent, mechanism.[60,61] In the middle cerebral artery of the rabbit, prostacyclin-induced relaxations involve both K_V and BK_{Ca}, but not K-ATP.[62] In the rat hepatic and in the guinea pig coronary artery, the potassium channels involved in the relaxations and hyperpolarizations to iloprost and to endogeneous prostaglandins have not been determined[63,64] and in the isolated coronary artery of the rat, prosta-cyclin and iloprost do not hyperpolarize the smooth muscle cells.[49]

In isolated smooth muscle cells of the rat portal vein, iloprost opens a potassium conductance that possesses the characteristics of BK_{Ca}.[65] Similarly, in smooth muscle cells of the rat pulmonary artery and of human coronary arteries, cell-permeable analogues of cyclic-AMP and prostaglandin E_2 activate BK_{Ca} via the cyclic-AMP cross-activation of PKG.[66,67] In the isolated smooth muscle cells of the bovine coronary artery, prostacyclin opens a 4-aminopyridine-sensitive K_V without affecting BK_{Ca}.[68]

The hyperpolarization produced by prostacyclin and the subtype of potassium channel that is activated may depend on the state of the vascular smooth muscle cells. In isolated guinea pig coronary arteries, the ability of endogenous or exogenous prostacyclin to hyperpolarize the cell membrane depends upon the stretch exerted on the smooth muscle. The amplitude of these responses reached a maximum when the tissues were stretched to the equivalent of that exerted by an intraluminal pressure of 50 mmHg. No hyperpolarization was observed in unstretched preparations. The resting membrane potential of the vascular smooth muscle cell was not affected by the stretching process.[49] Furthermore, the ability of prostacyclin to produce hyperpolarization of smooth muscle cells depends on the membrane potential of the latter. In blood vessels such as the uterine artery of the guinea pig, prostacyclin does not provoke hyperpolarization unless the tissue is depolarized.[49,50,69] The contribution of the hyperpolarizing mechanism in the relaxation process can be very significant in some tissues (rabbit coronary artery[48]; rat tail artery[59]) while in others, such as in the guinea pig coronary artery, the blockade of the hyperpolarization does not affect the relaxation produced by iloprost.[50] Finally, interactions between NO, prostacyclin, and potassium channels can confuse the issue. For instance, potassium channel activation can stimulate prostaglandin production.[70] In the porcine retinal and choroidal microcirculation, NO can release prostacyclin by opening endothelial BK_{Ca} channels,[71] while in the porcine coronary artery, prostacyclin releases NO from the endothelial cells.[72] Finally, in the guinea pig coronary artery, by a mechanism that is not understood, prostacyclin inhibits the effects of EDHF.[73]

Taken in conjunction, these findings indicate that prostacyclin, in numerous vascular beds, is an endothelium-derived hyperpolarizing substance that can activate diverse populations of potassium channels in smooth muscle cells.[74,75] However, prostacyclin can also evoke depolarization in the guinea pig carotid artery (Figures 29 and 30)[53] and contractile responses in preparations such as the rabbit aorta,[76] the human coronary and umbilical arteries,[77,78] and in the aorta of both normotensive and SHR.[79–81] These depolarizations and/or contractions are linked either to the activation of thromboxane receptors[53,81,82] or to the release of endothelium-derived contracting factors.[84]

These results shows that prostacyclin is not necessarily always a potent hyperpolarizing factor and/or vasodilator. Furthermore, EDHF-mediated responses are observed in the presence of inhibitors of cyclooxygenases, which, in most blood vessels, abolish the basal and stimulated release of prostacyclin (Figure 31) and other prostaglandins, indicating that this metabolite of arachidonic acid cannot account for most of the reported EDHF-mediated responses.

3.3 NITRIC OXIDE

The principal physiological action of NO is associated with the activation of cytosolic soluble guanylyl cyclase and the consequent formation of cyclic-GMP. The main targets of NO are the smooth muscle cells and the platelets. However, in the vascular wall, NO is also an autocrine factor that influences calcium handling in the endothelial cells and can also diffuse to the outer layers of the media and the adventitia to regulate neurotransmission.[84–86] Whereas the soluble guanylyl cyclase is undoubtedly the major effector of endothelial NO, this gaseous transmitter can directly regulate other targets, including ionic channels.[42]

In pulmonary and portal veins of the rabbit, nitrovasodilators (sodium nitroprusside or nitroglycerin) produce hyperpolarization of vascular smooth muscle as

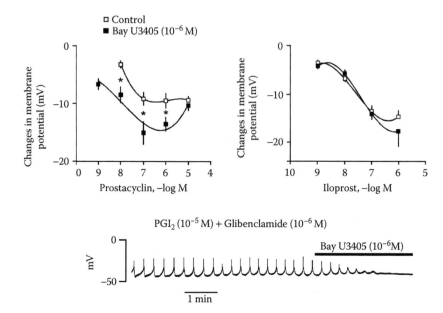

FIGURE 29 Differential effects of prostacyclin and iloprost on the membrane potential of the isolated carotid artery of the guinea pig (in the presence of inhibitors of NO synthases and cyclooxygenases). Top: The concentration-dependent response curve to prostacyclin is shifted to the right in the presence of a TP receptor antagonist (Bay U3405), while that to iloprost is not, indicating that, in addition to its effect on the IP receptors, prostacyclin also activates TP receptors. Bottom: The original trace shows that in the presence of the KATP blocker, glibenclamide, prostacyclin (PGI2) can elicit a spontaneous electrical activity that depends on TP receptor activation, since this activity is abolished by the addition of Bay U3405. (By permission from Elsevier Publishing Group; modified from Corriu et al.[53])

measured with intracellular microelectrodes.[87,88] This hyperpolarizing effect of nitro-vasodilators or of authentic NO was confirmed in other preparations from various species including mesenteric, coronary, and carotid arteries of the guinea pig, and in the aorta, tail, and mesenteric artery of the rat (Figure 32).[49,52,69,90] In carotid and femoral arteries of the rabbit, uterine arteries of the guinea pig, or mesenteric arteries of the dog, NO and nitrovasodilators do not produce hyperpolarization in resting tissue, but repolarize smooth muscle cells previously depolarized by an agonist (Figure 33).[69,90–92] In tissues such as the mesenteric artery of the rabbit, in order to observe hyperpolarization in response to nitrovasodilators, endogenous production of NO has to be suppressed either by removal of the endothelium or with an inhibitor of NO synthase.[93] Finally, in blood vessels such as the canine and porcine coronary arteries, the hepatic artery and the portal vein of the rat, and in the basilar artery of the rabbit, NO or/and nitrovasodilators do not influence at all the cell membrane potential.[63,89,94–100]

Electrophysiological (different configurations of the patch-clamp technique, intracellular microelectrodes) and functional experiments have characterized the

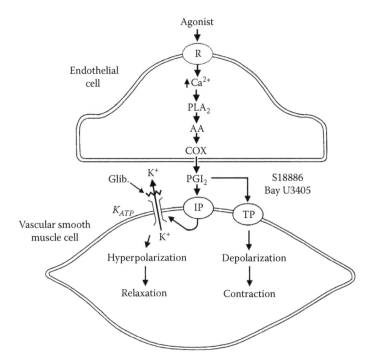

FIGURE 30 Schematic showing the potential relaxing and contracting effects of prostacyclin (PGI$_2$). R: receptor; IP: IP receptor; TP: TP receptor; PLA$_2$: phospholipase A2; AA: arachidonic acid; COX: cyclooxygenase; Glib: glibenclamide.

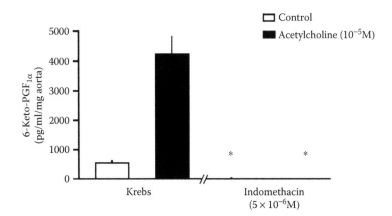

FIGURE 31 Cyclooxygenase inhibition and prostacyclin production in isolated blood vessels. Indomethacin abolishes both the spontaneous and the acetylcholine-stimulated release of prostacyclin from isolated arterial rings (aorta from spontaneously hypertensive rats, SHR). Thus, one may assume that, in experiments involving isolated blood vessels, the presence of a cycloxygenase inhibitor such as indomethacin, given at effective concentrations, rules out the involvement of cyclooxygenase derivatives.

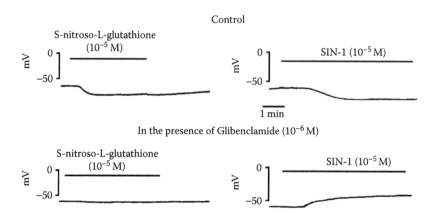

FIGURE 32 NO donors and cell membrane potential of the smooth muscle cells of the guinea pig carotid artery (in the presence of inhibitors of NO synthase and cyclooxygenase). Top: Both S-nitroso-L-glutathione and SIN-1 produce an endothelium-independent hyperpolarization of the vascular smooth muscle cells. Bottom: Glibenclamide prevents the hyperpolarization indicating that, in the guinea pig carotid artery, NO activates K_{ATP}. In the presence of glibenclamide SIN-1, but not S-nitroso-L-glutathione, produces a depolarization. Besides NO, the former, but not the latter, generates superoxide anion.

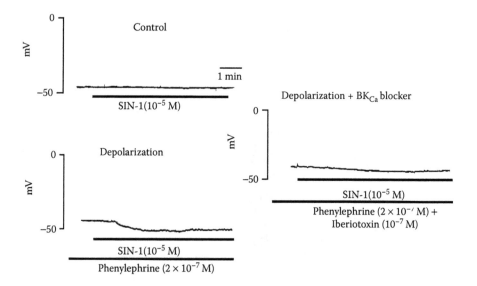

FIGURE 33 SIN-1 and cell membrane potential of smooth muscle cells of the rabbit carotid artery (in the presence of inhibitors of NO synthase and cyclooxygenase). In this artery, SIN-1 does not hyperpolarize the smooth muscle cell (control). However, in the presence of the depolarizing agonist phenylephrine, SIN-1 repolarizes the smooth muscle cell (Depolarization). In the presence of iberiotoxin, the repolarizing effect of SIN-1 is prevented (Depolarization + BK_{Ca} blocker), indicating that the repolarizing effect of SIN-1 is linked to the activation of BK_{Ca} channels. (By permission from Elsevier Publishing Group; modified from Quignard et al.[146])

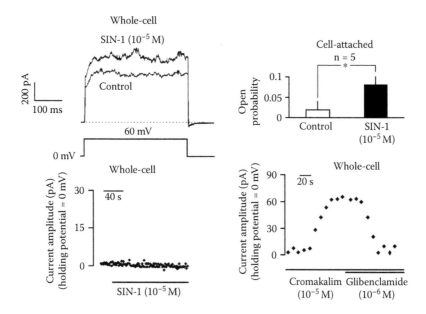

FIGURE 34 Effects of SIN-1 on outward K^+ currents in freshly isolated smooth muscle cells from the carotid artery of the rabbit. Top left: (whole-cell configuration of the patch-clamp technique, in the presence of an intracellular solution with a high concentration of Ca^{2+}: 0.5 µM): SIN-1 increases the outward and noisy K^+ current generated by a depolarizing step (from a potential of 0 mV to the test potential of 60 mV), suggestive of BK_{Ca} activation. Top right: SIN-1 increases the mean open probability of BK_{Ca} (cell attached configuration of the patch-clamp technique; patch potential, –20 mV). Bottom: (whole-cell configuration of the patch-clamp technique; holding potential, 0 mV, intracellular solution with a low concentration of ATP). SIN-1 does not generate an outward K^+ current (left) while cromakalim induces a glibenclamide-sensitive outward current under the same conditions (right). These results indicate that, in smooth muscle cells of the rabbit carotid artery, both BK_{Ca} and K_{ATP} are expressed and that in those cells SIN-1 activates the former but not the latter. (By permission from Elsevier Publishing Group; modified from Quignard et al.[146])

potassium channels activated by nitrovasodilators, exogenous authentic NO, and NO produced by either the constitutive NOS-3 or the inducible NOS-2.

3.3.1 BK_{Ca}

As shown with the patch-clamp technique, NO activates BK_{Ca} in isolated smooth muscle cells of cerebral and carotid arteries and aorta of the rabbit (Figure 34)[92,101–103] pulmonary, coronary, mesenteric, and cerebral arteries of the rat,[104–108] and human pulmonary and coronary arteries.[109,110] In most of the blood vessels, the activation of BK_{Ca} is mediated via a cyclic-GMP-dependent protein kinase phosphorylation of the channel.[104,106,109,111,112] The cyclic-GMP-dependent activation of BK_{Ca} by NO donors, as observed with the human channel expressed in Xenopus oocytes, has been attributed to the phosphorylation of two serine residues (855 and 869).[113]

However, NO also produces a direct, cyclic-GMP-independent activation of BK_{Ca} in smooth muscle cells from the rabbit aorta,[102] the mesenteric artery of the rat,[107] and in BK_{Ca} reconstituted into a planar lipid bilayer.[114] This activation may involve the binding of NO (or of one of its oxidized derivatives) to thiols, most likely cysteine residues located on the α subunit (*Slo1* α), in order to form S-nitrosothiols that may establish disulfide bridges if another reduced thiol is in close vicinity.[115–117]

In most of the functional studies performed on isolated or perfused blood vessels, the relaxation or the vasodilatation produced by NO and/or NO donors is more or less sensitive to BK_{Ca} blockers (tetraethylammonium, charybdotoxin, or iberiotoxin) suggesting that these potassium channels participate in the relaxation of vascular smooth muscle. This includes pulmonary, mesenteric, coronary and basilar arteries and the aorta of the rat,[104,105,118–123] mesenteric and carotid arteries and aorta of the rabbit,[102,124–126] pulmonary artery of the guinea pig,[127] coronary and cerebral arteries of the dog,[128–130] and human, porcine, and bovine coronary arteries.[110,112,131] In anesthetized pigs, the vasodilatation produced by NO donors is inhibited potently by iberiotoxin, suggesting that activation of vascular BK_{Ca} is an important component of the response.[132]

Some experiments involving the measurement of membrane potential with an intracellular microelectrode, also suggest a contributing effect of BK_{Ca}. Thus, in the rabbit carotid artery and the rat mesenteric artery, depolarized with phenylephrine, the repolarization of the smooth muscle cells in response to NO, is sensitive to charybdotoxin or iberiotoxin indicating the involvement of these channels.[103,126,133,134]

Differences in protocols may explain why, in most of the patch-clamp studies, NO activates BK_{Ca}, which is coherent with most of the functional studies where NO-induced relaxations are sensitive to BK_{Ca} blockers, while hardly any studies measuring changes in membrane potential with an intracellular microelectrode have shown such a phenomenon. In functional studies, tissues are contracted, i.e., depolarized, in order to observe relaxations. In contrast, most of the electrophysiological studies with an intracellular microelectrode involve tissues at rest. BK_{Ca} channels are both voltage- and calcium-regulated potassium channels. The approximate threshold activation potential of BK_{Ca}, with a low intracellular calcium concentration (approximately 100 nM), is –20 mV,[135] a value of potential much higher than the potential recorded in resting arteries (–50 to –65 mV). This may explain why, in the canine coronary artery for example, nitroglycerin, NO, or sodium nitroprusside do not produce hyperpolarization in resting tissue,[95,96,98] while in functional studies nitroglycerin induces a relaxation sensitive to iberiotoxin[128] and in isolated smooth muscle cells of the same artery, BK_{ca} are stimulated by a cyclic-GMP-dependent protein kinase, as observed in patch-clamp experiments.[111]

Several reports involving microelectrode and functional experiments suggest a cyclic-GMP-dependent activation of BK_{Ca}. However, in fairly good agreement with the patch-clamp data, in the rabbit isolated carotid artery NO induces hyperpolarization of the smooth muscle cells by means of both a cyclic GMP-dependent potassium conductance and a cyclic GMP-independent, charybdotoxin-sensitive conductance.[126] In the rat mesenteric and cerebral arteries, NO evokes a cyclic GMP-independent activation of BK_{Ca} channels.[133,134,136] In functional studies in the mesenteric and pulmonary artery of the rat[119,120,133,134] and in the carotid artery of the rabbit,[126] a cyclic GMP-independent activation of BK_{ca} is obtained, confirming the hypothesis that direct activation of BK_{ca} by NO might be an important mechanism leading to the relaxation of the smooth muscle cells.

NO can also indirectly activate BK_{Ca} by preventing the formation of an endogenous inhibitor of these channels. In renal and cerebral arteries, NO, by binding to the heme moiety of the cytochrome P450 monooxygenase, inhibits the enzymatic formation of 20-HETE, a potent inhibitor of BK_{Ca} activity.[137] The physiological relevance of this pathway has been demonstrated in functional experiments involving the measurement of the changes in diameter of renal arteries in response to nitro-vasodilators.[138,139]

By contrast, peroxynitrite, the product of the interaction between superoxide anion and NO that is formed under various circumstances, inhibits BK_{Ca} activity, possibly by a direct effect on the channel, as demonstrated in the smooth muscle of human coronary arterioles.[140]

3.3.2 OTHER POTASSIUM CHANNELS

BK_{Ca} is not the only population of K^+ channels activated by NO and this channel is not always activated by the endothelial mediator. As observed in patch-clamp experiments on bovine coronary artery, NO activates both BK_{Ca} and K_V.[68] In isolated smooth muscle cells of the porcine coronary artery it activates both BK_{Ca} and K-ATP.[141,142] In rat pulmonary arteries, NO promotes the opening of K_V by a mechanism that may be independent of soluble guanylyl cyclase.[143] In the rat tail artery, sodium nitroprusside activates K_{IR}, most likely Kir2.1.[144] In the carotid artery of the guinea pig, SIN-1 and a cell-permeable analogue of cyclic-GMP, dibutyril cyclic-GMP, activate K-ATP but do not modify significantly the open probability and the mean open time of BK_{Ca} (Figure 35).[92,145,146] Finally, NO can also activate, in a cyclic-GMP-dependent manner, the K_{2P} channel, TREK-1.[147]

In the rat mesenteric artery, depending on the experimental conditions, the hyperpolarization evoked by NO relies on BK_{Ca}, K_V, and/or K-ATP activation.[90,133,148] The hyperpolarization produced by NO and/or NO donors in the coronary, carotid, and cochlear arteries and mesenteric lymphatic vessels of the guinea pig,[49,52,146,149,150] as well as in the mesenteric and femoral artery of the rabbit,[91,93] are sensitive to glibenclamide, indicating the implication of K-ATP channels.

In functional studies, a 4-aminopyridine-sensitive relaxation is observed with NO donors in the isolated pulmonary and mesenteric arteries of the rat.[119,133] In anesthetized rats, the sodium nitroprusside-induced hypotension is also inhibited by 4-aminopyridine, but not by iberiotoxin, suggesting the activation of K_V.[151] In the canine femoral vein, the relaxation to NO involves both K-ATP and BK_{Ca} channels, with the latter, but not the former, depending on a cyclic-GMP-dependent pathway.[152] An apamin-sensitive effect, suggestive of SK_{Ca} involvement, is observed in the rat mesenteric and the lamb coronary arteries.[120,153] In the isolated aorta and carotid artery of the guinea pig, as well as in the anesthetized rat, the dilatation caused by NO donors does not involve BK_{ca}, as blockers of this channel do not inhibit the response.[127,151]

Some of the disparity observed in the published literature could be explained by the fact that endogenous endothelial NO (or exogenous authentic NO) and

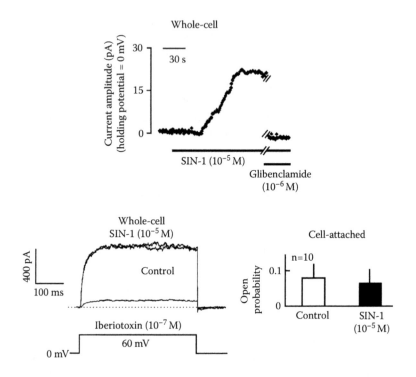

FIGURE 35 Effects of SIN-1 on outward K^+ currents in freshly isolated smooth muscle cells from the carotid artery of the guinea pig. Top: Whole-cell configuration of the patch-clamp technique; holding potential, 0 mV, intracellular solution with a low concentration of ATP. SIN-1 induces a glibenclamide-sensitive outward K^+ current. Bottom left: Whole-cell configuration of the patch-clamp technique, in the presence of an intracellular solution with a high concentration of Ca^{2+}: 0.5 μM. The outward K^+ currents generated by a depolarizing step from a holding potential of 0 mV to the test potential of 60 mV was unaffected by the presence of SIN-1 (10^{-5} M) but was inhibited by iberiotoxin. Bottom right: SIN-1 does not influence the mean open probability of BK_{Ca} (cell attached configuration, patch potential –20 mV). These results indicate that, in the smooth muscle cells of the guinea pig carotid artery, both BK_{Ca} and K_{ATP} are expressed and that in those cells SIN-1 activates the latter but not the former. (By permission from Elsevier Publishing Group; modified from Quignard et al. [146])

nitrovasodilators do not necessarily activate the same population of potassium channels. In the isolated rabbit carotid artery, endogenous endothelial NO (or exogenous authentic NO) activates a cyclic-GMP-dependent potassium conductance and a cyclic-GMP-independent, charybdotoxin-sensitive conductance, while nitrovasodilators activate only the former.[126] In the rat mesenteric artery, authentic NO and nitrovasodilators induce the cyclic-GMP-independent activation of both K_V and K_{Ca} channels, the contribution of each pathway being different depending on the source of NO.[133,134]

Furthermore, although hyperpolarization of the smooth muscle cells is a potent mechanism for dilatation, in some instances it may not be the predominant one responsible for the relaxation of a given artery. In the coronary artery of the guinea pig,[50] as well as in the basilar and femoral artery of the rabbit,[91,154] NO evoked glibenclamide-sensitive hyperpolarizations, which do not appear to contribute to the relaxation. This could be explained by the presence of other contributing pathways. For instance, in the mesenteric artery of the rat, in order to observe the contribution of the hyperpolarizing mechanism to the relaxation, prior inhibition of the other pathways is required.[120]

Taken in conjunction, these reports highlight the heterogeneity of the vascular smooth muscle cells in their response to NO and nitrovasodilators, as well as the diversity of the population of potassium channels expressed in these cells. This heterogeneity can be observed in the same blood vessel. In the pulmonary arteries of the rat, various subpopulations of smooth muscle cells can be identified according to the relative distribution of K_{Ca} and K_V. NO relaxes conduit pulmonary arteries with a predominant subpopulation of smooth muscle cells expressing mostly K_{Ca}, while resistance arteries that express mostly K_V are minimally affected by the endothelial mediator.[155]

3.3.3 OTHER IONIC CHANNELS

Beyond K^+ channels, NO also interacts with other ionic channels and can therefore further influence the membrane potential of the smooth muscle cells.

NO inhibits Cl^- currents and prevents depolarization of the smooth muscle cells. In the smooth muscle cells of the rabbit portal vein, NO donors can, depending on the cells, either enhance $I_{Cl,Swell}$ via a cyclic-GMP-dependent phosphorylation, or inhibit it in a cyclic-GMP-independent manner.[156] In the smooth muscle cells of the rat cerebral arterioles, endothelial NO and NO donors inhibit CLCA.[157] NO also can indirectly regulate chloride channel activity by decreasing calcium release from internal stores[158] or by decreasing calcium entry.[159] In functional studies, on coronary arteries and aorta of the rat, NO tonically suppresses a contractile mechanism that involves Cl^- conductance.[160,161]

Additionally, in smooth muscle cells, NO produces cyclic-GMP-dependent inhibition of a Ca^{2+}-permeable nonselective cation channel activated by agonists[162] and of a Ca^{2+} store depletion-activated, nonselective cation current.[163]

3.3.4 INDIRECT ENDOTHELIAL EFFECTS OF NO

NO can also indirectly affect the cell membrane potential of the underlying smooth muscle by influencing the ionic channels expressed at the surface of the endothelial cells or by regulating the calcium homeostasis of the latter. In porcine renal arteries, endothelial cells express BK_{Ca}, which are strongly activated by NO.[164] This could represent an autocrine regulation of endothelial function. As the hyperpolarization increases the electrochemical gradient for Ca^{2+} ions, the activation of BK_{Ca} by endothelial NO could constitute a positive feedback mechanism to further activate NO-synthase. In addition, the endothelial hyperpolarization could be transferred to the smooth muscle cells provided these cells express myoendothelial gap junctions.[165] Most of the endothelial cells preferentially express SK_{Ca} and IK_{Ca}.

NO activates these two K+ channel subtypes in gastrointestinal smooth muscle cells,[166] but whether this is the case in endothelial cells is not known. In Eahy926, a cell line derived from human umbilical vein endothelial cells, NO inhibits a Ca^{2+}-activated cationic current, possibly TRPM4. Preventing the depolarization caused by the activation of this current would also constitute a positive feedback mechanism to further activate NO-synthase.[167] However, in the same cell line NO does not activate BK_{Ca}.[168] In aortic endothelial cells, store-operated calcium entry is inhibited by NO donors.[169]

Thus, under certain circumstances, an autocrine role of NO in the reinforcement of endothelium-dependent hyperpolarization of the smooth muscle cells can be suspected. Additional experiments are required to evaluate the contribution of this mechanism as well as where and when this could play a role in the control of vascular tone.

3.3.5 NO AS AN EDHF

3.3.5.1 Incomplete Blockade of NO-Synthase

Taken collectively, these reports show that NO can affect, directly or indirectly, the activity of many different ionic channels and therefore can influence the membrane potential of vascular smooth muscle cells in various ways. As EDHF-mediated responses were observed in the presence of NO-synthase inhibitors, NO was originally excluded as a putative EDHF. Nevertheless, in the presence of these inhibitors, residual NO can be produced by the endothelial cells because of an incomplete blockade of the enzyme. In the rabbit carotid artery, this residual NO, measured with a porphyrinic-based microprobe[170] may account for the EDHF-mediated responses (Figure 36).[90,126] In the superior mesenteric artery of the rat, a NOS inhibitor did not abolish the endothelial production of NO and the residual NO release contributed to the L-nitro-arginine-resistant relaxation to acetylcholine. However, the addition of hemoglobin, a scavenger of NO, abolished the contribution of the residual NO produced and the related relaxations.[171] Indeed, in other arteries such as the porcine femoral and coronary arteries, NO synthase is not fully blocked by the inhibitors of NO synthase used. But if the residual NO accounted for the EDHF-like response in the femoral artery, it did not in the coronary artery.[172,173] Furthermore, in other arteries, such as human subcutaneous arteries, NO synthase inhibitors block totally the endothelial production of NO and, therefore, the observed EDHF-mediated responses cannot be attributed to residual production of the gaseous mediator.[174] In addition, the combination of a NO synthase inhibitor with a NO scavenger that consistently abolishes the access of NO to the smooth muscle does not prevent EDHF-mediated responses in small coronary and mammary arteries of humans, as well as in guinea pig carotid arteries.[92,175–177] Moreover, EDHF-mediated responses occur in NOS-3 knock-out mice, definitively ruling out the responsibility of residual endothelial-derived NO in EDHF-mediated responses.[178–182]

3.3.5.2 Stored NO

NO can contribute in another way to EDHF-mediated responses. It is generally thought that the endothelial production of NO by the NOS-3 is performed according to demand. However, NO can also be stored and released independently of a

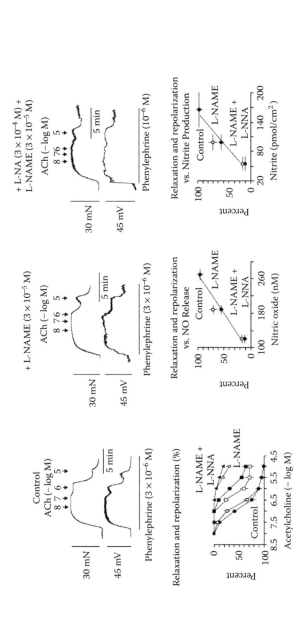

FIGURE 36 NO release and relaxation and hyperpolarization of the smooth muscle cells of the rabbit isolated carotid artery. Effects of inhibitors of NO-synthases (in the presence of cyclooxygenase inhibitors). Top: Original traces showing the changes in tone (top trace) and the associated changes in membrane potential of the vascular smooth muscle (lower trace). In contracted and depolarized arteries with phenylephrine, acetylcholine produces an endothelium-dependent relaxation and a parallel repolarization of the smooth muscle cells (left). In the presence of a NO-synthase inhibitor, L-NAME, both endothelium-dependent relaxations and repolarizations are partially inhibited (middle). The subsequent addition of another NO-synthase inhibitor, L-NA, produces a further inhibition and the virtual abolition of the endothelium-dependent relaxation and repolarization (right). Bottom left: Summary of the above observations showing the parallel inhibition produced by the NO-synthase inhibitors of the endothelium-dependent relaxation (open symbols) and repolarization (closed symbols). Bottom middle and right: Close correlation between the endothelium-dependent NO release evoked by acetylcholine (3×10^{-6}M), evaluated either directly with a porphyrinic microsensor (middle) or indirectly by measuring nitrite accumulation (right), and the maximal relaxation (filled symbols) or the maximal repolarization (open symbols). These results show that, in the rabbit carotid artery and under the tested experimental conditions, the endothelium-dependent relaxations and repolarizations produced by acetylcholine involve only the release of NO. In this artery, NO is an endothelium-derived hyperpolarizing factor. Furthermore, these data also indicate that in certain vascular preparations, the presence of a single inhibitor of NO-synthases may only partially inhibit NO synthase and NO release. (By permission of the National Academy of Sciences, USA; modified from Cohen et al.[90])

concomitant activation of the enzyme. In the blood, NO interacts with albumin to form a-nitroso-albumin, which possesses vasodilatator properties.[183,184] Similarly, NO is scavenged by hemoglobin, but in addition to this consumption pathway, NO can form S-nitroso-hemoglobin and be transported by the flowing blood to activate distant proteins by transnitrosation.[185] Although the possibility for NO to form long-lasting, slow-releasing complexes exists, definitive functional evidence of these complexes under physiological conditions has yet to be produced.[186]

In the blood vessel wall, two different putative stores have been described. The first one is the so-called photosensitive store, which involves the release of NO by ultraviolet light and the relaxation of vascular smooth muscle. This store is replenished during a recovery period in the dark.[187,188] NOS-3 plays a role in refilling this photosensitive store, but is not essential since the formation of the photorelaxation store is observed in NOS-3 knock-out mice.[189] The activation of K_V, in a soluble guanylyl cyclase-dependent manner, by the release of authentic NO or/and nitrosothiols, contributes to the mechanism of photorelaxation.[187,189,190] However, the mechanism of NO release by this store under physiological conditions, i.e., in the absence of ultraviolet light, is unknown. Nevertheless, in the rabbit and rat mesenteric arteries, the same pool of NO that is released during photoactivation may play a role in the endothelium-dependent relaxations caused by the release of a non-adrenergic-non-cholinergic mediator from capsaicin-sensitive sensory fibers or to EDHF-like responses.[191,192] The NO-dependent responses are sensitive to the guanylyl cyclase inhibitor, ODQ, and the NO scavenger hemoglobin, but not to NO-synthase inhibitors. In the rat mesenteric artery, the contribution of preformed vascular NO to the EDHF-mediated responses averages 30% of the whole hyperpolarization and involves a barium plus ouabain-sensitive mechanism.[192]

The second storage site is located in the adventitia and is generated by the formation of protein-bound dinitrosyl nonheme iron complexes and S-nitrosated proteins. Low-molecular-weight thiols displace NO from these stores and transfer it to various target membrane proteins including potassium channels.[193,194] These complexes are formed when elevated concentrations of NO are produced, such as when the expression of NOS-2 has been induced by lipopolysaccharides.[194] Whether these stores can play a role under physiological conditions in healthy blood vessels is unknown. In porcine coronary microvessels, L-nitrosocysteine can, in addition to its effects on the smooth muscle cells, activate, in a stereoselective manner, endothelial IK_{Ca} and SK_{Ca}, suggesting that this compound or other low-molecular-weight S-nitrosothiols may function as EDHF in these arteries.[195,196]

Considering all these results, it is obvious that the role of NO as an EDHF has certainly been underestimated on the assumption that the presence of an inhibitor of NO synthases rules out its contribution. This is clearly not the case. Whether or not the residual NO originates from an incomplete blockade of NO synthase or from preformed stores is difficult to assess. A "true" EDHF-mediated response, distinct from the production of vasodilator prostaglandins and NO, should be resistant not only to the combination of cyclooxygenase and NO synthase inhibitors, but to these inhibitors in the additional presence of a NO scavenger and/or an inhibitor of guanylyl cyclase (Figure 37). However, it should be kept in mind that pharmacological substances alone or in association can show nonselective effects. For instance,

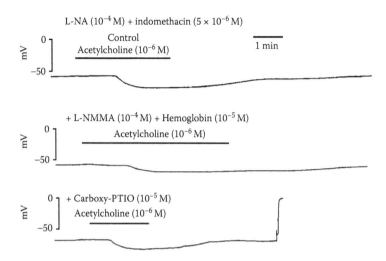

FIGURE 37 Endothelium-dependent hyperpolarizations in the guinea pig carotid artery. Top (control): Acetylcholine induces hyperpolarization in the presence of one inhibitor of NO-synthase, L-NA (10^{-4} M) and an inhibitor of cyclooxygenase, indomethacin (5×10^{-6} M). Middle: In the presence of L-NA plus indomethacin and after the addition of a second NO-synthase inhibitor, L-NMMA (10^{-4} M) plus a NO scavenger, hemoglobin (10^{-5} M), acetyl-choline still induces hyperpolarization. Bottom: In the presence of L-NA plus indomethacin and after the addition of a different NO scavenger, carboxy-PTIO (10^{-5} M), acetylcholine still evokes hyperpolarization. These results indicate that, in the guinea pig carotid artery, the endothelium-dependent hyperpolarization elicited by acetylcholine cannot be attributed to residual NO release (Chataigneau and Félétou, unpublished observations).

oxyhemoglobin can also inhibit BK_{Ca} activation and the associated spontaneous transient outward currents in a NO-independent manner, by a free-radical-mediated inhibition of calcium sparks.[197] This indicates that the further inhibition of EDHF-mediated responses produced by the addition of the NO scavenger may not necessarily be due to the scavenging of residual NO.

Despite these caveats, the presence of a third endothelial pathway, besides prostacyclin and NO, contributing to the endothelium-dependent relaxation of the smooth muscle cells has been demonstrated beyond reasonable doubt in various blood vessels (Figure 38).[198]

3.4 ENDOTHELIUM-DEPENDENT HYPERPOLARIZATIONS: INVOLVEMENT OF CALCIUM-ACTIVATED POTASSIUM CHANNELS

Initially, several lines of evidence suggested that the endothelium-dependent hyper-polarization of smooth muscle cells resulted from the opening of K^+ channels in the plasmalemma of vascular smooth muscle. First, the amplitude of the hyperpolariza-tion was inversely related to the extracellular concentration in K^+ and abolished by K^+ concentrations higher than 25 mM.[199-203] Second, agonists that produced

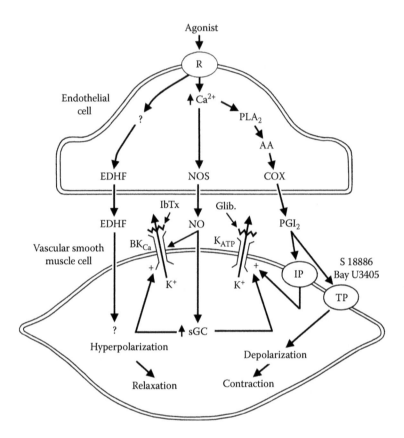

FIGURE 38 Schematic showing that besides nitric oxide (NO) and prostacyclin (PGI$_2$), a third pathway, the endothelium-derived hyperpolarizing factor (EDHF) pathway, is likely to be involved in the endothelium-dependent hyperpolarization and relaxation of some blood vessels. For the sake of clarity, only two of the multiple targets of NO have been shown: large conductance calcium-activated potassium channel (BK$_{Ca}$) and ATP-sensitive potassium channels (K$_{ATP}$). R: receptor; PLA$_2$: phospholipase A2; AA: arachidonic acid; COX: cyclooxygenase; Glib: glibenclamide; NOS: nitric oxide synthase; sGC: soluble guanylyl cyclase; IbTx: iberiotoxin.

endothelium-dependent hyperpolarizations also stimulated the efflux of ^{42}K (or ^{86}Rb) from pre-loaded arteries.[19,20] Third, a decrease in membrane resistance of the vascular smooth muscle cells was observed during endothelium-dependent hyperpolarizations.[15,199,200] suggesting that the hyperpolarization was due to the opening of a K$^+$ channel rather than the closing of a chloride or a nonspecific cationic channel (Figures 39 and 40).

EDHF-mediated responses, like the relaxations to endothelium-derived NO, are inhibited when the extracellular concentration of calcium is decreased (Figure 40).[204–206] The production of NO and endothelium-dependent hyperpolarization, in response to agonists that stimulate G-protein-coupled receptors is associated with an increase in the endothelial cell [Ca^{2+}]$_i$.[207,208] This is linked to the phospholipase C-dependent

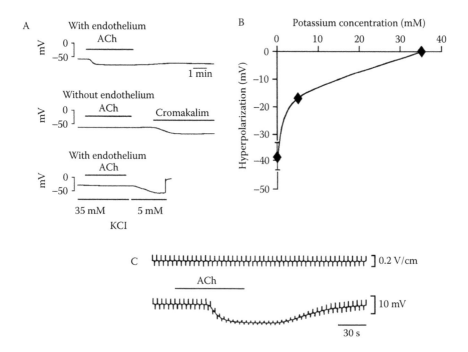

FIGURE 39 Characteristics of endothelium-dependent hyperpolarizations (1). (A) Acetylcholine-induced hyperpolarizations in the vascular smooth muscle of the guinea pig carotid artery (in the presence of inhibitors of NO-synthase and cyclooxygenase) are endothelium-dependent (top). In the absence of the endothelium acetylcholine no longer produces a hyperpolarization (middle). This is not due to damage following removal of the endothelium since the K_{ATP} opener, cromakalim, still hyperpolarizes the smooth muscle cells. In the presence of the endothelial cells, raising the extracellular potassium concentration from 5 to 35 mM depolarizes the smooth muscle cells and prevents the hyperpolarizing effect of acetylcholine (bottom; by permission from the Nature Publishing Group; modified from Corriu et al.[202]. (B) The amplitude of the endothelium-dependent hyperolarization to acetylcholine is inversely correlated to the extracellular concentration in potassium (guinea pig carotid artery, in the presence of inhibitors of NO-synthase and cyclooxygenase; Quignard and Félétou, unpublished observations). (C) Acetylcholine-induced endothelium-dependent hyperpolarization in the rat intrapulmonary vein. The amplitude of the electrotonic potentials produced by alternate application of current stimuli is decreased during acetylcholine-induced hyperpolarization, showing that this hyperpolarization is accompanied by a decrease in cell membrane resistance. (By Permission of the Physiological Society of Japan; modified from Chen and Suzuki[200]) Taken in conjunction, these observations indicate that endothelium-dependent hyperpolarizations are associated with the opening of a potassium conductance.

formation of inositol-trisphosphate, which depletes endothelial calcium stores and the subsequent capacitive calcium entry.[209,210] Inhibitors of phospholipase C inhibit totally or partially agonist-elicited EDHF-mediated responses in numerous blood vessels and vascular beds.[176,211–215] Furthermore, substances (calcium ionophore, thapsigargin, cyclopiazonic acid) that increase endothelial $[Ca^{2+}]_i$ in a receptor-independent

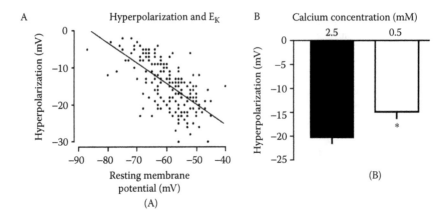

FIGURE 40 Characteristics of endothelium-dependent hyperpolarizations (2). (A) Hyperpolarizations to acetylcholine in the vascular smooth muscle cell of the guinea-pig carotid artery (in the presence of inhibitors of NO-synthase and cyclooxygenase). The amplitude of the hyperpolarization is linked to the membrane potential of the smooth muscle and tends to reach the equilibrium potential for potassium ions (E_K), supporting the involvement of a potassium conductance (by permission from the Nature Publishing Group; modified from Chataigneau et al.[230]). (B) Endothelium-dependent hyperpolarizations evoked by acetylcholine in the vascular smooth muscle of the guinea pig carotid artery (in the presence of inhibitors of NO-synthase and cyclooxygenase). The amplitude of the hyperpolarization is linked to the extracellular calcium concentration. (By permission from the Nature Publishing Group; modified from Gluais et al.[206])

manner also produce hyperpolarization of the endothelial cells[216–221] and endothelium-dependent hyperpolarization of the smooth muscle cells.[201,222]

Collectively, these observations suggested that the endothelium-dependent hyperpolarization of the smooth muscle cells involved an increase in smooth muscle K^+ conductance and that EDHF was an endothelium-derived K^+ channel opener released following an increase in endothelial $[Ca^{2+}]_i$ in a similar manner to NO or prostacyclin. However, this early interpretation had to be modified in the light of more recent experimental evidence.[165]

3.5 IDENTIFICATION OF THE POTASSIUM CHANNELS INVOLVED IN EDHF-MEDIATED RESPONSES

In all the species studied so far, EDHF-mediated responses are insensitive to glibenclamide (an inhibitor of ATP-sensitive potassium channels), with the possible exception of cerebral arteries such as the middle cerebral arteries of the rabbit and pial arteries of the pig.[223,224] However, they are blocked by nonselective inhibitors of K_{Ca}, including tetraethylammonium or tetrabutylammonium.[202,225,226] Depending on the tissue studied, apamin, a specific blocker of SK_{Ca}, has no effect or produces partial and even in some cases complete inhibition of the EDHF-mediated responses (Table 2). Similarly, a wide range of effects is obtained with charybdotoxin, a nonspecific inhibitor of BK_{Ca}, IK_{Ca}, and some K_V. However, in virtually every blood vessel studied, the combination of these two toxins abolishes EDHF-mediated responses (Table 2). Iberiotoxin (a specific inhibitor of BK_{Ca}) alone or in combination with apamin is in most (but not all) blood vessels ineffective or poorly effective, indicating that in most (but not all) cases iberiotoxin cannot substitute for charybdotoxin (Table 2). This crucial observation demonstrated that BK_{Ca} are not involved in most EDHF-mediated responses (Figure 41).[227–231]

The effects of apamin can be mimicked by peptidic and nonpeptidic, structurally different SK_{Ca} inhibitors, scyllatoxin, tubocurarine, and UCL 1684, demonstrating that SK_{Ca} is involved in endothelium-dependent hyperpolarizations.[202,206,232,233] Since BK_{Ca} were excluded, the target of charybdotoxin could have been either K_V or IK_{Ca}. A first series of experiments performed in isolated smooth muscle cells with the patch-clamp technique showed that the expressed KDR currents were not sensitive to charybdotoxin.[92,146,228,234] These observations were not supportive of K_V being the target of charybdotoxin, but could not definitively rule out this hypothesis. More conclusive data were obtained with 1-ethyl-2-benzimidazolinone (1-EBIO), an activator of SK_{Ca} and/or IK_{Ca} but not BK_{Ca}.[235–243] In various blood vessels of the rat, pig, and guinea pig and in human umbilical vein endothelial cells this compound can be regarded as a reasonably specific activator of IK_{Ca}.[236,240,243–247] In these arteries,

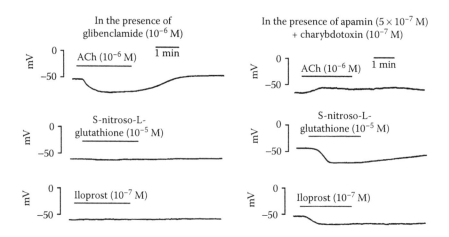

FIGURE 41 Potassium channel blockers and hyperpolarizations of the vascular smooth muscle of the guinea pig carotid artery (in the presence of inhibitors of NO-synthase and cyclooxygenase). Left: Glibenclamide, the K_{ATP} blocker, does not affect the endothelium-dependent hyperpolarization to acetylcholine (ACh) but abolishes the endothelium-independent hyperpolarizations to the NO donor, S-nitroso-L-glutathione, and that to the prostacyclin analogue, iloprost. Right: The combination of two blockers of calcium-activated potassium channels, apamin, a selective blocker of SK_{Ca}, and charybdotoxin, a nonselective blocker of IK_{Ca}, BK_{Ca}, and some K_V, inhibits the acetylcholine-induced endothelium-dependent hyperpolarization, while the endothelium-independent hyperpolarizations to either S-nitroso-L-glutathione or to iloprost are unaffected by the combination. (By permission from the Nature Publishing Group; modified from Corriu et al.[52])

TABLE 2
EDHF-mediated responses and blockers of potassium channels (presence of inhibitors of cyclooxygenases and nitric oxide synthases)

Species Blood Vessels	Agonist	Apamin	ChTX	IbTX	Apamin + ChTX	Apamin + IbTX	References
RAT							
Mesenteric Art.	ACh	Partial inh.	Ineffective	ND	Block	ND	80
	ACh	Partial inh.	Partial inh.	ND	Block	ND	772
	ACh	Partial inh.	Ineffective	ND	Block	ND	545
Hepatic Art.	ACh	Ineffective	Ineffective	Ineffective	Block	Ineffective	227, 228
Aff. Renal Art.	ACh	Partial inh.	Partial inh.	ND	Block	ND	775
Cerebral Art.	UTP	Partial inh.	Block	Partial inh.	ND	ND	776
GUINEA PIG							
Carotid Art.	ACh	Ineffective	Ineffective	ND	Block	ND	52
	ACh	Partial inh.	Ineffective	Ineffective	Block	Partial inh.	230
Submucosal Art.	ACh, SP	Partial inh.	Partial inh.	ND	Block	ND	773
	ACh	Partial inh.	Partial inh.		Block		343
Basilar Art.	ACh, A23187	Ineffective	Partial inh.	Ineffective	Block	Ineffective	229
		Ineffective	Ineffective	Ineffective	Block	ND	
Cerebral Art.	ACh	Ineffective	Block	Partial inh.	ND	Partial inh.	399
Coronary Art.	ACh	Ineffective	Partial inh.	Ineffective	Block	Ineffective	231
Mesenteric Art.	ACh	Partial inh.	Partial inh.	Partial inh.	ND	Partial inh.	399
RABBIT							
Carotid Art.	ACh	Partial inh	Block	Partial inh.	ND	ND	777
Mesenteric Art.	ACh	Ineffective	Partial inh.	Ineffective	Block	Ineffective	483

	Agonist						Ref.
MURINE							
Hindlimb	ACh	Ineffective	Partial inh.	Ineffective	Block	Ineffective	179
Mesenteric Art.	SLIGRL-NH$_2$	Partial inh.	Partial inh.	Ineffective	Block	Ineffective	383
Saphenous Art.	ACh	Ineffective	Ineffective	Ineffective	Block	Ineffective	180
PORCINE							
Coronary Art.	BK	Ineffective	Partial inh.	Partial inh.	Block	ND	173
	SP	Ineffective	Partial inh.	Ineffective	Block	Ineffective	248
Renal Art.	BK	Ineffective	Partial inh.	Partial inh.	Block	ND	387
CANINE							
Coronary Art.	ACh.	Ineffective	Partial inh.	Ineffective	ND	ND	778
	ACh	Partial inh.	Ineffective	No effect	Partial Inh.	ND	779
Corpus Cavernus	ACh	Block	Ineffective	Ineffective	ND	ND	774
BOVINE							
Ciliary Art.	ACh	Ineffective	Partial inh.	Ineffective	Block	Ineffective	780
Pulmonary Art.	BK	Ineffective	Ineffective	Ineffective	Block	Partial inh.	781
EQUINE							
Penile Art.	ACh, BK	Partial inh.	Partial inh.	ND	Block	ND	782
MONKEY							
Lingual Art.	ACh	Ineffective	Partial inh.	Ineffective	Block	ND	783
Coronary Art.	ACh	Partial inh.	Partial inh.	Ineffective	Block	ND	779
HUMAN							
Pial Art.	SP	Partial inh.	Partial inh.	ND	Block	ND	229
Gastroepiploic Art.	BK	Ineffective	Partial inh.	Ineffective	ND	ND	480
Coronary Art.	BK	Partial inh.	Partial inh.	ND	ND	ND	312
Radial Art.	Carbachol	ND	Partial inh.	Partial inh.	Partial inh.	ND.	757
Renal Art.	ACh	ND	ND	ND	Block	ND	388

Note: ChTX: charybdotoxin; IbTX: iberiotoxin; Art: artery; ACh: acetylcholine; BK: bradykinin; SP: substance P; SLIGRL-NH$_2$: agonist of protease-activated receptor-2; ND: not determined; partial inh.: partial inhibition.

1-EBIO produced an endothelium-dependent hyperpolarization that was blocked by charybdotoxin. The definitive identification of IK_{Ca} as the target of charybdotoxin in EDHF-mediated responses was made with TRAM-39 and TRAM-34, two specific inhibitors of IK_{Ca}.[249] Indeed, these nonpeptidic, potent, and specific blockers, devoid of BK_{Ca} channel inhibitory activity, fully mimic the action of charybdotoxin either alone or in combination with apamin or UCL 1684.[206,232,245,250,251]

In many EDHF-mediated responses each toxin alone produces no or minor inhibition, while the combination abolishes the responses (Table 2). There are two possible explanations for these observations. Either both SK_{Ca} and IK_{Ca} are activated similarly during EDHF-mediated responses and the two systems are redundant or, alternatively, the EDHF-channel is a "new channel," possibly a heteromultimer composed of IK1 and SK α subunits, which requires the presence of the two toxins

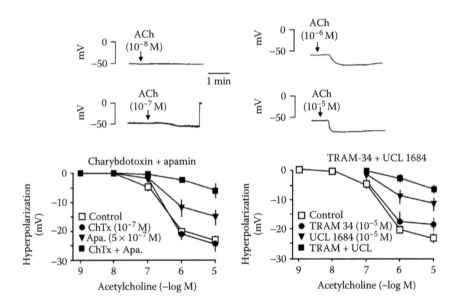

FIGURE 42 Involvement of SK_{Ca} and IK_{Ca} in the endothelium-dependent hyperpolarizations to acetylcholine in vascular smooth muscle of the guinea pig carotid artery (in the presence of inhibitors of NO-synthase and cyclooxygenase). Top: Original recording of the concentration-dependent hyperpolarizing effect of acetylcholine (ACh). Bottom: The concentration- and endothelium-dependent hyperpolarizations produced by acetylcholine are not affected by the presence of charybdotoxin (ChTx), partially inhibited by that of apamin (Apa), and virtually abolished by the combination of the two toxins (left). Similarly, TRAM-34, a nonpeptidic and selective blocker of IK_{Ca}, does not affect the hyperpolarization elicited by acetylcholine, while UCL 1684, a selective blocker of SK_{Ca}, produces a partial inhibition. The combination of the two blockers prevents the hyperpolarizing effect of acetylcholine. TRAM-34, UCL 1684, and their combination fully mimic the effects of charybdotoxin, apamin, and the association of the two toxins, respectively. These results show that both SK_{Ca} and IK_{Ca} are involved in EDHF-mediated responses. (By permission from the Nature Publishing Group; modified from Gluais et al.[206])

in order to be inhibited.[228,233,252] In favor of the latter hypothesis is the observation that in the rat brain cortex, the binding of [[125I]-charybdotoxin is increased in a dose-dependent manner by apamin,[228] suggesting that a novel channel with the two binding sites can be expressed. Indeed, in a heterologous expression system, SK3 can interact with the other α subunits SK1 and SK2 to form heteromeric channels.[253,254] Similarly, IK1 could theoretically form heteromultimers with SK1, SK2, or SK3,[255] although it is not known whether these heteromeric channels are formed in native cells. However, in the guinea pig carotid artery, by manipulating the calcium concentration it is possible to produce acetylcholine-induced endothelium-dependent hyperpolarizations that are preferentially inhibited either by apamin alone or by charybdotoxin (or TRAM 34) alone, indicating that either SK_{Ca} or IK_{Ca} can be activated and are able to produce endothelium-dependent hyperpolarizations.[206] These data are consistent with the hypothesis that, at least in the guinea pig carotid artery, the targets of charybdotoxin (or TRAM 34) and apamin are homomeric IK_{Ca} and SK_{Ca}, respectively, and, thus, do not support the existence of SK and IK heterotetramers. This interpretation is supported by the demonstration that IK1 does not form heteromeric entity with SK1, although the latter co-assemble with either SK2 or SK3.[254]

Taken in conjunction, these results indicate that, in most blood vessels, the activation of both SK_{Ca} and IK_{Ca} is required in order to observe endothelium-dependent hyperpolarizations. The subsequent step was to identify the cellular location of these two potassium channels, as two very different mechanisms could be hypothesized depending on the cell type expressing SK_{Ca} and IK_{Ca}. If the K_{Ca} channels, activated during EDHF-mediated responses, are located on the smooth muscle cells, this implies the release by the endothelial cells of a factor that diffuses and subsequently hyperpolarizes the smooth muscle cells. If the K_{Ca} channels are located on the endothelial cells, this implies that hyperpolarization of the endothelial cells is an important step in order to observe EDHF-mediated responses.

3.6 ENDOTHELIUM-DEPENDENT HYPERPOLARIZATION: LOCALIZATION OF THE POTASSIUM CHANNELS

In freshly isolated smooth muscle cells, an apamin-sensitive current is observed.[92,256] This current is voltage-sensitive, with an activation threshhold of approximately −30 mV, indicating that this channel is likely to be different from the classic SK_{Ca} channel, which is voltage-insensitive.[257] The sensitivity of this current to other blockers (scyllatoxin, tubocurarin, UCL 1684) is unknown. Whether the activation of this current could produce true hyperpolarization of the smooth muscle cells, i.e., whether its activation can be shifted to more negative potential by a putative EDHF, has not been determined. In freshly isolated vascular smooth muscle cells, IK_{Ca} is not or is very poorly expressed (Figures 7 and 8),[92,258] while in proliferating cells, as seen in cell culture or after vascular injury, its expression increases.[258,259] Furthermore, 1-EBIO, the activator of IK_{Ca}, does not affect directly the cell membrane potential of vascular smooth muscle cells but produces endothelium-dependent hyperpolarizations of these cells (Figure 43).[236,240,242,244,245,247,248]

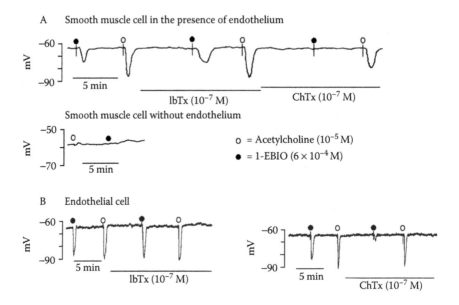

FIGURE 43 Effect of the opener of calcium activated potassium channels, 1-EBIO, on the cell membrane potential of vascular smooth muscle and endothelial cells of the rat hepatic artery (in the presence of inhibitors of NO-synthase and cyclooxygenase). (A) Membrane potential of vascular smooth muscle cells in the presence (top) and the absence (bottom) of the endothelium. Acetylcholine and 1-EBIO produce hyperpolarization of the smooth muscle cells only if the endothelium is present. These endothelium-dependent hyperpolarizations of the smooth muscle cells are unaffected by the presence of the BK_{Ca} blocker, iberiotoxin (IbTx), while charybdotoxin (ChTx) blocks the hyperpolarizing effect of 1-EBIO, leaving the response to acetylcholine minimally affected. (B) Similar results were obtained when measuring the membrane potential of the endothelial cells. These results show that 1-EBIO opens endothelial IK_{Ca} and that the opening of this endothelial channel leads to the hyperpolarization of the underlying smooth muscle cells. (By permission from the Nature Publishing Group; modified from Edwards et al.[236])

By contrast, both SK_{Ca} and IK_{Ca} channels are expressed in freshly isolated endothelial cells (Figures 10–13).[233,243,250,260–264] For instance, in freshly isolated endothelial cells from the porcine coronary artery, single-channel recordings of outside-out patches reveal calcium-sensitive K^+ conductances of 17.1 and 6.8 pS (and, very occasionally, a 2.7 pS conductance), the largest being reduced by charybdotoxin and the other by apamin, while the two were insensitive to iberiotoxin. Whole-cell perforated-patch analysis indicated that substance P and bradykinin, two substances that initiate EDHF-mediated responses, activate an outward current and that this current can be blocked partially by charybdotoxin or apamin and abolished by the combination of the two toxins. 1-EBIO evokes a charybdotoxin-sensitive, iberiotoxin-insensitive current. Messenger RNA encoding IK1, as well as SK2 and SK3, but not SK1, subunits of SK_{Ca} are detected by RT-PCR in samples of the endothelium. Western blot analysis indicates that SK3 protein is abundant in samples prepared from endothelium but not from whole arteries. SK2 protein is present in the nuclear fractions of whole arteries and also in endothelial samples. Immunofluorescent labeling confirms that IK1 and SK3 are expressed highly at the plasmalemma of endothelial cells but not in smooth muscle. SK2 is restricted to the perinuclear regions of both endothelium and smooth muscle.[243,264,265] Therefore, the endothelial cells of the porcine coronary artery express an intermediate- and a low-conductance Ca^{2+}-activated K^+ channel and the respective IK1 and SK3 gene products and these channels are opened by activators of the EDHF pathway. Similar conclusions have been reached in rat and murine arteries.[233,250] Together, these results suggest that endothelial IK_{Ca} and SK_{Ca} channels mediate the charybdotoxin- and apamin-sensitive components of the EDHF responses, respectively. An indirect piece of evidence strengthens this hypothesis. Indeed, in the isolated and perfused mesenteric artery of the rat, the combination of apamin plus charybdotoxin blocks EDHF-mediated responses if selectively applied to the lumen, i.e., in direct contact with the endothelium, but not when added to the solution bathing the adventitial side, which would have targeted preferentially the smooth muscle cells.[266]

The blockade of these two channels inhibits the hyperpolarization of the endothelial cells produced, for instance, by acetylcholine or bradykinin.[221,248,267] The question remains whether the hyperpolarization of the endothelial cells is essential to obtain EDHF-mediated responses?

3.7 IK_{CA}, SK_{CA}, AND ENDOTHELIAL HYPERPOLARIZATION

Stimulation of G-protein-coupled receptors, calcium ionophores, thapsigargin, and cyclopiazonic acid all increase endothelial $[Ca^{2+}]_i$ and produce hyperpolarization of the endothelial cells and endothelium-dependent hyperpolarization of the smooth muscle cells.[201,207,208,216–222,268] The hyperpolarization of the endothelial cells in turn favors the entry of calcium by increasing the driving force for this ion.[269,270] However, the principal mechanism that sustains the opening of endothelial K_{Ca} channels, following stimulation, is the capacitive calcium entry (TRP channels) elicited by the depletion of calcium stores.[210,271–273] It is therefore legitimate to wonder whether the intracellular increase in calcium *per se* is the essential contributor in eliciting EDHF-mediated responses or whether it is the hyperpolarization of the endothelial cells. Depolarization with KCl and inhibition of the two endothelial conductances (SK_{Ca} and IK_{Ca}) are two different maneuvers that fully prevent the hyperpolarization of the endothelial cells and the subsequent endothelium-dependent hyperpolarization of the vascular smooth muscle. However, these two maneuvers produce no or only partial inhibition in the sustained phase of the rise in endothelial intracellular calcium concentration (Figure 44).[231,270,274,275] Conversely, 1-EBIO, the activator of endothelial K_{Ca}, evokes EDHF-mediated responses without increasing endothelial $[Ca^{2+}]_i$.[247] The primary role of the increase in the endothelial calcium in EDHF-mediated responses is therefore to stimulate endothelial K_{Ca} channels. Thus, hyperpolarization of the endothelial cell is a prerequisite in order to observe the endothelium-dependent hyperpolarization of the underlying vascular smooth muscle.[92]

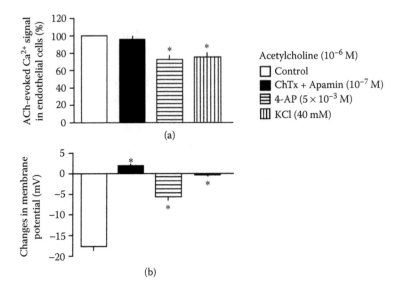

(a)

(b)

FIGURE 44 Endothelium-dependent hyperpolarization and calcium signaling in endothelial cells of the rat mesenteric artery (in the presence of inhibitors of NO-synthase and cyclooxygenase). (A) Acetylcholine induces an increase in endothelial intracellular calcium concentration that is unaffected by the presence of the combination of charybdotoxin (ChTx) plus apamin but partially inhibited by 4-aminopyridine (4-AP) or by an increase in extracellular potassium concentration. (B) By contrast, the acetylcholine-induced hyperpolarization is abolished by the combination of charybdotoxin (ChTx) plus apamin and by the increase in extracellular potassium concentration. The endothelium-dependent hyperpolarization also is partially inhibited by 4-aminopyridine. These results indicate that the primary role of the increase in the endothelial calcium concentration is to stimulate the opening of endothelial K_{Ca} channels. The hyperpolarization of the endothelial cell is a prerequisite in order to observe the subsequent endothelium-dependent hyperpolarization of the smooth muscle cells. (By permission from the Nature Publishing Group; modified from Ghisdal and Morel.[274])

Once the mechanism of endothelial hyperpolarization and its obligatory association with EDHF-mediated responses was determined, it was easier to understand the effects of various blockers of these responses. The calcium sensitivity of SK_{Ca} and IK_{Ca} channels is provided by the association of the calcium-binding protein calmodulin with the α subunit of these channels.[276–278] Although calmodulin is constitutively associated with the α subunits of K_{Ca}, calmodulin antagonists can inhibit IK_{Ca},[279] SK_{Ca},[280] and greatly reduce EDHF-mediated responses.[30,281] In some arteries, the nonspecific K_V blocker, 4-aminopyridine, inhibits EDHF-mediated responses. However, this effect should not be attributed to the blockade of plasmalemmal K_V but more likely to the prevention of calcium release from intracellular stores and the subsequent activation of endothelial IK_{Ca} and SK_{Ca} (Figure 44).[92,274,282]

In conclusion, hyperpolarization of the endothelial cells is a key step for the generation of endothelium-dependent hyperpolarization of the underlying smooth muscle cells (Figure 45).

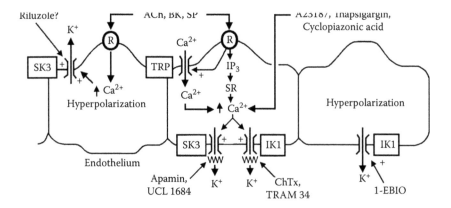

FIGURE 45 Schematic showing the mechanism underlying endothelial hyperpolarization, a key step in the generation of EDHF-mediated responses. R: receptor; IP3: inositol trisphosphate; SR: sarcoplasmic reticulum; TRP: Transient Receptor Potential Channel, which is likely to be the molecular identity for either the receptor-operated calcium channel activated by agonists, and store-operated calcium channel, which is activated following the calcium depletion of the internal stores (sarcoplasmic reticulum); SK3: small conductance calcium-activated potassium channel formed by SK3 α-subunits; IK1: intermediate conductance calcium-activated potassium channel formed by IK1 α-subunits; ChTx: charybdotoxin.

3.8 BEYOND ENDOTHELIAL HYPERPOLARIZATION

Increasing $[Ca^{2+}]_i$ in endothelial cells opens not only SK_{Ca} and IK_{Ca}, which results in the efflux and accumulation of K^+ in the myoendothelial space, but also leads to the activation of various enzymes including phospholipases and the subsequent metabolism of arachidonic acid by cyclooxygenases, cytochrome P450 epoxygenases, and lipoxygenase. The experimental evidence linking the activation of endothelial SK_{Ca} and IK_{Ca} to the hyperpolarization of the smooth muscle cells favors two explanations: (1) the hyperpolarization of the endothelial cell is transmitted directly to the vascular smooth muscle by means of gap junctions; and (2) the accumulation of K^+ ions, which are released from the endothelial cells through K_{Ca}, induces hyperpolarization of the smooth muscle by activating K^+ channels and/or Na^+/K^+-ATPase on the latter.

The increase in endothelial $[Ca^{2+}]_i$ triggering the synthesis of a cytochrome P450 metabolite, which is essential for the subsequent EDHF-mediated responses, has been the subject of many studies. However, this phenomenon may not rely fully on the activation of endothelial SK_{Ca} and IK_{Ca} and will be treated separately along with other explanations for the EDHF phenomenon.[283,284]

3.8.1 GAP JUNCTIONS

Docking of adjacent connexons leads to the creation of gap-junction channels that permit the transfer of ions and polar molecules (see Section 1.3.3). This provides an electrical continuity allowing a uniform membrane potential among coupled cells. Thus, both electrotonic spread of current and diffusion of signal molecules via gap junctions are likely to coordinate a functional syncytium such as vascular smooth muscle.[285–287] In the vascular wall, gap junctions not only link the same cell type, smooth muscle with other smooth muscle cells, or endothelial with other endothelial cells (homocellular gap junctions), but also smooth muscle with endothelial cells (heterocellular or myoendothelial gap junctions).[288]

3.8.1.1 Gap Junctions vs. a Diffusible Factor

The existence of a diffusible factor in EDHF-mediated responses has been reported in only a handful of studies.[17,91,289–291] However, a critical analysis of these studies shows that, in some of them, the involvement of either NO or prostacyclin cannot be ruled out, casting doubt on the existence of a diffusible EDHF.[17,92,289–291] In other studies performed in canine, porcine, and bovine large coronary arteries, the involvement of an EET is proposed[292,293] or can be suspected.[17,289,290,294] The production of EETs may not rely fully on the activation of endothelial SK_{Ca} and IK_{Ca} and will be discussed later (see section 3.8.3.1.1). On the other hand, several studies designed to demonstrate the existence of a diffusible EDHF failed to do so.[214,296–298]

From the passive electrical properties of single endothelial cells it can be calculated that the endothelium *in situ* should have a large electrical space constant of up to 500 μm, suggesting the aptitude of the endothelial layer to conduct electrical signals.[299] Taking these elements into consideration, it is not surprising that the involvement of gap junctions in EDHF-mediated responses was suggested very early on,[268,299–301] but this hypothesis was difficult to assess because of the lack of adequate tools. The inhibitors of gap junctions available at that time, such as heptanol and halothane, were difficult to manipulate and poorly specific and led to contradictory results.[227,302,303]

3.8.1.2 Evidence for the Involvement of Gap Junctions

There is now an impressive amount of evidence demonstrating the involvement of myoendothelial gap junctions in numerous EDHF-mediated responses. The endothelium-dependent hyperpolarization of vascular smooth muscle cells and the hyperpolarization of the endothelial cells show the same time course (Figure 46).[304] The number of myoendothelial gap junctions increases with the diminution of the size of the artery,[288,305,306] a phenomenon that parallels the contribution of the EDHF-mediated responses in endothelium-dependent relaxations.[31,307–312] Few studies have been dedicated to the specific identification of myoendothelial gap junctions, but they consistently show that the presence of these heterocellular gap junctions is associated with EDHF-mediated responses.[313–315] For instance, in the rat mesenteric artery, acetylcholine-induced hyperpolarizations of both endothelial and smooth muscle cells are associated with the characteristic pentalaminar myoendothelial gap junctions connecting the two cell types. By contrast, in the femoral artery of the same species, myoendothelial gap junctions cannot be detected and acetylcholine produces only the hyperpolarization of the endothelial cells without affecting the membrane potential of the smooth muscle cells.[313] Similarly, in the rat saphenous artery,

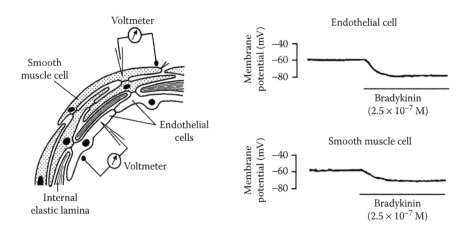

FIGURE 46 Simultaneous recording of the membrane potential of smooth muscle and endothelial cells in the porcine coronary artery (in the presence of inhibitors of NO-synthase and cyclooxygenase). Left: Schematic describing the experimental procedure. Right: The resting membrane potential is similar in the endothelium and smooth muscle cells. Bradykinin induces hyperpolarization of both the endothelial and the smooth muscle cells. The responses of the endothelial and smooth muscle cells shows a similar time course and amplitude. These experiments suggest that, in the porcine coronary artery, endothelial and smooth muscle cells are tightly coupled. (By permission of the American Physiological Society; modified from Beny and Pacicca.[302])

EDHF-mediated responses are observed in juvenile animals but disappear in adult rats. Numerous myoendothelial gap junctions are expressed in the young, while only very few of them can be observed in the adult.[314] These studies not only show a close correlation between the expression of myoendothelial gap junctions and the occurrence of EDHF-mediated responses, but also that the expression of these junctions is a plastic phenomenon involved in development and also possibly in the adaptation to diseases such as hypertension.[316]

Endothelium and smooth muscle cells communicate via these myoendothelial gap junctions as calcium can diffuse from one cell type to another.[317,318] The conduction of depolarization and hyperpolarization from smooth muscle cells to endothelial cells and from endothelial cells to smooth muscle cells occurs in various arteries.[240,302,319–323] More specific blockers of gap junctions include glycyrrhetinic acid, extracted from the liquorice root, and some of its derivative, HEPES buffer, and other taurine-based buffers, as well as Gap peptides. Glycyrrhetinic acid derivatives (18α-glycyrrhetinic acid, 18β-glycyrrhetinic acid, and the hemisuccinate of the α isoform carbenoxolone; Figure 17) are poorly soluble compounds with limited selectivity,[240,324] as they also target 11β-hydroxysteroid-dehydrogenase.[325] They are, nevertheless, better tools than heptanol or halothane as, in the high micromolar range, they inhibit gap junctions[326,327] and, in some arteries, EDHF-mediated responses.[244,321,328–333] HEPES and taurine-based buffer inhibit gap junctions by a pH-independent mechanism that may involve the binding of the buffer amino-sulphonate moiety to the gap junction protein.[334–336] Gap26 or Gap27 peptides are connexin mimetics that possess conserved sequence homology with the first and second extracellular loop domains, respectively, of the connexins thought to be involved in gap junction communication in the vascular wall (especially Cx 37, 40, and 43). The mechanism underlying the inhibitory action of these Gap peptides is not completely understood. The rapid and reversible inhibitory effect appears to be more consistent with a modulation of channel gating than prevention or disruption of connexon docking.[297,337] Gap peptides alone or in combination abolish or partially inhibit EDHF-like responses in many isolated arteries of the rabbit, guinea pig, rat, and pig,[214,244,248,287,297,329,337–341] as well as in the rat renal artery studied *in vivo*.[342]

3.8.1.3 Microcirculation

In peripheral vascular beds, the stronger electrical coupling between endothelial and smooth muscle cells, the increased relative proportion of the endothelial vs. smooth muscle mass, and the augmented dependence of smooth muscle cells on extracellular calcium (activation of Ca_V) to activate the contractile process concur to make the transmission of endothelial hyperpolarization toward smooth muscle cells an efficient mechanism to achieve relaxation. Indeed, in various microcirculatory beds, EDHF-mediated responses can be explained by the spreading of an electrotonic current from the endothelial to the smooth muscle cells via myoendothelial gap junctions. For instance, in submucosal arterioles of the small intestine of the guinea pig, acetylcholine induces outward currents in endothelial and smooth muscle cells that are blocked by the combination of charybdotoxin plus apamin. After the administration of blockers of gap junctions, acetylcholine elicits an outward current in endothelial cells but no longer in the smooth muscle cells, suggesting that the two cell types are connected electrically and form a functional syncytium.[240,321,322,343] In arterioles and feed arteries from the retractor muscle of the hamster, simultaneous measurements of the membrane potential in endothelial and smooth muscle cells show that electrical signals travel freely and bidirectionally between the two layers.[323,344] These electrical signals are associated with vasomotor responses since electrical events originating from a single endothelial cell can drive the relaxation of smooth muscle cells throughout an entire arteriolar segment.[323,344] The efficient coupling between endothelial cells provides the pathway for conducting hyperpolarization and vasodilatation along the arteriole.[345]

3.8.1.4 Larger Blood Vessels

In larger blood vessels, results are not necessarily as clear-cut. Inhibitors of gap junctions produce complete or near complete inhibition of the endothelium-dependent hyperpolarizations or relaxations attributed to EDHF in the aorta and mesenteric artery of the rabbit,[297,339] as well as in carotid arteries of the guinea pig (Figure 47).[244] However, only partial inhibition is observed in the rat mesenteric artery[244] and, depending on the study, minimal or complete inhibition in the rat hepatic artery.[244,338] The differences in the contribution of gap junctions to EDHF-mediated responses, as observed between vascular beds and between species, may

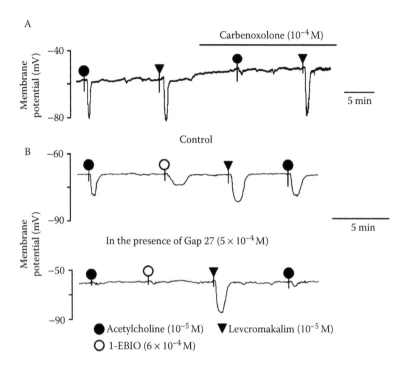

FIGURE 47 Inhibitors of gap junctions and endothelium-dependent hyperpolarization in response to acetylcholine in the guinea pig carotid artery (in the presence of inhibitors of NO-synthases and cyclooxygenases). (A) The addition of the inhibitor of gap junction, carbenoxolone, a water-soluble salt of 18β-glycyrrhetinic acid, inhibits the acetylcholine-induced, endothelium-dependent hyperpolarization of the vascular smooth muscle cells but does not affect the hyperpolarization produced by the K_{ATP} opener, levcromakalim. (B) Similar effect of another inhibitor of gap junction, Gap 27, a peptide with a sequence identical to a portion of an extracellular loop of connexin 43. Both acetylcholine and 1-EBIO produced an endothelium-dependent hyperpolarization (top) that is prevented by the presence of Gap 27 (bottom). By contrast, the endothelium-independent hyperpolarization in response to the K_{ATP} opener, levcromakalim, is unaffected by the gap junction inhibitor. These results indicate that in the guinea pig carotid artery, the hyperpolarization of the endothelial cells is transmitted to the smooth muscle cells via myoendothelial gap junctional communication. (By permission from the Nature Publishing Group; modified from Edwards et al.[244])

be linked to the differential expression of connexin subtypes in the vascular wall. For instance, in the hepatic artery of the rat, the three connexins 37, 40, and 43 are expressed.[346] In this artery, [37,43]Gap27 (an inhibitor of connexins 37 and 43), [40]Gap27 (inhibitor of connexin 40), or [43]Gap26 (inhibitor of connexin 43) individually do not affect EDHF-mediated responses.[244,338] However, the combination of three Gap peptides in order to target the three connexin subtypes inhibits them.[338]

3.8.1.5 Limitations in the Interpretation of the Results with Gap Junction Blockers

Although Gap peptides are supposed to be specific and their effects fully revers-ible,[347] the elevated concentration required to produce inhibition should temper the interpretation of the results obtained. Furthermore, connexins in addition to homo-typic gap junction channels (composed of a single connexin) can form heterotypic and heteromeric gap junctions that may show different conductances and are likely to be regulated differently.[286,348] Whether Gap peptides (or other gap junction blockers) similarly affect homotypic, heterotypic, and heteromeric gap junctions is unknown.

Gap junction inhibitors prevent the spread of endothelium-dependent hyperpo-larization, but the gap junctions inhibited are not necessarily myoendothelial gap junctions. In the small arteries of the retractor muscle of the hamster or the intestine of the guinea pig, these myoendothelial junctions play an essential role.[240,321–323,343–345] By contrast, in the hamster cheek pouch, a NO wave and the possible additional release of an EET at the stimulation site followed by homocellular electrotonic transmission (smooth muscle-to-smooth muscle and parallel endothelial-to-endo-thelial communications) underpins the conducted vasodilatation in response to ace-tylcholine, without involvement of myoendothelial gap junctions.[349–351] However, the expression of heterocellular gap junctions is similar in the arterioles of the hamster retractor muscle and those of the cheek pouch.[332] This indicates that the absence of a contribution of myoendothelial gap junctions in the cellular conduction pathways of the cheek pouch arterioles is not related to the absence of such myoendothelial gap junctions, as observed in the saphenous and femoral arteries of the adult rat.[313,314] It is due to the differential regulation of the gating of these myoendothelial gap junctions, illustrating the plasticity of intercellular communication in the blood vessel wall. The differences in the sympathetic and vasodilatator autonomic inner-vations of these two vascular beds of the hamster (feeding artery of the retractor muscle and arterioles of the cheek pouch), once thought to underlie the differential regulation mentioned above,[352] has been ruled out by the demonstration that sym-pathetic denervation does not alter either the profile of connexin expression nor the conduction of vasomotor responses in the hamster retractor muscle.[353]

In the porcine coronary artery, endothelium-dependent hyperpolarizations to substance P can be recorded with sharp microelectrodes in the smooth muscle cells situated close to the intimal layer or in those close to the adventitia. Gap junction blockers inhibit the hyperpolarization in the outer layer while the response remains unaffected in the intimal layer.[248] Similarly, relaxation and hyperpolarization can be observed in porcine coronary smooth muscle cells in segments without endothelial lining (separated by a partition chamber from the rest of the artery) only if this area is not physically separated from the intact segment.[354,355] These observations suggest that in large blood vessels, such as the porcine coronary artery, muscle-muscle gap junctions play a preponderant role in the spread of endothelium-dependent hyper-polarizations (Figure 48). In the rabbit iliac artery, the use of connexin-selective Gap peptides has allowed the discrimination of the role of myoendothelial gap junctions and that of homocellular smooth muscle communication.[356]

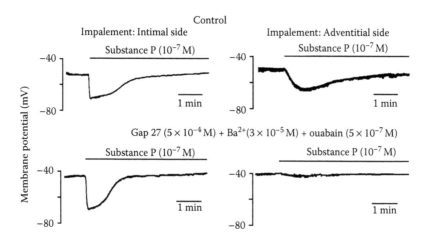

FIGURE 48 Endothelium-dependent hyperpolarization and gap junctions in isolated porcine coronary arteries (in the presence of inhibitors of NO-synthase and cyclooxygenase). Impalement of the smooth muscle cells via the intimal side (left) leads to the recording of membrane potential in smooth muscle cells situated in the inner layer of the artery and therefore in close proximity to the endothelial cells, while impalement of the smooth muscle cells via the adventitial side (right) produces recordings of membrane potential in smooth muscle cells situated in the outer layer of the artery and therefore distant from the endothelial cells. Under control conditions (top) the endothelium-dependent hyperpolarizations elicited by substance P are of similar amplitude in the two populations of smooth muscle cells studied, although the time course of the hyperpolarization appears faster in the smooth muscle cells situated near the intima when compared to that localized near the adventitia. In the presence of the gap junction inhibitor Gap 27 (and barium plus ouabain in order to prevent the hyperpolarizing effect of potassium; see Chapter 4; bottom), the hyperpolarization is fully blocked in the smooth muscle cells of the adventitial layer, while in those of the intimal layer, it is not affected. Therefore, under the tested experimental conditions, the gap junction inhibitor Gap 27 is preventing the transmission of the hyperpolarization between smooth muscle cells. The hyperpolarization observed in smooth muscle from the inner layer could be explained by two different phenomena not necessarily mutually exclusive: (1) the presence of myoendothelial gap junctions involving different connexins not inhibited by Gap 27 or (2) the release of a labile transferable endothelial factor that hyperpolarizes the innermost layer of smooth muscle cells (see section 3.8.3.1.1). These data show that the inhibition of an EDHF-mediated response by an inhibitor of gap junction does not necessarily imply that the endothelium-dependent hyperpolarization involves the transmission of the hyperpolarization via myoendothelial gap junctions. (By permission from the Nature Publishing Group; modified from Edwards et al.[248])

Overall, the data available show that gap junctions play an important role in EDHF-mediated responses. However, inhibition of a given EDHF-mediated response by gap junction blocker(s) does not necessarily exclude the presence of a diffusible factor and does not necessarily indicate that the mechanism of this response is the direct transmission of the endothelial hyperpolarization to smooth muscle cells through myoendothelial gap junctions.

The question remains whether gap junctions are a privileged site for the transfer of electrical charges or of messenger molecules (EDHF). Indeed, in some microcirculatory beds, such as the retractor muscle of the hamster, an electronic current spread from the endothelial to the smooth muscle seems to be the most likely explanation. However, from relaxation studies performed in various isolated arteries of the rabbit, a different hypothesis can be proposed. In these blood vessels, relaxations attributed to EDHF are sensitive to gap junction blockers when the response is receptor-mediated but not when the response is evoked by the calcium ionophore, A23187. This suggests that the calcium ionophore releases a factor across the endothelial surface while agonists provoke the preferential diffusion of this substance through gap junctions.[214]

3.8.1.6 Regulation of Gap Junction Communications

In arteries of rats and rabbits, EDHF-mediated responses are associated with a small but significant early and transient endothelium-dependent increase in cyclic-AMP content of the smooth muscle.[357–359] This phenomenon had not been detected in earlier studies since it was thought that EDHF-mediated responses occur independently of changes in the levels of cyclic nucleotides. Acetylcholine stimulates an efflux of cyclic-AMP from the endothelium. Inhibitors of adenylyl cyclase and of protein kinase A inhibit both the production of cyclic-AMP and the endothelium-dependent hyperpolarization of the subintimal smooth muscle cells, while a phosphodiesterase inhibitor amplifies both phenomena.[357,358] These results suggest that cyclic-AMP facilitates the electrotonic transmission of the endothelial hyperpolarization to the smooth muscle layers by increasing the conductance of both myoendothelial and smooth muscle-smooth muscle gap junctions.[357] Capacitive calcium influx promotes not only endothelial hyperpolarization by activating K_{Ca} but also stimulates the calcium-sensitive adenylyl cyclase isoform, resulting in an increase in the production of cyclic-AMP with subsequent enhancement of gap junction communication.[360]

Metabolites of arachidonic acid formed by cytochrome P450 monooxygenase are also involved in the regulation of gap junction communication (see section 3.8.3.1.1.4).

3.8.1.7 Genetically Modified Animals

Connexins 37 and 40 are the predominant gap junction proteins in the endothelial cells of the mouse.[361] The deletion of endothelial connexin 40 is associated with a diminution in dye transfer between adjacent endothelial cells, while that of connexin 37 does not produce a measurable alteration of this parameter. The endothelial cells taken from double knock-out mice (connexin 37 and 40) show a complete loss of endothelial dye transfer.[361,362]

Mice deficient for connexin 40 are hypertensive and exhibit a reduced spread of dilatation in response to endothelium-dependent vasodilators and electrical stimulation, as well as irregular arteriolar vasomotion.[363–365] Whether the hypertension is linked to the deletion of connexin 40 in the vascular wall itself is uncertain. The cardiovascular phenotype of the double knock-out mice with deletion of connexin 37 and 40 is unknown since these deletions are lethal, as a result of gross vascular structural abnormality.[362] Mice subjected to specific deletion of endothelial connexin 43 do not present any alteration in blood pressure,[366,367] possibly because this connexin is not the major one expressed in murine endothelial cells.

The interpretation of these data is complicated by concomitant changes in the expression of the other connexins. For instance, endothelial cells taken from connexin 40 (–/–) mice show an upregulation and redistribution of connexin 37 (as well as connexin 43 in the smooth muscle cells).[368] However, in another study, the deletion of either connexin 37 or 40 resulted in a significant post-transcriptional-dependent drop in the expression of the other nonablated connexins in the endothelial cells, while the overexpression of connexins 37 and 43 was still observed in the smooth muscle cells.[361]

3.8.1.8 Conclusion

Taken together these data indicate that the transmission of endothelial hyperpolarization to the underlying smooth muscle cells is likely to play a predominant role in the EDHF-mediated responses in many blood vessels (Figure 49).

3.8.2 Potassium Ions

An additional mechanism to achieve hyperpolarization and relaxation of the underlying vascular smooth muscle cells is also directly linked to the hyperpolarization of the endothelial cells. Indeed, the activation of endothelial IK_{Ca} and SK_{Ca} causes an efflux of potassium ions from the intracellular compartment toward the extracellular space. Potassium has long been recognized to be an endogenous metabolic vasodilator involved in exercise hyperemia or in the so-called active hyperemia occurring in the brain during neuronal activity.[369–374] A moderate increase in the extracellular potassium concentration (1 to 15 mM) can provoke the relaxation of vascular smooth muscle cells.[375,376] This observation is counter-intuitive since the Nernst equation would predict a depolarization of the smooth muscle cells as a result of such an increase in the extracellular potassium ion concentration. The depolarization of vascular smooth muscle cells produces the opening of Ca_V and the contraction of these cells. However, small increases in the extracellular concentration of potassium ions can also activate a specific population of potassium channels, K_{IR}[377] and the Na^+/K^+ pump.[378] The activation of these two systems overcomes the small depolarizing effects linked to the increase in potassium ions *per se* and the net result is hyperpolarization and thus relaxation of the smooth muscle cells. The endothelium is a cell monolayer and it would be expected that an efflux of potassium in the lumen of the blood vessel from this small cell mass would be washed away by the flowing blood and most likely be without any physiological consequences. However, an efflux of potassium toward the abluminal side could accumulate in the intercellular space between endothelial and smooth muscle cells and reach sufficient levels to activate K_{IR} and the Na^+/K^+ pump on the smooth muscle cells in the immediate vicinity of the endothelial cells releasing K^+. Therefore, potassium ions could be an EDHF or contribute to the mechanism of EDHF-mediated responses.

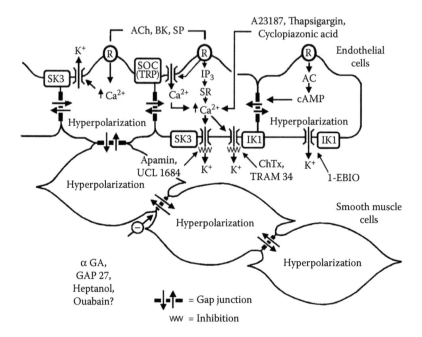

FIGURE 49 EDHF-mediated responses and gap junction communications. The endothelial hyperpolarization is conducted along the axis of the blood vessel via homocellular gap junctions and transmitted to the adjacent smooth muscle cells via myoendothelial gap junctions. The hyperpolarization of the smooth muscle cells of the inner layer of the media can also be transmitted to the entire vascular wall via homocellular gap junctions. Cyclic-AMP (cAMP) facilitates the electrotonic transmission of the endothelial hyperpolarization to the smooth muscle layers by increasing the conductance of both myoendothelial and smooth muscle-smooth muscle gap junctions. R: receptor; IP3: inositol trisphosphate; SR: sarcoplasmic reticulum; TRP: Transient Receptor Potential Channel; AC: adenylyl cyclase; αGA: glycyrrhetinic acid derivatives; SK3: small conductance calcium-activated potassium channel formed by SK3 α-subunits; IK1: intermediate conductance calcium-activated potassium channel formed by IK1 α-subunits; ChTx: charybdotoxin; SOC: store-operated channels.

3.8.2.1 Potassium Ions as EDHF

This hypothesis was demonstrated successfully in the hepatic and mesenteric arteries of the rat by Edwards et al. (Figures 50 and 51).[267] These authors provided evidence that the targets of apamin and charybdotoxin are located on the endothelial cells, that potassium ions accumulate in the intercellular space between endothelial and smooth muscle cells, and that the potassium efflux associated with the activation of endothelial IK_{Ca} and SK_{Ca} channels produced hyperpolarization by activating both K_{IR} and the Na^+/K^+ pump on the smooth muscle cells.[267] The contribution of K^+ in EDHF-mediated responses was confirmed in the same two arteries[236,379,380] and suggested in many other blood vessels including the femoral and skeletal muscle arteries of the rat,[381,382] murine mesenteric arteries,[252,383] porcine and bovine coronary

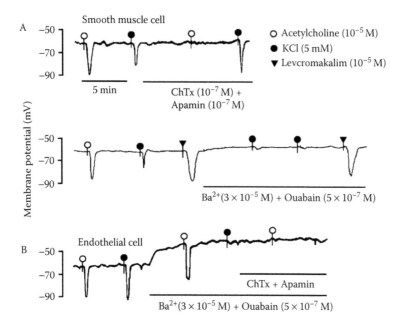

FIGURE 50 Membrane potential of vascular smooth muscle and endothelial cells in the isolated rat hepatic artery (in the presence of inhibitors of NO-synthases and cyclooxygenases). (A) The endothelium-dependent hyperpolarization evoked by acetylcholine is blocked by the combination of the two toxins charybdotoxin (ChTx) plus apamin. Raising the extracellular concentration of potassium by 5 mM produces a hyperpolarization that is not affected by the presence of the two toxins (top). However, the combination of the blocker of inward rectifying potassium channel (K_{IR}), barium (Ba^{2+}), plus the inhibitor of Na^+/K^+-ATPase, ouabain, blocks the hyperpolarization elicited by either acetylcholine or the small increase in potassium concentration. In contrast, the endothelium-independent hyperpolarization in response to the K_{ATP} opener, levcromakalim, is unaffected by the combination of barium plus ouabain. (B) Raising the extracellular concentration of potassium by 5 mM produces a hyperpolarization of the endothelial cells, which is blocked by combination of barium plus ouabain, while the hyperpolarization elicited by acetylcholine is unaffected by this combination. However, the effect of acetylcholine is blocked by the combination of the two toxins, charybdotoxin plus apamin, blockers of IK_{Ca} and SK_{Ca}, respectively. These results indicate that acetylcholine, by opening endothelial SK_{Ca} and IK_{Ca}, generates a potassium efflux. The accumulation of potassium ions in the intercellular space hyperpolarizes the smooth muscle cells by activating K_{IR} and Na^+/K^+-ATPase, which are inhibited by barium and oubain, respectively. In this artery, the potassium ion is, in essence, an EDHF. (By permission from the Nature Publishing Group; modified from Edwards et al.[267])

arteries,[384–386] porcine and human renal interlobar arteries,[387,388] and human thyroid arteries.[389]

From RT-PCR and immunohistochemistry studies, the K_{IR} channel most likely involved in potassium-ion-induced hyperpolarization in rat arteries, is composed of the Kir2.1 α subunits.[390,391] This conclusion is supported by the disappearance of

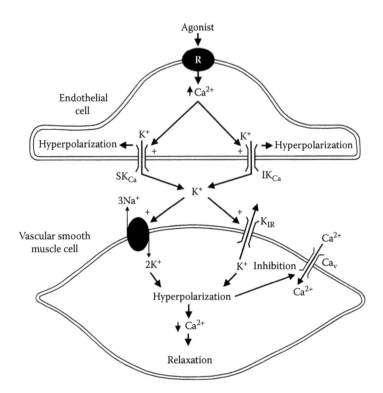

FIGURE 51 Schematic showing potassium ion acting as an EDHF. R: receptor; SK_{Ca}: small conductance calcium-activated potassium channel; IK_{Ca}: intermediate conductance calcium-activated potassium channel; ChTx: charybdotoxin; Ca_V: voltage-dependent calcium channel.

potassium-induced relaxations in mice knock-out for Kir2.1 while these responses are preserved in mice knock-out for Kir2.2.[392] The Na^+/K^+ pump activated during EDHF-mediated responses is unlikely to be composed of $\alpha 1$ subunits since this isoform is nearly fully activated at the physiological concentration of extracellular potassium (5 mM) and is poorly sensitive to ouabain ($IC_{50} > 10^{-5}$ M) in smooth muscle cells of the rat. However, the vascular endothelial and smooth muscle cells, including those of the rat mesenteric artery, can express the $\alpha 2$ and/or $\alpha 3$ isoforms. These isoforms are activated by a rise in extracellular concentrations of potassium from 1 to 15 mM, a window of concentrations compatible with the augmented interstitial concentration of potassium (6–8 mM) observed during an EDHF-mediated response in rat arteries. Furthermore, in the rat, these isoforms are markedly more sensitive to the inhibitory action of ouabain, a property that is also consistent with the observation that a reasonably low concentration of the cardiac glycoside (500 nM) inhibits endothelium-dependent hyperpolarizations in arteries of that species.[236,267,393–395]

In peripheral vascular beds, but also in autoregulatory vascular beds such as the coronary and cerebral circulations,[396,397] potassium can induce dilatation and therefore could function as a putative EDHF. However, in some blood vessels, such as the guinea pig carotid, mesenteric, coronary, submucosal, and cerebral arteries,[145,240,244,398,399] rat gastric and renal arteries,[400,401] porcine and bovine coronary arteries,[145,248,402,403] and human subcutaneous arteries[240] potassium does not evoke or inconsistently produces relaxations and hyperpolarizations. Therefore, in these blood vessels the contribution of K^+ ions in EDHF-mediated responses must be, if anything, minimal. Additionally, in some preparations, such as bovine coronary arteries,[404] rat renal juxtamedullary efferent arterioles,[405] and human subcutaneous arteries,[406,407] potassium induces hyperpolarization and/or relaxation, but this cannot account for EDHF-mediated responses, since the latter is insensitive to barium and/or ouabain. Finally, in mouse aorta and human umbilical vein endothelial cells, an increase in extracellular potassium ions within the physiological range reduces the cell membrane potential and inhibits the increase in $[Ca^{2+}]_i$ in the endothelial cells as well as the associated endothelium-dependent relaxation.[408] This suggests that in some arteries, K^+ ions not only are not vasodilators but could even inhibit endothelium-dependent relaxations.

3.8.2.2 Potassium Cloud

The elegant hypothesis suggesting K^+ as EDHF generated immediate controversy. Several reports were published soon after the princeps paper contesting the K^+ ion hypothesis in the very tissues where it was first described, i.e., the rat hepatic and mesenteric arteries.[232,340,400,409–412] As often in science, the inability to repeat experiments, in this case to attribute a role for K^+ ion as a vasodilator agent was due to differences in experimental conditions. Indeed the studies questioning the K^+ hypothesis were performed in experiments measuring relaxation or vasodilatation in isolated contracted or pressurized arterial segments. These maneuvers depolarize the vascular smooth muscle. Depending on the level of depolarization, the contribution of K_{IR} in the hyperpolarizing mechanism will be reduced since the outward potassium current flowing through this channel is observed only for a narrow window of membrane potentials. Furthermore, the depolarization of the smooth muscle cells increases the intracellular calcium concentration that in turn activates, as a braking mechanism, voltage-dependent potassium channels (sensitive to 4-aminopyridine) and BK_{Ca} (sensitive to iberiotoxin). This potassium efflux generates a local increase in extracellular K^+, which surrounds the myocytes ("potassium cloud"), preventing activation of the Na^+/K^+ pump and/or K_{IR} by any subsequent increases in potassium caused by the opening of the endothelial SK_{Ca} and IK_{Ca}. Thus, in the rat mesenteric artery, K^+-induced hyperpolarization is inhibited by the presence of phenylephrine,[380,413] but is restored by the addition of iberiotoxin, although the smooth muscles are even more depolarized in the presence of the BK_{Ca} blocker than in the presence of phenylephrine alone.[394,413] This "potassium cloud" hypothesis can explain some of the controversies concerning the involvement of K^+ ions in EDHF-mediated responses (Figures 52 and 53).

In quiescent rat mesenteric arteries, endothelium-dependent hyperpolarizations are relatively insensitive to gap-junction inhibitors under control conditions and are blocked by the combination of Ba^{2+} plus ouabain. However, they become sensitive to Gap27 in the presence of the $\alpha1$-adrenoceptor agonist, phenylephrine, unmasking the contribution of gap junctions.[413] The obligatory use of contractile agonists in the studies where the involvement of EDHF is implied from measuring relaxations may account

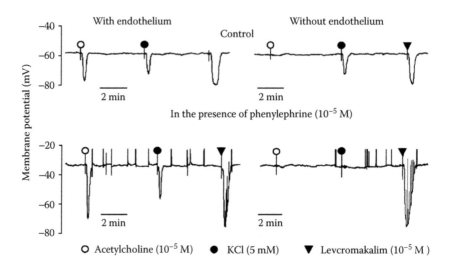

FIGURE 52 Acetylcholine and K⁺-induced hyperpolarizations in vascular smooth muscle cells of the isolated rat mesenteric artery either under resting conditions or after contraction and depolarization with phenylephrine (in the presence of inhibitors of NO-synthase and cyclooxygenase). Top: Resting conditions. Acetylcholine induces an endothelium-dependent hyperpolarization of the smooth muscle cells while K⁺ and levcromakalim produces endothelium-independent hyperpolarizations. Bottom: Phenylephrine-induced depolarization. In smooth muscle cells depolarized with phenylephrine, acetylcholine still produces endothelium-dependent hyperpolarization and levcromakalim endothelium-independent hyperpolarization. However, K⁺ no longer hyperpolarizes the arterial smooth muscle cells in strips without endothelium. These results indicate that the presence of a contractile and depolarizing agent (phenylephrine) impedes the response to K⁺. As most of the electrophysiological experiments are performed in resting tissues, while functional experiments are performed in contracted arterial rings or strips, this observation explains while under the former experimental conditions K⁺ ions may play a major role in EDHF-mediated responses and under the latter experimental conditions the role of K⁺ ions may be negligible. (By permission from the Nature Publishing Group; modified from Richards et al.[413])

for the divergent conclusions that gap junctions play a major role in EDHF-mediated responses, and that K⁺ is not an EDHF. *In vivo*, nearly all arteries are contracted to some extent depending on the degree of sympathetic nerve activation and the levels of circulating vasoconstrictor hormones. However, the magnitude of the contraction in the various vascular beds is likely to be markedly different. In the human forearm, the infusion of K⁺ produces a vasodilatation that is inhibited by Ba²⁺ and ouabain and almost abolished by the combination of the two inhibitors, suggesting a role for K⁺ ions as an endogenous vasodilatator under physiological conditions.[414,415] In addition, vasodilatation to bradykinin is mediated by an ouabain-sensitive pathway as a compensatory mechanism for impaired nitric oxide availability in essential hypertensive patients, suggesting that K⁺ can act as an EDHF in humans.[416]

Taken in conjunction, these findings indicate that gap junctions and K⁺ ions are not necessarily mutually exclusive and that, in a given blood vessel, for instance, in

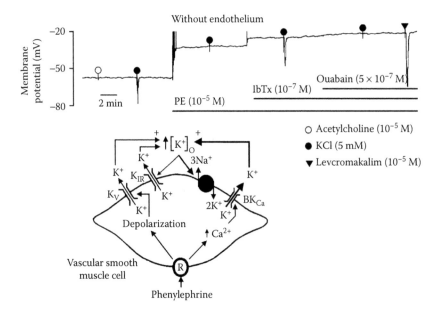

FIGURE 53 The potassium cloud hypothesis. Top: Membrane potential recording in smooth muscle of an isolated rat mesenteric artery without endothelium (in the presence of inhibitors of NO-synthase and cyclooxygenase). Acetylcholine does not hyperpolarize the smooth muscle cells since the arterial strip has been denuded. K^+ produces a hyperpolarization, which is prevented by the addition of the depolarizing agent phenylephrine (PE). The subsequent addition of iberiotoxin (IbTx), the specific blocker of large-conductance calcium-activated potassium channels (BK_{Ca}), produces a further depolarization and restores the hyperpolarizing response to K^+. This restored hyperpolarization is blocked by the presence of ouabain. (By permission from the Nature Publishing Group; modified from Weston et al.[394]) These results are explained in the schematic shown in the lower part of the figure. Phenylephrine interacts with its receptor (R) on vascular smooth muscle cells increasing intracellular calcium (Ca^{2+}) concentration and depolarizing the vascular smooth muscle cells. The rise in intracellular calcium activates BK_{Ca}, while the depolarization activates K_V as a brake mechanism. Potassium effluxes through these two channels and accumulates in the intercellular space (potassium cloud). Potassium activates the inward rectifying potassium channel (K_{IR}) and Na^+/K^+-ATPase, preventing the effects of any subsequent rise (addition) of potassium. Blocking BK_{Ca} and, to a lesser extent K_V, prevents the formation of the potassium cloud and restores a hyperpolarization to K^+. The impact on the increase in intercellular K^+ (width of the arrows) is dependent on the predominance of the mechanism involved.

the spiral modiolar artery of the guinea pig,[417] they can occur simultaneously or sequentially and also may act synergistically. The relative proportion of each mechanism almost certainly depends on numerous parameters including the state of activation of the vascular smooth muscle, the density of myoendothelial gap junctions, and the level of the expression of the appropriate isoforms of Na^+/K^+-ATPase and/or K_{IR} on the vascular smooth muscle (Figure 54).

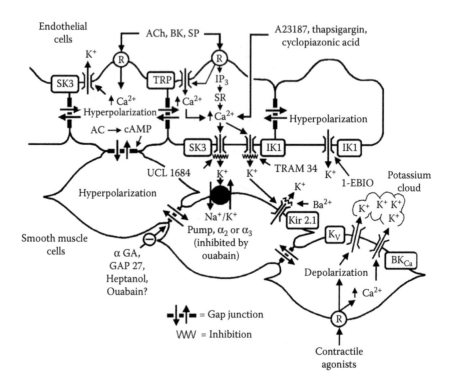

FIGURE 54 EDHF-mediated responses. The activation of endothelial potassium channels produces not only endothelial hyperpolarization conducted through myoendothelial gap junctions to the underlying vascular smooth muscle, but also accumulation of potassium ions in the intercellular space. These mechanisms are not necessarily mutually exclusive and, in a given blood vessel, they can occur simultaneously, sequentially, or even synergistically. The relative proportion of each mechanism almost certainly depends on numerous parameters, including the state of activation of the vascular smooth muscle, the density of myoendothelial gap junctions, and the level of the expression of the appropriate isoforms of Na$^+$/K$^+$-ATPase and/or K$_{IR}$. The presence or not of a "potassium cloud" can markedly affect the contribution of K$^+$ ions to EDHF-mediated responses. R: receptor; Bk: bradykinin; SP: substance P; IP3: inositol trisphosphate; SR: sarcoplasmic reticulum; TRP: Transient Receptor Potential Channel; AC: adenylyl cyclase; cAMP: cyclic-AMP; αGA: glycyrrhetinic acid derivatives; SK3: small conductance calcium-activated potassium channel formed by SK3 α-subunits; IK1: intermediate conductance calcium-activated potassium channel formed by IK1 α-subunits; Kir2.1: Inward rectifying potassium channel constituted of Kir2.1 α-subunits; K$_V$: voltage-gated potassium channels; BK$_{Ca}$: large conductance calcium-activated potassium channels.

3.8.3 OTHER IDENTIFIED ENDOTHELIUM-DERIVED HYPERPOLARIZING SUBSTANCES

Besides the production of NO and prostaglandins and the opening of IK$_{Ca}$ and SK$_{Ca}$, the endothelial cells can release other relaxing factors. The contribution of these depends on the species and/or the vascular bed studied. The hyperpolarization of the smooth muscle cells evoked by these other endothelium-derived hyperpolarizing factors is the major mechanism underlying the relaxation or contributes only partially to it. Because of the poor specificity of the tools that are available and/or because the various endothelial pathways involved in the regulation of vascular tone are not completely independent, the action of these other relaxing factors has frequently been mixed up with the pathway that involves the opening of endothelial IK_{Ca} and SK_{Ca}, generating intense confusion in the field. This subsection of the monograph will attempt to characterize these various relaxing factors and summarize the most appropriate means for their proper identification. Once these mediators are identified and their involvement is confirmed in a given vascular bed, they should be referred by their proper name (e.g., H_2O_2, EETs, CNP, adenosine, CO). The term EDHF-mediated responses should be restricted to the pathway that requires the activation of endothelial SK_{Ca} and IK_{Ca}. Implementation of this simple nomenclature could dramatically simplify the interpretation of future studies.

3.8.3.1 Metabolites of Arachidonic Acid

Following the demonstration of the contribution of metabolites of arachidonic acid to certain endothelium-dependent relaxations,[1] the involvement of a short-lived metabolite of arachidonic acid produced through either the cytochrome P450 monooxygenase or the lipoxygenase pathway was proposed to explain EDHF-mediated responses.[4,41,418] In several arteries, EDHF-mediated responses are blocked by inhibitors of phospholipase A2.[4,214,419,420] In porcine coronary arteries, EDHF signaling involves the association of the calcium-activated $cPLA_2$ with caveolin-1. Disintegration of caveolae is linked to the supression of EDHF-mediated responses.[421]

3.8.3.1.1 Cytochrome P450 Monooxygenase

EETs, derived from cytochrome P450 2C or 2J epoxygenases, play an important role in endothelium-dependent hyperpolarizations and relaxations in various blood vessels.[422] This is particularly the case in bovine,[419,423] porcine,[292,419] canine,[424] and human large coronary arteries.[312]

3.8.3.1.1.1 EETs and Vascular Smooth Muscle

EETs relax many blood vessels including cat cerebral and guinea pig, rat, bovine, porcine, and canine coronary arteries.[423,425–430] EETs hyperpolarize coronary arterial smooth muscle cells (Figure 55)[423,427,430] and increase the open-state probability of large-conductance calcium-activated potassium channels sensitive to tetraethyl ammonium, charybdotoxin, and iberiotoxin.[38,68,293,423,425,430,431] EETs can interact directly with BK_{Ca} to enhance the mean open time of this channel.[432] However, in vascular smooth muscle cells, EETs appear to interact with other sites. Thus, in isolated smooth muscle cells of the bovine coronary artery, as well as in human embryonic kidney cells, 11,12-EET activates BK_{Ca} in a cyclic-AMP- and cyclic-GMP-independent manner, but only in the presence of GTP and the guanine nucleotide binding protein, $Gs\alpha$.[433] The subsequent activation of the *Slo1* α subunit increases the activity of the channel without affecting the unitary conductance or

A   Porcine coronary artery

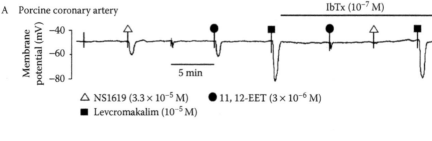

△ NS1619 (3.3×10^{-5} M)   ● 11, 12-EET (3×10^{-6} M)
■ Levcromakalim (10^{-5} M)

B   Guinea-pig carotid artery

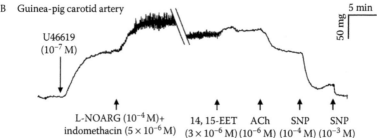

L-NOARG (10^{-4} M)+   14, 15-EET   ACh   SNP   SNP
indomethacin (5×10^{-6} M) (3×10^{-6} M) (10^{-6} M) (10^{-4} M) (10^{-3} M)

FIGURE 55 Species-specific effects of epoxyeicosatrienoic acids (EETs). (A) Membrane potential recording in smooth muscle cells of the isolated porcine coronary artery (in the presence of inhibitors of NO-synthase and cyclooxygenase). 11,12-EET and the opener of BK_{Ca}, NS 1619, produce a hyperpolarization blocked by the presence of iberiotoxin, while the hyperpolarization to the K_{ATP} opener levcromakalim is unaffected by the toxin. These results indicate that, in the porcine coronary artery, 11,12-EET activates BK_{Ca} of the smooth muscle. (By permission from the Nature Publishing Group; modified from Edwards et al.[248]) (B) Isometric tension recording in an isolated guinea pig carotid artery with endothelium. The thromboxane A_2 analogue, U46619, elicits a contraction. The addition of inhibitors of NO-synthases and cyclooxygenases produces a further increase in tension. The administration of 14,15-EET (as well as the other EETs, data not shown) does not affect either tone or the membrane potential (data not shown), while acetylcholine (ACh) produces a partial relaxation and the subsequent addition of sodium nitroprusside (SNP) a complete relaxation. These results indicate that in the guinea pig carotid artery, the release of EETs cannot account for the acetylcholine-induced EDHF-mediated relaxation. (By permission from the Nature Publishing Group; modified from Chataigneau et al.[230])

the calcium sensitivity.[434] Furthermore, in isolated bovine coronary arterial smooth muscle cells, 11,12-EET stimulates the production of ADP-ribose and the ADP ribosylation of Gsα, an important signaling molecule in BK_{Ca} activation and relaxation of smooth muscle cells.[435,436] Additionally, EETs activate K_{ATP} in cardiac myocytes[437] and in freshly isolated rat mesenteric artery myocytes via a protein kinase A-dependent mechanism.[438]

Although EETs activate BK_{Ca} through a G-protein-signaling cascade, the existence of a specific receptor(s) for EETs has not yet been proven in vascular cells.

Such a high-affinity binding site has been shown in guinea pig mononuclear cells.[439] The presence of such a receptor(s)/binding site(s) for EETs in vascular smooth muscle cells is suggested by the observation that some EET analogues [14,15-epoxyeicosa-5(Z)-enoic acid (14,15-EEZE) and 14,15-epoxyeicosa-5(Z)-enoic-methylsulfonylimide (14,15-EEZE-mSI)] act as inhibitors of the action of EETs.[440–443]

3.8.3.1.1.2 EETs as Endothelium-Derived Mediators

Inhibitors of cytochrome P450 monooxygenases can inhibit endothelium-dependent vasodilator responses to acetylcholine, bradykinin, or arachidonic acid, which are resistant to inhibitors of NO synthase and cyclooxygenase. This is the case *in vivo* in the hamster cheek pouch microcirculation, in the canine coronary circulation, and in the human forearm, as well as the perfused heart and kidney of the rat and in isolated arteries including canine, porcine, and bovine coronary arteries, small bovine adrenal cortical arteries, murine skeletal muscle arterioles, and human coronary, renal, internal mammary, and subcutaneous arteries (Figure 56).[181,349,406,419,426,444–454] The inhibition of a given EDHF-mediated response by a cytochrome P450 inhibitor has been frequently interpreted as the signature for the endothelial release of EETs. This may not necessarily be the case. First, cytochrome P450 inhibitors, such as clotrimazole, miconazole, and 17-ODYA, especially when studied at high concentrations, are notoriously unspecific and, among other effects, interfere with the activity of potassium channels.[230,455–458] They inhibit hyperpolarizations and relaxations of vascular smooth muscle induced by mechanisms unrelated to that of EDHF-mediated responses, including those evoked by levcromakalim, an activator of K_{ATP}.[456,459–462] The lack of selectivity of the available tools has long made it difficult to determine whether the activation of the two endothelial potassium channels IK_{Ca} and SK_{Ca} or that of EETs was involved in the mechanism underlying an EDHF-mediated response. For instance, clotrimazole, the reference inhibitor of cytochrome P450 epoxygenases, is also a reference blocker of IK_{Ca} (Figure 9).[463,464] Charybdotoxin, the other reference blocker of endothelial cell IK_{Ca}, also blocks the smooth muscle BK_{Ca} that are opened by EETs.

However, more selective inhibitors of cytochrome P450[465] and of the cellular targets of EETs [receptor antagonist or partial agonist (14,15-EEZE and 14,15-EEZE-mSi)],[440–442,466] as well as specific blockers of IK_{Ca} that are devoid of activity toward either BK_{Ca} or cytochrome P450 (TRAM 34 and TRAM 39),[249,467] allow the proper definition of the contribution of EET to EDHF-mediated responses. Porcine coronary arterial endothelial cells express several cytochrome P450 epoxygenases and in this artery, as well as in hamster resistance arteries, endothelium-dependent hyperpolarizations and relaxations can be inhibited by antisense oligonucleotides against cytochrome P450-2C8-9 (Figure 56).[430,468] In isolated bovine and porcine coronary arteries and in cultured endothelial cells, receptor-dependent and/or -independent agonists and pulsatile stretch release EETs from the endothelial cells, an effect blocked by inhibitors of cytochrome P450.[292,293,423,430,469,470] Bradykinin stimulates the release of a transferable factor from isolated bovine and porcine coronary arteries as well as from cultures of human umbilical vein endothelial cells,

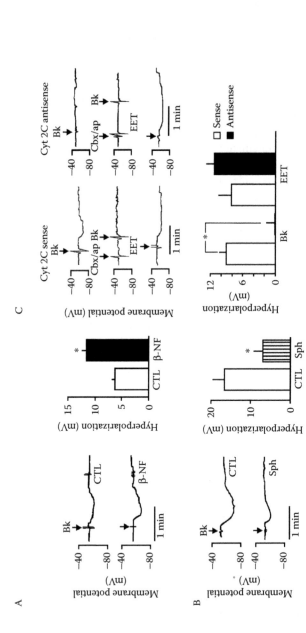

FIGURE 56 EETs as EDHF in isolated porcine coronary arteries (in the presence of inhibitors of NO-synthases and cyclooxygenases). (A) Membrane potential of smooth muscle cells and cytochrome P450 induction. Bradykinin (Bk) induces an endothelium-dependent hyperpolarization (CTL: control), which is enhanced by the induction of cytochrome P450 2C expression. The induction was achieved by a 48-h incubation of the isolated artery with β-naphtoflavone. Left: original traces; right summary bar graph. (B) Membrane potential of smooth muscle cells and cytochrome P450 inhibition. The bradykinin-induced hyperpolarization is suppressed by the cytochrome P450 inhibitor sulfaphenazole (Sph). Left: original traces; right summary bar graph. (C) Membrane potential of smooth muscle cells and antisense oligonucleotides directed against cytochrome P450 2C (Cyt 2C). The bradykinin-induced hyperpolarization is also prevented by a treatment with cytochrome 2C antisense oligonucleotides. The original traces (left) were control experiments performed with sense oligonucleotides. The hyperpolarization (top) is blocked by the combination of charybdotoxin (Cbx) plus apamin (ap: middle). 11,12-EET (EET) also hyperpolarizes the smooth muscle cells (bottom). In arteries treated with cytochrome 2C antisense oligonucleotides (right), the effects of 11,12-EET are preserved. The summary data are shown in the lower portion of the graph. These results indicate that, under the tested experimental conditions, the activation of cytochrome P450 2C is required to observe bradykinin-induced EDHF-mediated responses. Whether 11,12-EETs is released from the endothelial cells cannot be determined from these experiments. (By permission from the Nature Publishing Group; modified from Fissltthaler et al.[430])

that activates BK_{Ca} and hyperpolarizes vascular smooth muscle cells.[38,292,393,471] Furthermore, in coronary arteries, EDHF-mediated responses are increased by agents that enhance the endothelial expression of cytochrome P450 (Figure 56).[292,430] Taken in conjunction, these observations support the concept that, in some vascular beds, and particularly but not exclusively in the coronary circulation, EETs act as an EDHF, eliciting the relaxation of smooth muscle by opening BK_{Ca}.

In porcine and bovine coronary arteries the EDHF-mediated hyperpolarizations are inhibited by both 14,15-EEZE, a nonspecific antagonist of EETs, and by 14,15-EEZE-mSi, a preferential antagonist of 5,6-EET and 14,15-EET that does not inhibit the hyperpolarization evoked by 11,12-EET.[441,471,472] Both 5,6-EET and 14,15-EET relax the coronary smooth muscle cells by activating BK_{Ca}.[441,471–473] In bovine coronary arteries, bradykinin releases 14,15-EET from the endothelial cells, strongly suggesting that, as least in this artery, it is an endothelium-derived hyperpolarizing factor.[471]

The fact that in bovine and porcine coronary arteries, an EET, most likely 14,15-EET, is an endothelium-derived hyperpolarizing factor does not imply that the release of this EET explain all the EDHF-mediated responses in those arteries. For instance, in the porcine coronary artery, the endothelium-dependent hyperpolarization evoked by substance P involve exclusively the activation of the endothelial potassium channels SK_{Ca} and IK_{Ca} and the subsequent propagation of the hyperpolarization by myoendothelial gap junctions, while that to bradykinin involves two different mechanisms: the activation of endothelial potassium channels and, additionally and independently of the activation of the endothelial potassium channels, the release of 14,15-EET with the subsequent activation of BK_{Ca} on the smooth muscle.[336,472]

However, in other vascular preparations such as the guinea pig carotid and basilar arteries or in the rat hepatic, mesenteric, and gracilis arteries, EETs evoke no or minor relaxations and/or hyperpolarization (Figure 55). Furthermore, involvement of EETs in EDHF-mediated responses appears unlikely in these[52,202,206,233,474–477] and many other preparations, including the human omental, gastroepiploic, subcutaneous, mesenteric, renal interlobar, and submucosal intestinal arteries,[174,388,478–482] guinea pig coronary and mesenteric arteries,[231,399] rabbit renal and mesenteric arteries,[296,483] rat renal and carotid arteries,[250,484] murine hindlimb vascular bed,[179] and porcine coronary microvessels.[403]

In conclusion, in some vascular beds and following appropriate stimulation, the endothelial cells can release EETs, which act as diffusible endothelium-derived hyperpolarizing factors. In order to dissipate the confusion associated with the involvement, or not, of EETs in EDHF-mediated responses, most of those responses need to be revisited with the specific tools available in order to properly determine the exact mechanism(s) underlying each endothelium-dependent hyperpolarization in any given artery.

3.8.3.1.1.3   *EETs as Intracellular Second Messengers*
EETs can be involved in EDHF-mediated responses without being a diffusible factor *per se*. Indeed, the hyperpolarization of the endothelial cells may be partly regulated by the activation of cytochrome P450. For example, EETs can regulate endothelial calcium homeostasis, by modulating store-operated Ca^{2+} channels (SOC) in response

to calcium-store depletion,[485–488] and thus can be identified as the elusive calcium influx factor (CIF).[489] Store depletion causes mobilization of arachidonic acid from the endoplasmic reticulum membrane[490–492] and activation of the nearby cytochrome P450.[493,494] EETs, possibly the 5-6 isomer, would then diffuse from the endoplasmic reticulum and activate SOC (Figure 16). Although still controversial, TRP channels have been proposed as the molecular identity of SOC and receptor-operated channels (ROC).[495] In vascular endothelial cells, the 5,6-EET isomer activates TRPV4 channels (Figure 57),[496] further supporting a possible role for EETs in endothelial calcium homeostasis. The regulation of endothelial $[Ca^{2+}]i$ controls the activation of endothelial K^+ channels. In addition, independent of their role in calcium homeostasis, EETs could, as in smooth muscle cells, regulate the activity of K_{Ca} (Figure 57).[497] Finally, EETs also cause a biphasic change in gap junction communication

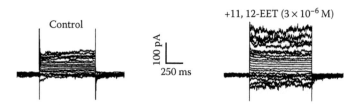

FIGURE 57 EETs as intracellular messengers and autocrine factors. (A) Murine aortic endothelial cells (inside-out configuration of the patch-clamp technique). Single channel activation of a TRPV4 (Transient Receptor Channel Vanilloid-4) by 5,6-EET causes calcium entry. Left: original traces; right: open probability. (By permission from the Nature Publishing Group; modified from Watanabe et al.[496]) (B) Freshly dissociated porcine coronary arterial endothelial cells (whole-cell configuration of the patch-clamp technique). 11,12-EET increases the outward current carried by the activation of calcium activated potassium channels (Bychkov and Félétou, unpublished observations). These two examples show that EETs have multiple targets on endothelial cells, which can be involved in the initiation and/or regulation of EDHF-mediated responses.

between endothelial cells,[498] suggesting that these arachidonic acid metabolites, provided they produce similar effects on myoendothelial gap junctions, could favor transmission of the endothelial hyperpolarization toward the smooth muscle cells (Figure 58).

EETs and other products of cytochrome P450 can therefore also be classified as intracellular second messengers that in some vascular beds could be crucial for

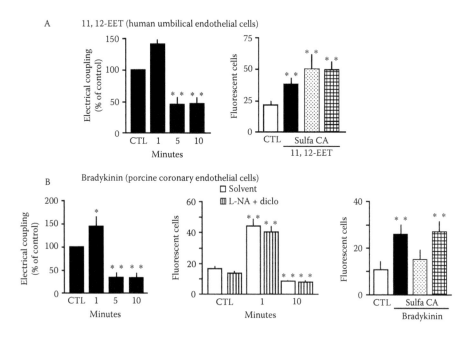

FIGURE 58 EETs and gap-junctions. (A) Effects of 11,12-EET on gap junctional commu- nication between endothelial cells of the human umbilical vein (in the presence of inhibitors of NO-synthase and cyclooxygenase). Left: Time course of the biphasic changes in electrical coupling produced by 11,12-EET, assessed by measuring electrical capacity with the patch- clamp technique (cell-attached configuration). Right: Increase in dye transfer (Lucifer yellow) between endothelial cells produced by 11,12-EET. The effects of 11,12-EET are not altered by an inhibitor of cytochrome P450, sulfaphenazole (sulfa), or by the presence of the combination of charybdotoxin plus apamin (CA). (B) Effects of bradykinin on gap junctional communi- cation between porcine coronary artery endothelial cells. Left: Time course of the biphasic changes in electrical coupling produced by bradykinin. Middle: The effects of bradykinin are not affected by the presence of inhibitors of NO-synthases (L-NA) and cyclooxygenases (diclofenac: diclo). Right: Increase in dye transfer between endothelial cells produced by bradykinin. The effects of bradykinin are inhibited by sulfaphenazole (sulfa) but unaffected by the presence of the combination of charybdotoxin and apamin (CA). These results show that bradykinin produces an early increase in gap-junctional communications via the activation of cytochrome P450 and the generation of EETs. This effect is independent of the activation of endothelial SK_{Ca} and IK_{Ca}. (By permission of the American Heart Association, Lippincott Williams & Wilkins; modified from Popp et al.[498])

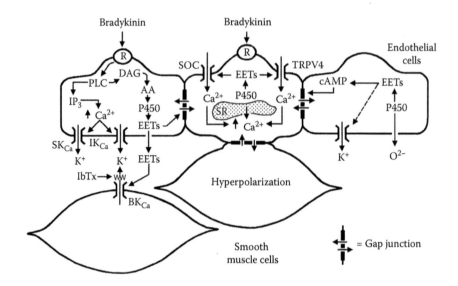

FIGURE 59 Schematic summarizing the multiple effects of EETs in the vascular wall. EETs can interact at various levels regulating EDHF-mediated responses by regulating the intracellular calcium concentration, potassium channel activity, and gap-junctional communications or by acting as an active diffusible hyperpolarizing factor. R: receptor; PLC: phospholipase C; DAG: diacylglycerol; AA: arachidonic acid; SR: sarcoplasmic reticulum; TRP: Transient Receptor Potential Channel; SOC: store-operated channel; AC: adenylyl cyclase; cAMP: cyclic-AMP; P450: cytochrome P450 monooxygenase; EETs: epoxyeicosatrienoic acids; SK_{Ca}: small conductance calcium-activated potassium channel; IK_{Ca}: intermediate conductance calcium-activated potassium channel; BK_{Ca}: large conductance calcium-activated potassium channels.

the initiation and transmission of endothelial cell hyperpolarization and, as a consequence, for EDHF-mediated hyperpolarization and relaxation of vascular smooth muscle cells. However, more work is needed to determine precisely in which blood vessels and under which conditions EETs contribute to EDHF-mediated responses and whether EETs are released and act as diffusible hyperpolarizing factors or mainly as intracellular second messengers (Figure 59).

3.8.3.1.1.4   Other Effects of EETs

EETs can also hyperpolarize platelets and inactivate them by inhibiting the expression of adhesion molecules and platelet adhesion to endothelial cells. This effect is directly linked to the membrane potential of the platelets.[499] EETs also possess fibrinolytic properties through the induction of t-PA in vascular endothelial cells.[500] They prevent leukocyte adhesion to the vascular wall by decreasing the leukocyte-induced expression of adhesion molecule by endothelial cell through a mechanism involving inhibition of the transcription factor NF-kappaB, independently of changes in endothelial cell membrane potential.[501] This effect, which in human endothelial cells has been attributed to the activation of cytochrome P450 2J2, suggests that EETs can modulate thrombosis and inflammation. Additionally, eicosanoids stimulate tyrosine kinase activity in endothelial and smooth muscle cells, including the extracellular signal regulated kinase 1 and 2 (Erk1/2), p38 MAP and protein kinase B/Akt.[502–504] The further activation of a complex cascade of phosphorylation stimulates endothelial cell proliferation and angiogenesis by a mechanism involving activation of EGF receptors.[505–510] Furthermore, EETs inhibit migration of the smooth muscle cells through a cyclic-AMP-PKA pathway,[511] indicating that these cytochrome P450 epoxygenase derivatives play a role not only in angiogenesis but also in vascular remodeling.

These positive effects of cytochrome P450 activation could be blunted by the fact that cytochrome P450 2C9, the isoform most commonly associated with EDHF-mediated responses, is also a significant source of reactive oxygen species in coronary arteries.[512] This suggests that this enzyme could be involved in the generation of oxidative stress and vascular disease (Figure 59). However, different cytochrome P450 isoforms generate varying amounts of oxygen-derived free radicals and over-expression of cytochrome P450 2J2, in bovine aortic endothelial cells, protects against oxidative stress.[513,514] Furthermore, reactive oxygen species can no longer be regarded solely as an indicator of cellular damage or as byproducts of metabolism. They are mediators of physiological and pathophysiological events and are also involved in the regulation of vascular tone.[515]

3.8.3.1.2 Lipoxygenases

Under various physiological and/or pathophysiological conditions, endothelial cells express various lipoxygenases, including 5-lipoxygenase, platelet type 12-lipoxygenase, 12/15 lipoxygenase (rat, mouse, pig), and 15-lipoxygenase type I.[516–519] These enzymes metabolize arachidonic acid into various vasoactive derivatives (relaxing and contracting substances).

Some of the derivatives can be considered as endothelium-derived hyperpolarizing factors since they are released by endothelial cells in response to neurohumoral substances (e.g., acetylcholine, thrombin, arachidonic acid) or oxidative stress, and produce relaxation and hyperpolarization of vascular smooth muscle cells. In coronary arteries of the rat, the EDHF-mediated vasodilatation in response to protease-activated receptor-2 involves such a lipoxygenase-derived eicosanoid.[520] In the basilar artery of the same species, the arachidonic acid-induced vasodilatation is attributed to the release of the 12-(S)-HETE and the subsequent activation of BK_{Ca} on the smooth muscle.[521] Similarly, in porcine coronary microvessels and, to a much lesser extent, in large conduit coronary arteries, 12-(S)-HETE, released by the endothelial cells in response to oxidative stress, acts as a potent relaxing and hyperpolarizing factor by opening BK_{Ca} on the smooth muscle.[522] In these coronary arteries the relaxing and hyperpolarizing effect of H_2O_2 involves the lipoxygenase-dependent activation of BK_{Ca}.[523] Provided G-protein-regulated inward-rectifier potassium channels (GIRK) are expressed in vascular smooth muscle cells, 5-lipoxygenase metabolites could also theoretically stimulate directly G-proteins and activate this population of K_{IR} in a receptor-independent manner.[524]

The best characterization of a lipoxygenase derivative as an endothelium-derived hyperpolarizing factor was obtained in rabbit arteries. In the aorta, in response to acetylcholine, arachidonic acid, or interleukin-13, and in the mesenteric artery in response to acetylcholine, a NO-and PGI$_2$-independent component of the endothelium-dependent relaxation involves a lipoxygenase metabolite,[525–527] which relies on the activation of potassium channels of the smooth muscle.[528] Upon exposure to acetylcholine, arachidonic acid, mobilized following activation of phospholipase A$_2$, is metabolized by 15-lipoxygenase to 15-HPETE, which undergoes an enzymatic rearrangement to the hydroxy (H)-epoxyeicosatrienoic acid (11-H-14,15-EETA and/or 15-H-11,12-EETA). The hydrolysis of the epoxy group results in the formation of 11,12,15-trihydroxyeicosatrienoic acid (11,12,15-THETA) and 11,14,15-THETA, the former possessing relaxing and hyperpolarizing properties, while the latter is inactive.[529,530] In the smooth muscle cells of the rabbit aorta, 11,12,15-THETA produces relaxation by activating an apamin-sensitive but charybdotoxin-insensitive K$_{Ca}$ of "small" conductance (24 pS; Figure 60)[443] while in the mesenteric artery BK$_{Ca}$ are involved.[527]

However, although a strong case for lipoxygenase derivatives acting as endothelium-derived mediators can be built in specific arteries, most of the EDHF-mediated responses do not involve an arachidonic acid metabolite from these pathways. This conclusion has been reached, for instance, for guinea pig carotid and coronary arteries,[64,176,202,427] rat uterine artery,[531] porcine coronary artery,[532,533] or human resistance arteries.[174] One should keep in mind that inhibitors of lipoxygenase, like some cytochrome P450 monooxygenase inhibitors, are far from specific as they can affect guanylyl cyclase as well as ionic channels including potassium channels or voltage-dependent calcium channels.[176,534,535]

3.8.3.1.3 Endocannabinoids

Endocannabinoids are the endogenous ligands that bind and activate the cannabinoid receptors first identified from the action of (-)-Δ9-tetrahydrocannabinol, the active substance of the recreational drug marijuana, derived from the plant *Cannabis sativa*. To date, five lipid mediators have been identified as endogenous cannabinoids: anandamide (*N*-archidonoyl-ethanolamine), the first identified ligand and therefore the most studied, 2-arachidonoyl-glycerol (2-AG), noladin (2-arachidonyl-glyceryl-ether), virhodamine (*O*-arachidonoyl-ethanolamine), and *N*-arachidonoyl-dopamine (NADA). These endogenous mediators are agonists of either one or both of the cloned and characterized G-protein coupled cannabinoid receptor subtypes, CB$_1$ and CB$_2$. Anandamide is thought to be synthesized and released in a calcium-dependent manner following the enzymatic hydrolysis of the precursor *N*-acyl-phosphatidylethanolamine by a specific phospholipase D, and is a preferential activator of the CB$_1$ receptors (Figure 61).[536–538] Both in isolated blood vessels or in anesthetized animals, endogeneous and exogeneous cannabinoids usually have vasodilator properties.[539]

In isolated and perfused mesenteric and coronary arterial beds of the rat, anandamide induces a dilatation that mimics EDHF-mediated responses. Indeed, both responses are blocked by the combination of charybdotoxin plus apamin and also by high concentrations of the CB$_1$ receptor antagonist SR141716A (rimonabant).[540] These observations prompted the proposal that anandamide was EDHF in these vascular beds.[541–544] However, further analysis demonstrated that anandamide did not fulfill the criteria to be classified as such. In isolated blood vessels from various species (pig, guinea pig, rat, rabbit), anandamide either did not produce hyperpolarization

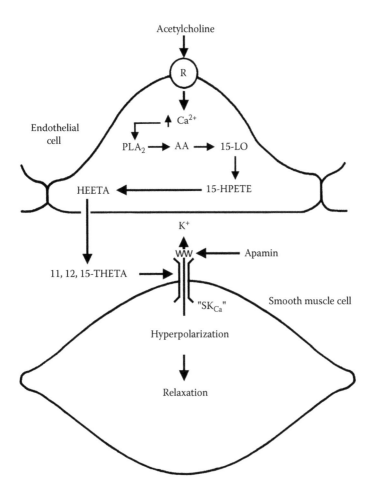

FIGURE 60 Schematic showing a metabolite of arachidonic acid produced by 15 lipoxyge-nase (15-LO) acting as an EDHF. This pathway has only been identified so far in the rabbit aorta. Whether or not the apamin-sensitive potassium channel on the smooth muscle is a true SK_{Ca} is not known. R: receptor; PLA_2: phospholipase A_2; AA: arachidonic acid; 15-HPETE: 15-hydroperoxy-5,8,11,13-eicosatetraenoic acid; HEETA: hydroxy (H)-epoxyeicosatrienoic acid; 11,12,15-THETA: 11,12,15-trihydroxyeicosatrienoic acid; "SK_{Ca}": small conductance calcium-activated-like potassium channel.

or produced a hyperpolarization that did not share the characteristics of the EDHF-mediated response.[234,458,545,546] For instance, in the isolated mesenteric artery of the rat, the acetylcholine-induced endothelium-dependent hyperpolarization involves the activation of both IK_{Ca} and SK_{Ca} and is reproducible over time, while the endothe-lium-independent hyperpolarization elicited by anandamide is tachyphylactic and involves the activation of K_{ATP}.[458] Furthermore, EDHF-mediated responses were not inhibited by pharmacologically relevant concentrations of the CB_1 receptor

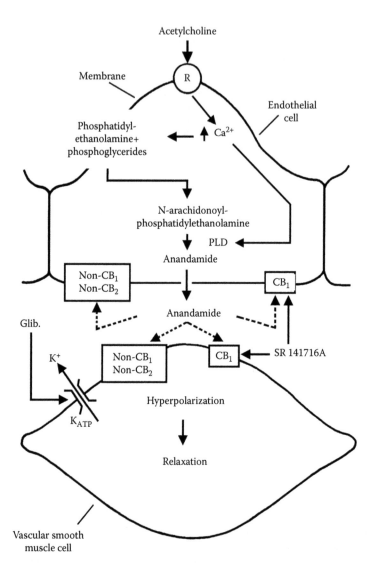

FIGURE 61 Theoretical pathway by which anandamide could act as an EDHF. No EDHF-mediated responses have actually been linked to this mechanism. R: receptor; PLD: phospholipase D; Glib.: glibenclamide; CB_1: CB_1 cannabinoid receptor subtype; Non-CB_1-non-CB_2: putative alternative cannabinoid receptor.

antagonist.[234,458,545,547] Finally, some of the effects of anandamide were endothelium-dependent, i.e., anandamide, depending on the vascular bed, releases NO and prostacyclin, and induces EDHF-mediated responses or conversely inhibits those responses.[234,458,547–550] These observations make it unlikely that an endocannabinoid mediates endothelium-dependent hyperpolarizations (Figure 62).

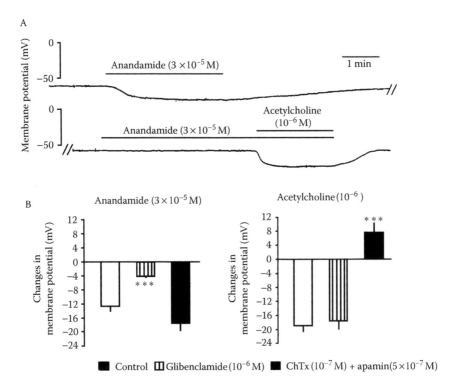

FIGURE 62 Acetylcholine-induced endothelium-dependent hyperpolarization in isolated mesenteric of the rat does not involve anandamide (in the presence of inhibitors of NO-synthase and cyclooxygenase). Top: Membrane potential of the vascular smooth muscle cells of isolated rat mesenteric artery. The original trace shows the hyperpolarization evoked by anandamide. However, the hyperpolarization in response to anandamide is tachyphylactic and a second application of anandamide no longer affects the membrane potential. Nevertheless, acetylcholine is still able to produce an endothelium-dependent hyperpolarization in the presence of anandamide. This experiment indicates that the mechanism of acetylcholine-induced hyperpolarization does not involve the release of anandamide. Bottom: Summary analysis of the mechanisms of anandamide- (left) and acetylcholine- (right) induced hyperpolarization. The hyperpolarization induced by anandamide involves the activation of K_{ATP} since it is blocked by glibenclamide but not that of K_{Ca}, since the combination of charybdotoxin (ChTx) plus apamin is ineffective. The converse is true for the hyperpolarization evoked by acetylcholine. (By permission from the Nature Publishing Group; modified from Chataigneau et al.[458])

Nevertheless, anandamide and possibly some of the other endocannibinoids, besides their actions in the central and peripheral nervous system, are likely to play an important role in the regulation of the cardiovascular system.[539,551–553] The analysis of the effects of these compounds is difficult to assess because they can activate multiple molecular targets. For instance, anandamide, a preferential CB_1 receptor agonist with antagonistic properties at the CB_2 receptor subtype, can be converted (and vice versa) to virhodamine, which, in contrast, is an antagonist of the CB_1 and an agonist of the CB_2 receptor subtypes.[538] Additionally, anandamide is also an agonist of the not-yet-cloned, elusive non-CB1, non-CB2, G-protein-coupled cannabinoid receptor that may underlie some of the endothelial effects of the lipid mediator.[554,555] Furthermore, anandamide stimulates and inhibits different populations of channels/receptors. Anandamide activates the vanilloid receptors V1 (also known as TRPV1) on sensory nerves and causes vasodilatation by releasing calcitonin gene-related peptide (CGRP), mimicking the effects of capsaicin, the active ingredient of "hot" red peppers.[556] Anandamide also inhibits the T-type calcium channel[557] and the K_{2P} potassium channel, TASK-1.[558]

Taken in conjunction, the available data indicate that, even if the release of endocannabinoid cannot account for EDHF-mediated responses, these lipid mediators can modulate vascular tone and blood pressure by many different mechanisms, not only in the periphery, via actions on endothelium and smooth muscle, sympathetic nerve endings, and sensory nerves, but also in the central nervous system.[539]

3.8.3.2 Hydrogen Peroxide (H_2O_2)

Superoxide anion, the one-electron reduction of molecular oxygen ($O_2^{\cdot-}$), can be generated by different enzymes (NADPH oxidase, xanthine oxidase, cyclooxygenases, NO synthases, cytochrome P450 monooxygenases, enzymes of the mitochondrial respiratory chain) in virtually all cell types including vascular smooth muscle and endothelial cells. Superoxide either spontaneously or enzymatically through dismutation by superoxide dismutase (SOD) is reduced to the uncharged H_2O_2. H_2O_2 in the presence of the enzyme catalase or glutathione peroxidase is then dismutated into water and oxygen. In the presence of transition metals (copper, iron) or superoxide anion, H_2O_2 generates, through the Fenton or Haber-Weiss reaction, respectively, the highly reactive hydroxyl radicals ($\cdot$OH), which can be scavenged by mannitol or dimethylthiourea. H_2O_2 can act locally close to its site of production or, since it is an uncharged molecule, diffuse through the cell membrane and act on neighboring cells (Figure 63).

3.8.3.2.1 H_2O_2 and Vascular Smooth Muscle

Both endothelial[403,559–563] and smooth muscle cells[564–568] generate significant amounts of reactive oxygen species, either spontaneously or in response to receptor-mediated and nonreceptor-mediated stimuli. H_2O_2 can, depending on the tissue, the experimental conditions, or the concentrations studied, possess dilator or constrictor properties.

H_2O_2 induces relaxations in porcine, canine, bovine, and human coronary arteries,[403,533,569–578] murine, rabbit, rat, and human mesenteric arteries,[481,579–583] rabbit iliofemoral artery,[563] feline and canine cerebral arteries,[584–586] and murine, rabbit, rat, and guinea pig aorta.[581,587–591] By contrast, contractions have been reported in the bovine coronary artery,[592] the canine cerebral artery,[593] the mesenteric artery of the rat,[582] the aorta of the rat and rabbit,[587,594–598] and in human submucosal microvessels.[482] *In vivo*, the administration of either norepinephrine or angiotensin II in mice produces a vasopressor response associated with the production of H_2O_2. In genetically modified mice overexpressing catalase, the basal arterial pressure and the basal

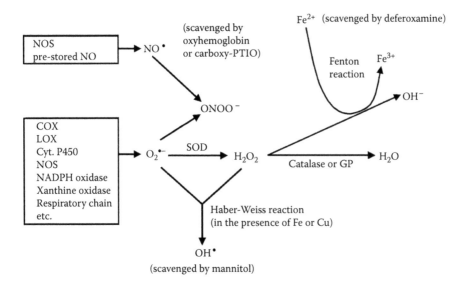

FIGURE 63 Production of hydrogen peroxide and other biologically active reactive oxygen species. Superoxide anion (O_2^-) is the product or byproduct of many cellular enzymatic reactions catalyzed by, e.g., cyclooxygenases (COX), lipoxygenases (LOX), cytochrome P450 monooxygenases (cyt. P450), nitric oxide synthases (NOS), nicotinamide adenine dinucleotide oxidases (NADPH oxidase), xanthine oxidase, or enzymes of the mitochondrial electron transport chain. Superoxide anions are either reduced enzymatically into hydrogen peroxide (H_2O_2) by superoxide dismutases (SOD), or spontaneously react with other reactive oxygen species. In the presence of nitric oxide peroxynitrite (ONOO-) is formed. Superoxide anions can react with H_2O_2 and traces of catalytic metals to generate the highly reactive hydroxyl radical (OH·) by the so-called Haber-Weiss reaction. H_2O_2 is metabolized into water by catalase or glutathione peroxidase (GP) or in presence of iron (Fe^{2+}) can also produce OH· (Fenton reaction). Reactive oxygen species are not regarded any longer as metabolic byproducts responsible for unspecific cellular damage. They can play physiological and pathophysiological roles by interacting with heme-containing proteins or thiol redox-regulated systems leading to the regulation of enzymes, ion channels, and contractile proteins.

level of H_2O_2 are similar to those of wild-type animals. However, the pressor response induced by the administration of either norepinephrine or angiotensin II is decreased in the transgenic animals and this is associated with a suppression of H_2O_2 production, indicating that endogenously produced H_2O_2 contributes to the vasopressor responses evoked by these cardiovascular hormones.[568,599]

In the isolated murine mesenteric artery, H_2O_2 (<50 µM) produces an endothelium-independent relaxation providing that K_{Ca} channels are operational, but, at the same concentrations, elicits a potent contractile response if the activity of these channels is compromised.[600] In some blood vessels, such as the rabbit aorta or the rat mesenteric artery, low concentrations of H_2O_2 (10–100 µM) evoke contractions, while higher concentrations (0.3–1 mM) produce transient relaxations followed by sustained contractions.[582,587]

In most blood vessels that do relax in response to exogenously added H_2O_2, the relaxation is endothelium-independent. However, in some arteries the relaxing effect of H_2O_2 can be partially endothelium-dependent, such as in the canine and porcine coronary artery.[571,578] In the latter the endothelium-dependent relaxation elicited by H_2O_2 involves the COX-1-dependent release of PGE_2.[578] The vasodilator effects of H_2O_2 are totally endothelium- and NO-dependent in the canine basilar artery,[585] in the rabbit and rat aorta,[587,589] and in human submucosal microvessels.[482]

The endothelium-independent relaxations of arterial myocytes to H_2O_2 involve multiple pathways. Direct and indirect evidence for the activation of guanylyl cyclase and the involvement of cyclic GMP has been obtained in porcine, bovine, and human coronary arteries,[575,577,588] guinea pig aorta,[590] and rabbit mesenteric artery.[580] The involvement of the metabolism of arachidonic acid either through cyclooxygenases (canine cerebral artery),[586] lipoxygenases (porcine coronary artery),[523] or both (rabbit mesenteric artery)[580,583] is also possible.

3.8.3.2.2 H_2O_2 and the Membrane Potential of Vascular Smooth Muscle and Endothelial Cells

H_2O_2 can produce hyperpolarization or depolarization of the vascular smooth muscle cells in many arteries. In patch-clamp experiments on isolated vascular smooth muscle cells of the porcine coronary artery, H_2O_2 activates BK_{Ca}.[523,574,575] In studies on freshly isolated cells, the effects of H_2O_2 are mediated by a lipoxygenase derivative of arachidonic acid,[523] while in cultured smooth muscle cells, H_2O_2 evokes direct as well as soluble guanylyl cyclase-dependent activation of BK_{Ca}.[575] In addition, in freshly isolated smooth muscle cells of the rat mesenteric artery, H_2O_2 activates K_V.[582] By contrast, H_2O_2 can also inactivate BK_{Ca} of vascular smooth muscle by altering the cystein-mediated calcium sensing of the *Slo1* α subunit.[601] Lastly, H_2O_2 can depolarize vascular smooth muscle cells, for instance, by inhibiting K_V channels, as seen in the human ductus arteriosus[602] or activating ICl_{Ca}, as observed in the rabbit portal vein.[603]

H_2O_2 also modulates the ionic channel activity of the endothelial cells, which, in turn, can influence the membrane potential and thus the contractile activity of the underlying smooth muscle cells. In cultured endothelial cells of the human umbilical vein, H_2O_2 elicits both depolarization and hyperpolarization. In the micromolar range, it inhibits K_{IR}, while at 1 mM it activates BK_{Ca}.[604] The activation of BK_{Ca} is due to an increase in $[Ca^{2+}]_i$ since, in concentrations higher than 100 μM, H_2O_2 activates in those cells a calcium-permeable, nonselective cation current.[605] In human aortic endothelial cells, H_2O_2 causes the release of Ca^{2+} from the endoplasmic reticulum.[606] By contrast, in bovine aortic endothelial cells, within a similar range of concentrations (high micromolar-low millimolar), H_2O_2 reduces the activity of IK_{Ca}, without affecting the unitary conductance, because of oxidation of sulfhydril groups involved in the gating of the channel.[607] In addition, in endothelial cells of the porcine renal artery, patched *in situ*, H_2O_2 inhibits the stimulation of BK_{Ca} elicited by NO,[164] suggesting that reactive oxygen species may impair endothelium-dependent vasodilatations.

The hyperpolarization caused by H_2O_2, or at least the component of the relaxation it causes attributed to hyperpolarization of the vascular smooth muscle cells, is associated with the opening of different potassium channels. Thus, the activation of H_2O_2 probably opens K_{Ca} in cultured aortic smooth muscle cells and isolated cerebral artery of the rat[608,609] and in murine mesenteric[579] and human and porcine coronary[403,576,578] arteries. K_{ATP} are activated by H_2O_2 in feline cerebral and rabbit mesenteric arteries,[583,610] K_V in rat mesenteric arteries,[582,586] and K_{Ca}, K_V, and K_{IR} in canine cerebral arteries.[586]

However, H_2O_2 does not always hyperpolarize vascular smooth muscle cells (guinea pig carotid artery)[611] and its relaxing effect is not necessarily associated with hyperpolarization of the vascular smooth muscle cells or/and opening of potassium channels (rabbit ilio-femoral artery).[563]

Taken in conjunction, these results show that H_2O_2 has multiple effects on vascular cells. Nevertheless, as this reactive oxygen species is produced by the endothelial cells and in some arteries evokes hyperpolarization and relaxation of smooth muscle cells, H_2O_2 is a potential candidate as an endothelium-derived hyperpolarizing mediator.[612]

3.8.3.2.3 H_2O_2 and Endothelium-Dependent Hyperpolarizations

Most of the evidence in favor of H_2O_2 being an endothelium-derived hyperpolarizing factor comes from the observation that, in certain blood vessels, the agonist- and flow-induced responses attributed to EDHF are partially or totally sensitive to catalase. This is the case in murine, human, and rat mesenteric arteries,[481,579,613,614] porcine pial arteries,[224] and in canine, porcine, and human coronary arteries.[403,576,615] In most of these blood vessels, with the exception of the porcine pial arteries, exogenously added H_2O_2 appears to mimic EDHF-mediated responses since the peroxide produces endothelium-independent relaxation and/or hyperpolarization of the vascular smooth muscle cells by a mechanism involving K_{Ca} activation. In porcine pial arteries, the bradykinin-induced, EDHF-mediated vasodilatation is atypical since it involves K_{ATP} instead of K_{Ca}.[224] In murine mesenteric[579,614] and porcine coronary arteries,[403] the agonist-induced EDHF-mediated response is associated with a catalase-sensitive endothelial production of H_2O_2. Similarly, in human coronary arteries,[576] flow-induced dilatation increases the production of H_2O_2 from the endothelial cell layer.

Based on studies performed in NOS-3 knock-out mice, Matoba et al.[579] suggested that the enzymatic source of superoxide anion, precursor of H_2O_2, is the NOS-3 itself. However, this may not necessarily be the case in all blood vessels since several other endothelial enzymes such as cyclooxygenases, lipoxygenases, cytochrome P450 epoxygenases, and NADPH oxidases can generate superoxide anions.[612] The endothelial Cu, Zn-superoxide dismutase probably plays a major role, at least in mice, in producing H_2O_2 in order to elicit endothelium-dependent hyperpolarizations.[614]

Not all available data support the hypothesis that EDHF-mediated responses can be attributed to the generation of H_2O_2. Indeed, catalase does not inhibit such responses in the porcine, canine, and rat coronary,[453,533,572,616] human radial,[617] murine mesenteric,[581] guinea pig carotid,[611] rabbit ileo-femoral and mesenteric[563,567] arteries, or mouse aorta.[581] In the rabbit ileo-femoral arteries, although the calcium ionophore A23187 (and to a lesser extent acetylcholine) induces the endothelial production of

H_2O_2, which causes relaxation of smooth muscle, there is virtually no associated hyperpolarization.[563] Furthermore, in human submucosal microvessels,[482] although acetylcholine-induced EDHF-mediated vasodilatation is inhibited by catalase and is associated with the generation of H_2O_2, H_2O_2 *per se* does not evoke relaxation of the smooth muscle cells, indicating that, in this artery, H_2O_2 might be a permissive factor but definitively is not a diffusible hyperpolarizing substance.

These apparently controversial results have been explained tentatively by differences in experimental protocols and, in particular, those involving the use of catalase. Catalase, which is an enzyme with a high molecular weight when compared to most pharmacological inhibitors, may require a prolonged incubation time. This would be necessary to allow transcytosis and accumulation of this large molecule in the myo-endothelial intercellular space in order to inactivate the diffusing H_2O_2.[612] In addition, commercially available catalases may contain contaminants and do more than dismutating H_2O_2 into water and oxygen.[618] The results, whether positive or negative, of experiments involving catalase should therefore be interpreted with caution.

A critical appraisal of the available evidence linking H_2O_2 to EDHF-mediated responses requires a word of caution. H_2O_2 is a rather weak relaxing and hyperpolarizing substance. Quantification of the production of this reactive oxygen species was attempted by electron spin resonance spectroscopy in porcine coronary microvessels and a micromolar production was observed in response to bradykinin and substance P, two agonists that elicit EDHF-mediated responses.[403] However, in the same artery, H_2O_2 at 10 μM produces only a minor hyperpolarization, while at the high concentration of 100 μM, it hyperpolarizes the smooth muscle cells by only approximately 7 mV. In contrast, bradykinin induces a hyperpolarization of up to 20 mV. Furthermore, as mentioned above, one of the major achievements that has improved the understanding of EDHF-mediated responses has been the identification and localization of the potassium channels involved in these responses. In the various arteries where the hyperpolarizations in response to H_2O_2 and EDHF were compared, none of the specific potassium channels blockers have been studied and it is not yet known which population(s) of potassium channels are activated by H_2O_2 or whether they are situated on the endothelial or the smooth muscle cells. The absence of this information undermines any conclusion as to the potential role of H_2O_2 in EDHF-mediated responses. In particular, the question remains whether H_2O_2 is a diffusible factor activating potassium channels on the smooth muscle or is an intracellular messenger involved in the activation of endothelial potassium channels. Likewise, although H_2O_2 is produced in response to an increase in endothelial $[Ca^{2+}]i$[612] it is uncertain whether this is truly a pathway linked to activation of endothelial IK_{Ca} and SK_{Ca} or is an independent (epi)phenomenon (Figure 64).

3.8.3.3 Endothelium-Derived Peptides

Endothelial cells can synthesize numerous peptides, including endothelin, vasoactive intestinal peptide (VIP), substance P, calcitonin gene-related peptide (CGRP), adrenomedullin, C-type natriuretic peptide (CNP), and arginine-vasopressin.[619–621] Among these peptides, CNP, VIP CGRP, and adrenomedullin cause relaxation of vascular smooth muscle cells.

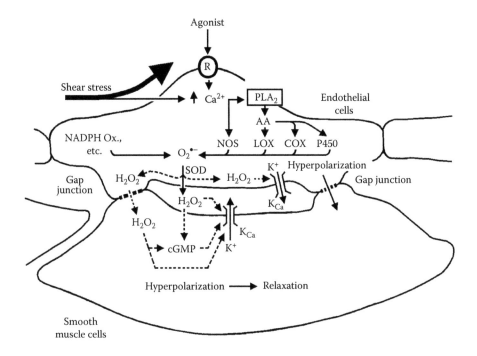

FIGURE 64 Multiple putative pathways by which H_2O_2 can act as an EDHF. The forces exerted by the flowing blood and/or receptor stimulation increase the endothelial intracellular calcium concentration, which activates many enzymes capable of generating superoxide anion ($O_2^{\bullet-}$). They include cyclooxygenases (COX), lipoxygenases (LOX), cytochrome P450 monooxygenases (P450), and nitric oxide synthases (NOS). Superoxide anions are reduced enzymatically into hydrogen peroxide (H_2O_2) by superoxide dismutases (SOD). H_2O_2 is cell permeable and can diffuse toward the smooth muscle cells directly through the intercellular space, but also can transit through myoendothelial gap junctions. In the vascular smooth muscle, H_2O_2 can open calcium-activated potassium channels (K_{Ca}) either directly or by stimulation of soluble guanylyl cyclase and the production of cyclic-GMP (cGMP). H_2O_2 can also act as an intracellular messenger. Thus, by activating endothelial K_{Ca}, H_2O_2 could hyperpolarize the endothelial cells and evoke and/or reinforce endothelium-dependent hyperpolarizations of the smooth muscle cells. NADPH Ox: nicotinamide adenine dinucleotide oxidases; PLA_2: phospholipase A_2; AA: arachidonic acid.

3.8.3.3.1 C-Type Natriuretic Peptide

The natriuretic peptide family includes four distinct peptides, atrial natriuretic peptide (ANP), brain natriuretic peptide (BNP), C-type natriuretic peptide (CNP), and urodilatin, which are involved in the control of body fluid homeostasis and blood pressure. CNP, a 22-amino-acid peptide, was originally isolated from the porcine brain. In contrast to ANP and BNP, which are synthesized predominantly in the heart, CNP is produced in the central nervous system, but has also been detected in peripheral endothelial cells.[622,623] Endothelial cells express the transcript of the CNP receptor gene, synthesize, and constitutively release CNP. The CNP gene expression and release of CNP are regulated by various cytokines.[620,624,625]

CNP evokes relaxation and hyperpolarization of arterial and venous smooth muscle cells.[620,624,626–630] In porcine coronary arteries and canine femoral veins, CNP induces the accumulation of cyclic-GMP and opens BK_{Ca}.[626,628] This endothelium-derived natriuretic peptide has therefore been proposed as a potential EDHF.[626,628] This hypothesis may seem at variance with the observation that, at least in the porcine coronary artery, the characteristics and the amplitude of the hyperpolarizations evoked by exogenous CNP are by no means comparable to those observed during EDHF-mediated responses.[631] By contrast, experiments performed in the rat mesenteric artery suggest that acetylcholine releases CNP from the endothelial cells, thus activating the smooth muscle NPR-C receptor subtype. Hyperpolarization of the smooth muscle cell is evoked via the cyclic-GMP-independent activation of a G-protein-regulated inward-rectifier potassium channel (GIRK; Figure 65).[632]

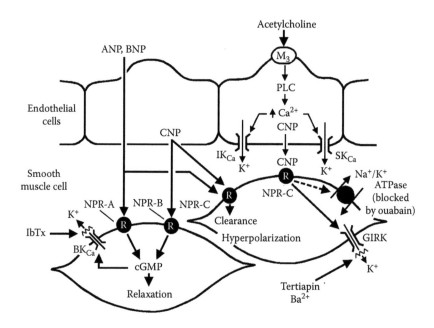

FIGURE 65 Schematic representation of how C-type natriuretic peptide (CNP) could act as an EDHF. Acetylcholine interacts with endothelial muscarinic M_3 receptors, activates phospholipase C (PLC), and increases the intracellular calcium concentration. The stimulation of endothelial calcium-activated potassium channels (IK_{Ca} and SK_{Ca}) could contribute to the endothelial release of CNP and the subsequent activation of the C-type natriuretic peptide receptor (NPR-C) on the vascular smooth muscle. Stimulation of NPR-C, which is also a clearance receptor for CNP and the other natriuretic peptides [atrial natriuretic peptide (ANP) and brain natriuretic peptide (BNP)], activates G-protein-gated inwardly rectifying potassium channel (GIRK) and possibly the Na^+/K^+-ATPase. CNP, as well as circulating ANP and BNP, can also activate NPR-A and NPR-B, two guanylyl cyclase-coupled receptors located on the plasma membrane of the smooth muscle. This produces the cyclic-GMP-dependent (cGMP) hyperpolarization and relaxation of the smooth muscle cells by opening large-conductance calcium-activated potassium channels (BK_{Ca}). IbTx: iberiotoxin.

The hypothesis that CNP could be an endothelium-derived hyperpolarizing substance should be examined with caution since it requires the validation of various concepts. First, there are three identified receptors for the natriuretic peptide family. The NPR-A subtype is preferentially activated by ANP/BNP and the NPR-B subtype is activated by CNP.[622] Most of the cardiovascular effects of CNP are probably mediated by the NPR-B receptors linked to particulate guanylyl cyclase, the activation of which results in an augmented generation of cyclic GMP.[633,634] By contrast, the NPR-C receptor is a truncated receptor with an extracellular natriuretic binding domain and a short intracellular portion that includes a G-protein activating sequence.[635] The NPR-C receptor is thought to be a clearance receptor, removing natriuretic peptides from the circulation.[636] However, this receptor, besides its clearance role, may be functional and mediate G-protein-dependent intracellular events such as the inhibition of adenylyl cyclase activity.[637] Whether the activation of this receptor by CNP in vascular smooth muscle cells produces such a cyclic-GMP-independent, pertussis toxin-sensitive signaling remains is unknown. The second critical point concerns the functional expression of GIRK in vascular smooth muscle cells. The expression and activity of GIRK has been well characterized in neurons and cardiac myocytes.[638–640] The constitutive activity of such channel cells (Kir3.1 and Kir3.2) has been reported in gastrointestinal but not in vascular smooth muscle cells,[641] although the mRNA expression of Kir3.1 has been detected in rat aortic smooth muscle cells.[642,643] This expression and the complete characterization of this channel in vascular smooth muscle cells await full demonstration. Finally, there is no evidence to date, in any cell type, that CNP can activate GIRK. In atrial myocytes, one of the few cell types in which NPR-C may be more than a clearance receptor, its activation was linked to an inhibition of calcium current without any significant effect on the inward rectifier potassium current.[644]

3.8.3.3.2   Other Peptides

Calcitonin gene-related peptides (αCGRP and βCGRP) and adrenomedullin are extremely potent vasodilator peptides belonging to the family that also includes calcitonin and amylin.[645–647] Adrenomedullin was originally identified in the extracts of human pheochromocytoma tissue, but is expressed constitutively in various other cells including epithelium and cardiac myocytes, fibroblasts, vascular smooth muscle, and, in particular, endothelial cells with a 20- to 40-fold higher expression than in pheochromocytoma cells.[621,645,648] CGRP is expressed mainly in neurons and endocrine cells,[647] but can also be expressed in endothelial cells where the peptide is stored in Weibel-Palade bodies.[649,650] These peptides activate a family of CGRP receptors consisting of a calcitonin receptor-like protein linked to one of the three receptor activity-modifying proteins (RAMPs) that are required for the functional activity of the receptor.[646,647]

The vasodilator effects of adrenomedullin are partially or totally endothelium-dependent in, for example, the human coronary artery,[651] rat kidney, mesenteric, and pulmonary vascular beds,[652–654] porcine ciliary arteries,[655] and bovine retinal arteries.[656] Both release of NO and hyperpolarization of the smooth muscle cells (mediated by NO and/or an EDHF-like mechanism) contribute to the endothelium-dependent relaxations elicited by adrenomedullin. Endothelium-independent relaxations have been reported only in a few preparations such as the basilar artery of the rat.[657] Adrenomedullin stimulates the formation of cyclic-AMP in vascular smooth muscle cells.[658,659] This could explain the endothelium-independent relaxations. Thus adrenomedullin is a potent endothelium-dependent vasodilator, but is unlikely *per se* to be an endothelium-derived hyperpolarizing factor.

Similarly, CGRP is both an endothelium-dependent and -independent relaxing substance. However, the contribution of the endothelium-independent component is more significant than that of adrenomedullin. Thus, in rabbit ophthalmic and mesenteric,[660,661] human uterine, mammary and gastroepiploic,[662,663] and rat basilar and mesenteric[653,664–666] arteries, the relaxations induced by CGRP are mostly endothelium-independent. In most, but not all, of these blood vessels the relaxations evoked by CGRP are associated with the opening K_{ATP}. However, in porcine coronary artery, CGRP activates the adenylyl cyclase-cyclic-AMP-protein kinase A pathway, which induces K-ATP and BK_{Ca} activation.[667,668] Similarly, endogenous CGRP, released by transmural nerve stimulation or by capsaicin analogues, evokes endothelium-independent relaxations that involve activation of K_{ATP}. This is the case in the canine lingual artery,[669] guinea pig aorta and coronary circulation,[670] or the rat mesenteric artery.[671] Since CGRP activates a different population of potassium channels than those involved in EDHF-mediated responses, it is unlikely that CGRP acts as an endothelium-derived hyperpolarizing substance in most blood vessels. Furthermore, CGRP receptor antagonists do not inhibit EDHF-mediated responses in any of the studies conducted to date. Acetylcholine, in the hepatic artery [672] and possibly also in the mesenteric artery[673] of the rat, releases CGRP from sensory nerves that produce an endothelium-independent relaxation and hyperpolarization. Thus CGRP should be considered as a nerve-derived hyperpolarizing factor.[672]

Vasoactive intestinal peptide (VIP) and pituitary adenylyl cyclase-activating polypeptide (PACAP) belong to the superfamily of structurally related peptides that includes glucagon, glucagon-like peptides, secretin, and growth hormone releasing factor. Three receptors for PACAP have been identified, two of which share a high affinity for VIP.[674] VIP produces mostly endothelium-dependent relaxations by releasing NO and/or prostacyclin. This has been observed in the bovine pulmonary artery,[675] the human uterine artery,[662,676] and the murine aorta.[677] VIP also directly relaxes the smooth muscle cells of bovine coronary arteries.[678] This peptide stimulates adenylyl cyclase of the mesenteric artery and aorta of the rat[679] and activates BK_{Ca} in the former.[680] In rabbit cerebral arteries, it produces hyperpolarization by opening K_{ATP}.[681] In smooth muscle cells of the porcine coronary artery, VIP activates both BK_{Ca} and Kv.[682] PACAP is also an endothelium-independent vasodilator in human pulmonary arteries,[683] as well as in human and porcine coronary arteries.[684] In the same human arteries, PACAP opens both K_{ATP} and BK_{Ca} and this contributes to the relaxation it causes of the vascular smooth muscle cells.[683,685] However, whereas VIP and PACAP are well established as neuroendocrine hormones and neurotransmitters, there are only a few anecdotal reports of the expression of these peptides in endothelial cells.[686,687] Therefore, neither VIP nor PACAP can qualify as endothelium-derived hyperpolarizing mediators. Nevertheless, as both peptides are expressed in nerve endings surrounding arteries and veins, they can be considered, in a similar manner to CGRP, as nerve-derived hyperpolarizing factors.

The development of specific nonpeptidic antagonists of the various receptors involved in the vasodilatator action of these different peptides will make it possible to determine whether or not CGRP, adrenomedullin, VIP, or PACAP (and possibly other peptides to come) could mediate endothelium-dependent hyperpolarizations under certain circumstances in specific vascular beds or under given physiological or pathophysiological conditions.

3.8.3.4 Gaseous Mediators other than NO

The discovery and characterization of NO was much more than simply the identification of another intercellular messenger. NO is not involved like other neurohumoral mediators in classic ligand-membrane-bound receptor interactions. It is a gas that readily crosses lipid membranes and therefore diffuses homogeneously, in a nonpolarized manner, from its production site. NO can be considered as the first member of a new family of neurohumoral substances, the gaseous mediators. Two other low-molecular-weight compounds can be included in this family, carbon monoxide and hydrogen sulfide.[688]

3.8.3.4.1 *Carbon Monoxide (CO)*

The predominant biological source of CO is from the degradation of heme (hemoglobin, myoglobin, cytochrome P450, etc.) by heme-oxygenase (HO), which generates equimolar quantities of biliverdin, free ferrous iron, and carbon monoxide. Biliverdin is rapidly converted to bilirubin, while free iron is promptly sequestered by ferritin.[689] These bile pigments are not simple byproducts of heme catabolism; they have antioxidant properties that protect the vascular cells from oxidative stress. Three isoforms of heme oxygenase have been described HO-1 (or heat shock protein 32, inducible isoform), HO-2 (constitutive isoform), and HO-3, which is closely related to HO-2 but is nearly devoid of catalytic activity.[690,691] HO-1 is ubiquitously expressed in response to a variety of stimuli including its own substrate, heme, and several oxidants. HO-2 is also ubiquitously expressed with notably high levels in brain and testes.[688,689] Both HO-1 and HO-2 are expressed in vascular smooth muscle and endothelial cells.

CO relaxes and hyperpolarizes vascular and nonvascular smooth muscles[187,690,692] by activating soluble guanylyl cyclase.[187,690,692,693] Additionally, CO directly increases the open probability of BK_{Ca}.[694,695] The direct action of CO on BK_{Ca} differs from the direct effect of NO on this population of potassium channels. Indeed, CO interacts with the α-1 subunit *Slo*-1, while NO interacts with the β subunit.[696] CO increases the apparent calcium sensitivity of the α subunit, enhancing the probability of eliciting spontaneous transient outward currents (STOC) in response to calcium sparks.[697,698] CO causes cyclic-GMP-independent activation of BK_{Ca} in vascular smooth muscle cells of the tail and mesenteric arteries of the rat[692,695,696,699] and the porcine cerebral artery.[698] By contrast, in canine cerebral arteries, the activation of BK_{Ca} is cyclic-GMP-dependent.[700] It is not known whether CO influences the activity of the endothelial potassium channels IK_{Ca} and SK_{Ca}. Furthermore, the inhibition of cytochrome P450 by CO suppresses the synthesis of 20-HETE, an endogenous vasoconstrictor, which lifts the inhibition of BK_{Ca} produced by this cytochrome P450 derivative (Figure 66).[701–703]

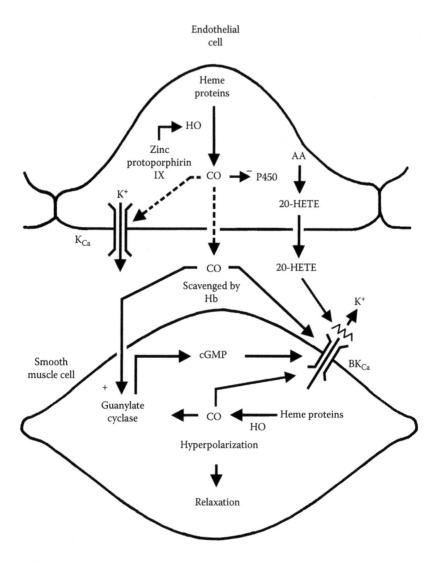

FIGURE 66 Schematic representation of how carbon monoxide (CO) could act as an EDHF. In response to stimuli such as oxidant stress, carbon monoxide is produced by heme oxygenase (HO) in both endothelial and vascular smooth muscle cells. Endothelium-derived CO can diffuse toward the smooth muscle cells and open large conductance calcium-activated potassium channels (BK_{Ca}) either directly or by stimulation of soluble guanylyl cyclase and the production of cyclic-GMP (cGMP). Carbon monoxide could also act as an intracellular messenger. Thus, by activating endothelial K_{Ca}, CO could hyperpolarize the endothelial cells and evoke and/or reinforce the endothelium-dependent hyperpolarization of the smooth muscle cells. Additionally, CO inhibits cytochrome P-450 monooxygenase (P450) and prevents the production of 20-hydroxyeicosatetrienoic (20-HETE), an endogenous blocker of BK_{Ca}. Hemoglobin (Hb) is a scavenger of CO. AA: arachidonic acid.

Endogenously produced CO attenuates the sensitivity to vasconstrictors and to pressure-induced vasoconstriction in the renal and gracilis arteries of the rat. The addition of substrate for HO produces vasodilatation in porcine cerebral and rat tail arteries.[692,704–706] However, CO also induces vasoconstriction in resistance arteries where NO is abundantly available. The systemic arterial blood pressure is elevated in transgenic mice overexpressing heme oxygenase-1 specifically in vascular smooth muscle cells.[707] Nevertheless, since CO can be produced by the endothelial cells and induces relaxation and hyperpolarization of the vascular smooth muscle cells, it may act as an endothelium-derived hyperpolarizing substance.

Heme oxygenase is inhibited by zinc protoporphyrin IX. Although this inhibitor lacks specificity as it can also inhibit guanylyl cyclase, NO-synthase and K_{ATP} channel activation,[690,708] it does not inhibit endothelium-dependent hyperpolarizations in the rat hepatic artery[227,708] Furthermore, CO is scavenged actively by oxyhemoglobin, which, in a variety of blood vessels, does not affect endothelium-dependent hyperpolarizations. Additionally, although CO can be produced by the endothelial cells, the smooth muscle cells are a more important source of CO production in the vascular wall.[688] Finally, there is no evidence showing that mediators eliciting EDHF-mediated responses do simultaneously evoke the endothelial release of CO. This conclusion does not exclude that this gaseous transmitter may play an important role in the regulation of vascular function under physiological conditions and in certain cardiovascular diseases.[689,691] As HO-1 is induced in response to a variety of stresses and especially oxidant stress, the endothelial release of CO could contribute to the regulation of vascular tone under pathological conditions such as atherosclerosis.[709] Under these conditions CO could be an endothelium-derived hyperpolarizing substance (Figure 66).

3.8.3.4.2 Hydrogen Sulfide (H_2S)

H_2S, like NO and CO, is a toxic gas. It is colorless with a strong smell of rotten eggs and is produced not only by bacteria but also by mammalian cells. In rats, the plasmatic concentration of H_2S is approximately 50 μM. Vascular tissues can produce H_2S.[688,710,711]

Two main enzymes are responsible for the production of H_2S, cystathionine β-synthase and cystathionine γ-lyase, which both use L-cysteine as substrate. The physiological cardiovascular effects of H_2S are linked to the latter enzyme, which, in the vascular wall, is expressed mostly in smooth muscle but is virtually absent in endothelial cells.[688] H_2S produces a decrease in blood pressure, vascular relaxation, and hyperpolarization of smooth muscle by activating K_{ATP}.[711] In the mesenteric arterial bed of the rat, the addition of L-cysteine enhances the production of H_2S and reduces vasoconstriction. In addition, H_2S may induce EDHF-mediated responses since a component of the vasodilatation is endothelium-dependent and sensitive to the combination of charybdotoxin plus apamin.[712,713] If exogenous H_2S given alone relaxes arterial smooth muscle cells, much lower concentrations of H_2S greatly enhance the relaxation induced by NO, suggesting a synergic relationship between H_2S and NO.[710] However, H_2S does not qualify as an endothelium-derived mediator since it is not generated by endothelial cells. Furthermore, the mechanism of H_2S-induced relaxation and hyperpolarization of smooth muscle is different from that involved in EDHF-mediated responses.

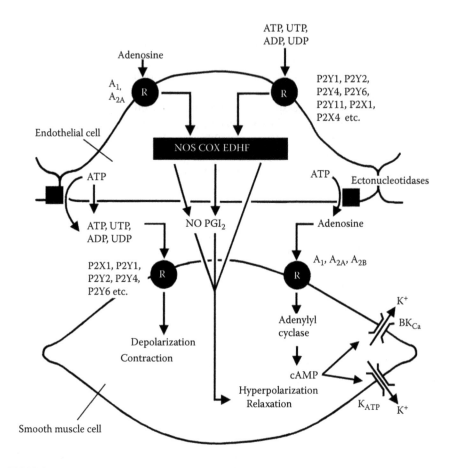

FIGURE 67 Schematic representation of how adenosine could act as an EDHF. Adenosine trisphosphate (ATP) released by various cells including endothelial cells is metabolized into adenosine diphosphate (ADP), adenosine monophosphate (AMP), and ultimately adenosine by membrane-bound ectonucleotidases. Adenosine, by interacting with A_1, A_{2A}, or A_{2B} receptor subtypes, produces hyperpolarization of the smooth muscle by cyclic-AMP- (cAMP) dependent activation of either ATP-sensitive potassium channels (K_{ATP}) or large-conductance calcium-activated potassium channels (BK_{Ca}). In addition, adenosine interacting with endothelial receptors (A_1 or A_{2A} receptor subtypes) and purines interacting with various purinoceptors (P2Y and P2X families) release endothelial factors such as nitric oxide (NO) and prostacyclin (PGI_2) and produce EDHF-mediated responses. The activation of purinoceptors on vascular smooth muscle usually elicits a contractile response. NOS: nitric oxide synthase; COX: cyclooxygenase; UTP: uridine trisphosphate.

3.8.3.5 Adenosine

Nucleotides and nucleosides are extracellular signaling molecules that are released by many different cells under various physiological and pathophysiological conditions. Adenosine interacts with P1 receptors, of which four types have been identified (A_1, A_{2A}, A_{2B}, and A_3), while nucleotides (ATP, ADP, UTP) interact with the P2 receptors divided into two main subgroups, the P2X ligand-gated ion channel (7 subtypes) and the P2Y G-protein-coupled receptor families (at least 6 subtypes in humans).[714,715] In the vascular system, nucleosides and nucleotides contribute to short-term signaling events associated with the moment-to-moment control of vascular tone. However, these molecules are also involved in the regulation of migration, proliferation, and apoptosis of vascular smooth muscle and endothelial cells and hence in the long-term normal control of the geometry of the vascular wall as well as in pathological events such as arteriosclerosis, restenosis, wound healing, and the genesis of collateral and tumoral vascularization.[716] In general, ATP and ADP produce endothelium-dependent relaxations and/or hyperpolarizations of smooth muscle cells or evoke direct contractions and/or depolarizations by activating various purinoceptors, P2X and/or P2Y subtypes located in the endothelial and smooth muscle cells (Figure 67).[717–719]

Adenosine is an endogenous vasodilator that is released from contracting skeletal muscle where it contributes to exercise hyperemia, and from the myocardium where it is an important regulator of coronary blood flow especially during ischemia.[374,720,721] Adenine nucleotides and adenosine are also released by endothelial cells.[722,723] Endothelial cells release mainly ATP, but the nucleotide is rapidly transformed into ADP, AMP, and adenosine by ectonucleotidases. The release of adenosine by endothelial cells can be stimulated by chemical and physical stimuli including hypoxia and increases in shear stress,[724–727] by neuromediators such as ATP itself,[728,729] and by drugs such as α_1-adrenergic agonists.[723]

Adenosine can induce both endothelium-dependent relaxations and direct relaxations of vascular smooth muscle cells. Often both mechanisms are involved in the same artery. In the renal artery of the rat, the relaxation in response to adenosine is fully endothelium-dependent,[730] while in other arteries an endothelium-dependent component contributes to the relaxation, evoked by adenosine, or by pharmacological agonists at the P1 purinoceptor subtype, as is the case in the aorta and the femoral vascular bed of the rat,[731,732] renal arteries of the rabbit,[733,734] and in canine, porcine, rat, and human coronary arteries.[7,732,735–739] These endothelium-dependent relaxations evoked by adenosine are linked to the activation of either A_1 or A_{2A} receptor subtypes and the subsequent release of NO or the involvement of an EDHF-mediated component. Adenosine can produce hyperpolarization of endothelial cells by activating K-ATP and/or K_{Ca}.[731,740]

Adenosine also produces direct relaxation and hyperpolarization of nonvascular[741] and vascular smooth muscles including those of the rat femoral vascular bed,[731] rabbit carotid mesenteric and renal arcuate arteries,[225,734,742] or guinea pig, canine, and human coronary arteries.[737,743–746] In most of these blood vessels, the hyperpolarization and relaxation of the vascular smooth muscle cells involve the activation of the A_{2A} adenosine receptor subtype[747] and the subsequent activation of K_{ATP} channels. In the rabbit mesenteric and guinea-pig coronary arteries, the adenosine receptor is coupled to adenylyl cyclase producing a cyclic-AMP-dependent activation of protein kinase A and the subsequent opening of K_{ATP} channels.[742,745] However, in canine coronary, rat pulmonary, and porcine retinal arteries, the activation of K_{ATP} appears to be independent of the accumulation of cyclic-AMP.[744,748,749] In porcine and human coronary arteries, adenosine activates the A_1 and the A_{2B}

receptor subtype, respectively.[746,750] In the latter artery, a yet unidentified potassium channel different from K-ATP is involved. In canine epicardial arteries, the relaxation produced by adenosine has been attributed to BK_{Ca} (Figure 67).[746,751,752]

Taken in conjunction, the available data indicate that adenosine can be released by the endothelial cells and can act as an autocrine or a paracrine factor to produce hyperpolarization of endothelial and/or vascular smooth muscle cells. Adenosine can therefore act as an endothelium-derived mediator under many circumstances, especially during hypoxia and exercise. The putative role of adenosine in endothelium-dependent relaxations and hyperpolarizations may have been overlooked because of the lack of specific agonists and antagonists of the adenosine receptor subtypes. The availability of these compounds will help to clarify when and where adenosine truly contributes to EDHF-mediated responses.

3.8.3.6 L-Citrulline

L-citrulline, the byproduct of NO biosynthesis by the NO-synthase, can relax the rabbit aorta by a cyclic-GMP-dependent mechanism that may involve potassium channel activation.[753] However, this observation has not been confirmed by subsequent work involving the same and different arteries.[754]

3.9 CONCLUSION

The endothelium controls the tone of the underlying vascular smooth muscle cells by releasing numerous vasoactive substances: NO, reactive oxygen species, potassium ions, metabolites of arachidonic acid (e.g., prostacyclin, EETs, lipoxygenase derivatives), peptides, and adenosine. In addition, the monolayer of endothelial cells behaves as a conductive tissue propagating an electrical signal along the axis of the blood vessel by means of homocellular gap junctions and throughout the vascular wall itself by means of myoendothelial gap junctions. This latter pathway is also a powerful means of controlling smooth muscle tone, especially in resistance arteries (Figure 68).

3.10 EDHF-MEDIATED RESPONSES IN HUMAN BLOOD VESSELS

3.10.1 ISOLATED BLOOD VESSELS

In human isolated vascular preparations, EDHF-mediated responses are observed in coronary,[21,175,312,576,755,756] internal mammary,[177,452] radial,[177,757] renal,[388,448] and cerebral arteries,[22,229] as well as in resistance arteries from various vascular beds[174,333,343,389,406,407,450,478–482,758,759] and in forearm veins.[760] As in animal arteries,[28,31,308] the contribution of EDHF-mediated responses appears significantly more important in smaller than in large human arteries.[480,761]

The observed EDHF-mediated responses (hyperpolarizations or/and relaxations resistant to inhibitors of nitric oxide synthases and cyclooxygenases) can be attributed to the activation of potassium channels as they are blocked by elevated potassium

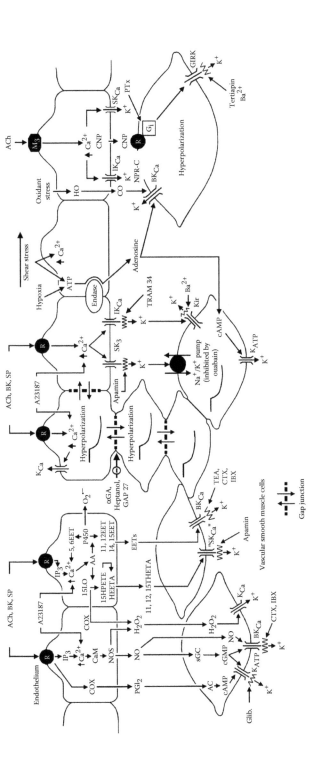

FIGURE 68 Summary of the various potential contributors to EDHF-mediated responses. Among the various proposed pathways, besides the release of NO and prostacyclin, the mechanism involving the hyperpolarization of endothelial cells via the activation of SK_{Ca} and IK_{Ca} is likely to be of major physiological significance in the regulation of vascular diameter and blood flow. The production of EETS may become preponderant in specific vascular beds (coronary circulation) and under specific conditions of activation (stimulation with bradykinin). The contribution of the other pathways in the overall endothelium-dependent regulation of vascular tone remains to be assessed properly. R: receptor; IP3: inositol trisphosphate; SR: sarcoplasmic reticulum; Trp: Transient Receptor Potential Channel; AC: adenylyl cyclase; αGA: glycyrrhetinic acid derivatives; IK1: intermediate conductance calcium-activated potassium channel formed by IK1 α-subunits; ChTx: charybdotoxin; ACh: acetylcholine; BK: bradykinin; SP: substance P; R: receptor; M_3: muscarinic receptor; NPR-C: natriuretic petide receptor; SK_{Ca}: small conductance calcium-activated potassium channel; SK3: small conductance calcium-activated potassium channel formed by SK3 α-subunits; IK_{Ca}: intermediate conductance calcium-activated potassium channel; BK_{Ca}: large conductance calcium-activated potassium channel; K_{ATP}: ATP-sensitive potassium channel; K_{IR}: inward rectifying potassium channel; GIRK: G-protein-gated inwardly rectifying potassium channel; CTX: charybdotoxin; IBX: iberiotoxin; Glib: glibenclamide; TEA: tetrethylammonium; Ba^{2+}: barium; AA: arachidonic acid; NOS: nitric oxide synthase; COX: cyclooxygenase; 15-LO: 15-lipoxygenase; P450: cytochrome P450; HO: heme oxygenase; PGI_2: prostacyclin; 15-HPETE: 15-hydroperoxy-5,8,11,13-eicosatetrienoic acid; HEETA: hydroxy (H)-epoxyeicosatrienoic acid; 11,12,15-THETA: 11,12,15-trihydroxy-eicosatrienoic acid; EETs: epoxyeicosatrienoic acids; NO: nitric oxide; CO: carbon monoxide; H_2O_2: hydrogen peroxide; IP3: inositol-trisphosphate; AC: adenylyl cyclase; sGC: soluble guanilyl cyclase; cAMP: cyclic-AMP; cGMP: cyclic-GMP; Gi: G-protein; ATP: adenosine triphosphate; CaM: calmodulin; ENdase: endonucleotidase; PTx: pertussis toxin; Gap27: connexin mimetics that possess conserved sequence homology with the second extracellular loop domains of the connexins; αGA: glycyrrhetinic acid derivatives.

concentrations. These responses are independent of K-ATP activation and in general are minimally or not at all affected by charybdotoxin, iberiotoxin, or apamin given alone, but are blocked either by nonspecific potassium channel inhibitors (tetrabutyl-ammonium or tetraethylammonium) or by the combination of charybdotoxin plus apamin.[21,229,389,448,478–480,756,758,759] The various hypotheses formulated for the identification of EDHF in animal blood vessels have also been proposed in human blood vessels: (1) epoxyeicosatrienoic acid derived from cytochrome P450 monooxygenase (EETs) in coronary, internal mammary, and subcutaneous arteries;[406,449,452,762] (2) myoendothelial

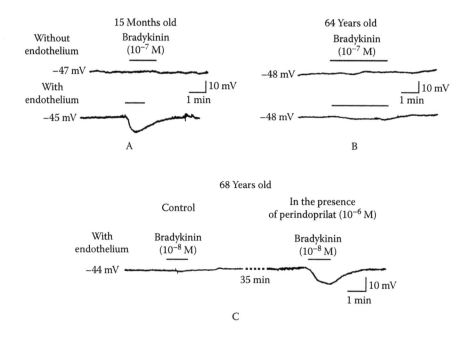

FIGURE 69 Endothelium-dependent hyperpolarizations in isolated human coronary arteries (in the presence of inhibitors of cyclooxygenases and NO-synthases). (A) In coronary arteries taken from a young patient (15 months old, congenital heart disease), 10^{-7} M bradykinin produces an endothelium-dependent hyperpolarization. (B) In coronary arteries taken from an older patient (64 years old, ischemic heart disease), the endothelium-dependent hyperpolarization to the same concentration of bradykinin is reduced markedly when compared to that observed in the juvenile patient. (C) In coronary arteries taken from a patient (68 years old, ischemic cardiomyopathy), the endothelium-dependent hyperpolarization to bradykinin is virtually absent. The presence of the angiotensin-converting enzyme inhibitor, perindoprilat, restores the endothelium-dependent hyperpolarization to the peptide. These experiments demonstrate that EDHF-mediated hyperpolarizations occur in human coronary arteries but that the response is altered by vascular diseases (and/or aging) and is susceptible to improvement by pharmacological intervention. This improvement may partially contribute to the beneficial therapeutic effects of some drugs such as angiotensin-converting enzyme inhibitors. Targeting the EDHF pathway could lead to exciting new drug discoveries for the treatment of cardiovascular diseases. (By permission of the *Journal of Clinical Investigation*; modified from Nakashima et al.[21])

gap junction communication in subcutaneous arteries;[333,343] (3) potassium ions in interlobar renal and thyroid arteries,[388,389] and (4) hydrogen peroxide in coronary and mesenteric arteries. [481,576]

Thus, the *in vitro* data available demonstrate that EDHF-mediated responses are observed in isolated human blood vessels and that the characteristics of these responses are similar to those observed in animals studies (Figure 69).

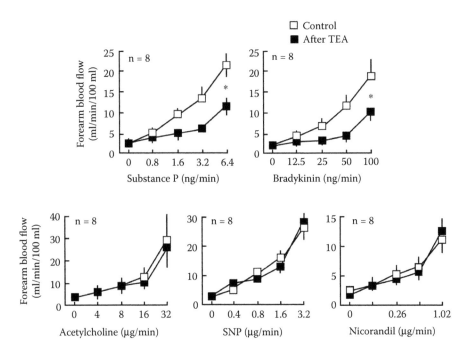

FIGURE 70 Contribution of EDHF-mediated responses as observed *in vivo* in humans. Top: Forearm blood flow was measured by strain-gauge plethysmography in healthy human treated with an inhibitor of cyclooxygenase, aspirin, and an inhibitor of NO-synthase, N-methyl-L-arginine. Under these conditions, the intra-arterial infusion of bradykinin or substance P (two endothelium-dependent vasodilators) produces an increase in forearm blood flow, which is partially inhibited by the administration of tetraethylammonium (TEA), a nonspecific blocker of calcium-activated potassium channel. Bottom: In contrast, the increase in forearm blood flow produced by another endothelium-dependent vasodilator, acetylcholine, and that to two endothelium-independent vasodilators, the NO-donor sodium nitroprusside (SNP) and the mixed NO-donor and activator of ATP-sensitive potassium channel, nicorandil, were not affected by the administration of TEA. These results indicate that bradykinin- and substance P-induced increases in forearm blood flow involve the contribution of a NO- and prostacyclin-independent mechanism, which relies on the activation of calcium-activated potassium channels. This strongly suggests that an EDHF-related mechanism contributes to the regulation of peripheral vascular tone and blood flow in healthy human. (By permission of the American Heart Association and Lippincott Williams & Wilkins; modified from Inokuchi et al.[768])

3.10.2 INTACT ORGANISM

In humans, the involvement of EDHF in the vasodilator responses is difficult to assess and few studies have been designed specifically to address this issue. An EDHF-mediated mechanism is often suggested to explain vasodilatations resistant to inhibitors of nitric oxide synthase. However, most of the human studies do not involve the administration of an inhibitor of cyclooxygenases. Furthermore, complete blockade of nitric oxide synthase is difficult to obtain (or to demonstrate) and/or a nonendothelial effect of the vasodilators, such as a direct effect on the smooth muscle cells or an inhibitory effect on the sympathetic nerve endings, cannot be excluded easily. In the human, apamin plus charybdotoxin cannot be administered for obvious ethical reasons. However, studies especially designed to study EDHF-mediated responses have shown convincingly that in the forearm vasculature, the increases in blood flow in response to bradykinin, substance P, and, to a lesser extent, acetyl-choline involve a non-prostanoid-, non-NO-mediated mechanism that depends on activation of potassium channels.[450,763–767] Thus, in the presence of inhibitors of both cyclooxygenase and NO synthase, the addition of the nonspecific K_{Ca} blocker tetra-ethylammonium reduces basal forearm blood flow and the increase in blood flow evoked by the intrabrachial administration of substance P and bradykinin, but not that of acetylcholine (Figure 70).[768] In the human forearm the EDHF-mediated dilation does not appear to be sensitive to feedback-inhibition from NO.[767]

These results suggest that, in humans, EDHF-mediated responses can occur *in vivo* and that they contribute to the regulation of basal blood flow as well as to the endothelium-dependent relaxations in response to certain mediators. The synthesis of nonpeptidic inhibitors of SK_{Ca} and IK_{Ca}[249,769,770] may in the future allow the selective blockade of EDHF-mediated responses and hence the proper determination of their physiological role in the human circulation. In any case, it should be emphasized that the EDHF-related studies performed in animals have been, so far, remarkably predictive for the characterization of the EDHF pathway in the human.

4 EDHF and the Physiological Control of Blood Flow

4.1 NOS-3 KNOCK-OUT MICE

In the absence of selective and user-friendly inhibitors of EDHF-mediated responses for *in vivo* studies, transgenic mice are important tools in the analysis of their contribution to cardiovascular physiology and pathology. Studies performed with NOS-3 knock-out mice have shown that both prostacyclin and especially EDHF-mediated responses play a compensatory role for the absence of endothelial NO in both agonist- and flow-induced endothelium-dependent vasodilatation, as well as in the basal regulation of myogenic tone (Figure 71).[1–7] In resistance arteries of NOS-3/COX-1 double knock-out female mice, endothelium-derived relaxations are preserved by an EDHF-mediated mechanism, while in arteries from the male, the endothelium-dependent relaxations are impaired severely. In female mice, the absence of NOS-3 and COX-1 has no effect on mean arterial blood pressure, while the corresponding males are hypertensive.[8] These results suggest that EDHF-mediated responses contribute to the overall regulation of arterial blood pressure.

However, the early adaptation of these genetically modified animals may overemphasize the role of EDHF-mediated responses. Experiments with selective conditional knock-out mice should be performed in order to properly assess the physiological role of EDHF.

4.2 EDHF AND ARTERIAL BLOOD PRESSURE

Blood flow is regulated by vascular tone, which is the result of various constrictor and dilator stimuli including myogenic tone, sympathetic activity, oxygen level, endothelium-derived vasoactive factors, circulating hormones, products of cell metabolism, and physical forces exerted by the flowing blood. It is generally accepted that arterial blood pressure and flow to the tissues are determined mainly by the degree of opening of resistance arteries with an internal diameter smaller than 300 µM. The presence of an EDHF component to endothelium-dependent relaxations (dilatations) becomes preponderant in resistance arteries[9,10] and is more pronounced in arteries than in veins.[11,12] However, whether or not this contributes to the regulation of arterial blood pressure is still a matter of debate.

Inhibition of NO-synthase *in vivo* leads to an increase in mean arterial blood pressure and to a reduction in basal blood flow in many species and all vascular beds. By contrast, in anesthetized rats, the inhibition of the EDHF-pathway by the combined administration of charybdotoxin plus apamin does not modify basal conductance in the mesenteric or the hindlimb, and does not affect arterial blood pressure, although

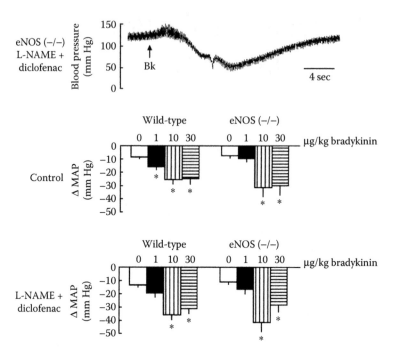

FIGURE 71 Bradykinin (Bk) induces changes in blood pressure in two strains of mice, wild-type (C57BL/6), and NOS-3 knock-out mice [eNOS(-/-)]. Top: Original trace showing the recording of arterial blood pressure in an anesthetized eNOS-(-/-) mouse treated with an inhibitor of cycloxygenase, diclofenac, and an inhibitor of NO-synthase, L-arginine-methyl-ester (L-NAME). The intravenous administration of the endothelium-dependent vasodilator bradykinin induces a transient decrease in arterial blood pressure. Bottom: Summary bar graphs showing the concentration-dependent changes in mean arterial pressure (Δ MAP) evoked by administration of bradykinin in wild-type and eNOS(-/-) mice under control conditions (top) or after treatment with inhibitors of cyclooxygenases and NO-synthases (bottom). The amplitude of the hypotensive response induced by bradykinin is similar in eNOS(-/-) and wild-type animals. In both strains, the effect of bradykinin is unaffected by the presence of inhibitors of cyclooxygenases and NO-synthases. These data strongly suggest that EDHF-mediated responses play an essential role in the hypotensive effect of bradykinin. (By permission of the National Academy of Sciences USA; modified from Brandes et al.[6])

the toxins inhibit the vasodilatation, in both vascular beds, in response to acetylcholine or bradykinin.[13] In the same species, inhibition of gap junction communication by the intra-renal administration of elevated concentrations of Gap peptide[14] produces a decrease in basal renal blood flow and an increase in arterial blood pressure.[15] In mice, the level of expression of SK3 channels on the endothelial cells is correlated with the value of cell membrane potential of both endothelial and vascular smooth muscle cells, the tone of isolated mesenteric arteries, the diameter of these arteries *in situ*, and the arterial blood pressure of the animals (Figure 72).[16] In NOS-3 knock-out mice, the hypotension evoked by the administration of bradykinin is resistant to the combined

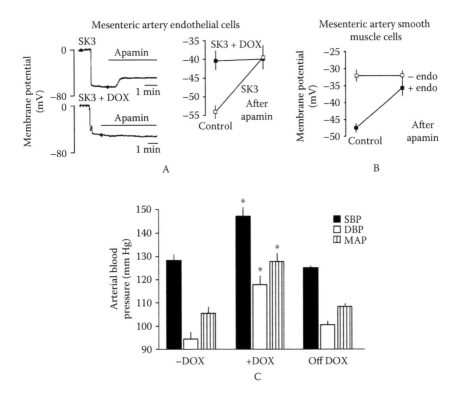

FIGURE 72 Membrane potential of endothelial and vascular smooth muscle cells in transgenic mice with doxocycline-sensitive overexpression of SK3. Effects on arterial blood pressure. (A) Original traces (left) and summary graph (right) of the membrane potential of endothelial cells of mesenteric arteries from transgenic mice overexpressing SK3 (threefold mainly in the endothelial cells and virtually not in the smooth muscle cells) fed with a standard diet (SK3 mice) or from transgenic mice subjected to dietary exposure to doxocycline (SK3 + DOX mice), which results in the virtual abolition of the SK3 expression. At rest, the membrane of the endothelial cells taken from SK3 mice is more polarized than that of those taken from SK3 + DOX mice. Apamin depolarizes the endothelial cells from SK3 mice, bringing the resting membrane potential to similar values as in endothelial cells of SK3 + DOX mice. Apamin is ineffective in the endothelial cells of the latter group of animals. (B) Summary graph of the effect of apamin on the membrane potential of smooth muscle cells of mesenteric artery (with and without endothelium) taken from SK3 mice. At rest, the membrane of the smooth muscle cells is more polarized in the presence of the endothelium (+ endo), than in its absence (– endo). In arteries without endothelium, apamin does not influence the resting membrane potential while, in the presence of the endothelium, the toxin depolarizes the smooth muscle cells, bringing the resting membrane potential to values similar to those recorded in smooth muscle cells from arteries without endothelium. These results show that the overexpression of SK3 in the endothelial cells hyperpolarizes the smooth muscle cells and indicate that the resting potential of the smooth muscle cells is driven by that of the endothelial cells. (C) Changes in arterial blood pressure produced by doxocycline (SBP: systolic blood pressure; DBP: diastolic blood pressure; MAP: mean arterial blood pressure). In transgenic animals overexpressing SK3, dietary exposure to doxocycline (+ DOX) increases both systolic and diastolic arterial blood pressure when compared to transgenic mice fed a standard diet (– DOX). The suppression of doxocycline from the

presence of inhibitors of NO-synthase and cyclooxygenase (Figure 71)[6] and in rats the hypotension elicited by agonists of protease-activated receptor-2 (PAR-2) is blocked by the combination of charybdotoxin plus apamin.[17]

These results indicate that the alteration in EDHF-mediated responses can affect mean arterial blood pressure. Whether or not they contribute tonically to the homeostatic regulation of arterial blood pressure is unknown. The issue to be addressed specifically is the role of EDHF-mediated responses in the control of peripheral resistance, when the activity of the NO-synthase is either inhibited or preserved, in conscious animals. Indeed, anesthetic drugs may inhibit some EDHF-mediated responses (see section 6.8).[18] The availability of new, nonpeptidic, and specific inhibitors of IK_{Ca} and SK_{Ca} should facilitate such studies.[19–21]

4.3 FLOW-INDUCED VASODILATATION

Efficient regulation of blood flow requires coordinated changes in the diameter of both arterioles and the feeding arteries that supply them. This involves flow-induced vasodilatation and a mechanism that is independent of flow or changes in internal pressure, spreading vasodilator responses. Flow-induced vasodilatation is an important physiological stimulus involved in the control of vascular tone as the blood vessel continuously adapts its diameter to minimize the shear forces exerted by the flowing blood on the endothelial surface.

Shear stress can activate stretch-activated cation channels on the endothelial cell surface and promote calcium influx and the release of calcium from intracellular stores, and thus increase intracellular calcium.[22,23] This is a key step not only for the activation of the L-arginine-NO-synthase pathway but also for the opening of endothelial potassium channels, the subsequent hyperpolarization of the endothelial cell membrane,[22,24] and thus EDHF-mediated responses.[25] Long-term exposure of human endothelial cells to laminar shear stress upregulates the expression and the function of IK_{Ca},[26] suggesting that this potassium channel, one of the key players in EDHF-mediated responses, contributes to the endothelium-dependent adaptation to hemodynamic changes. The involvement of a non-NO non-prostanoid mechanism in flow-induced dilatation has been demonstrated in various isolated arteries from different species including human coronary arteries (Figure 73).[4,27–32]

The involvement of a cytochrome P450 metabolite has been suggested in blood vessels from NOS-3 knock-out mice[4] and in isolated cremaster arterioles of the hamster.[32] Furthermore, in a perfusion system, the release of EETs from the endothelium of rat and murine gracilis arteries (donor vessels) subjected to changes in shear stress hyperpolarizes the smooth muscle cells of the downstream detector vessels

FIGURE 72 (Continued) diet (off DOX), and therefore the re-expression of SK3, restores the blood pressure to control values. These results strongly suggest that EDHF-mediated responses, which rely on the activation of endothelial calcium-activated potassium channels, including SK3, are a fundamental determinant of vascular tone and arterial blood pressure. (By permission of the American Heart Association and Lippincott Williams & Wilkins; modified from Taylor et al.[16])

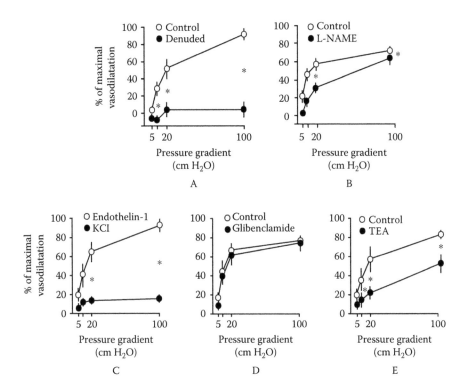

FIGURE 73 Flow-induced vasodilatation in isolated human coronary arteries. The experiments were performed in isolated human coronary arteries taken from patients without clinical evidence of coronary artery diseases. Changes in flow were produced by changes in the pressure gradient of the perfused coronary artery segments without changes in intraluminar pressure. (A) Flow-induced vasodilatation is endothelium-dependent (Control: with endothelium; Denuded: without endothelium). (B) Flow-induced vasodilatation involves NO production as the vasodilatation is inhibited partially by the presence of a NO-synthase inhibitor (L-nitro-arginine-methyl ester: L-NAME). However, a substantial portion of flow-induced vasodilatation relies on another mechanism. (C) Flow-induced vasodilatation is observed in vessels constricted by a receptor-dependent mechanism, stimulation of endothelin receptors (pharmaco-mechanical coupling), but is not seen in blood vessels constricted by a depolarizing solution (KCl: 40 mM). (D) Flow-induced vasodilatation is not affected by glibenclamide, a blocker of ATP-sensitive potassium channels. (E) Flow-induced vasodilatation is inhibited significantly by the nonselective blocker of calcium-activated potassium channels, tetraethylammonium (TEA). Altogether, these results suggest that in human coronary arteries flow-induced vasodilatation involves an EDHF-mediated response. (By permission of the American Heart Association and Lippincott Williams & Wilkins; Modified from Miura et al.[29])

(mesenteric arteries).[33] These results confirm that EETs can act as endothelium-derived hyperpolarizing factors and that gap junctional communication is not necessarily required for shear-stress-stimulated EDHF-mediated vasodilatation. Additionally, pulsatile stretch, another mechanical stimulus, promotes the expression of endothelial cytochrome P450 and the generation of epoxyeicosatrienoic acid, which causes hyperpolarization and relaxation of the smooth muscle cells.[34,35]

In human coronary arteries, both P450 metabolites and hydrogen peroxide may contribute to shear stress-induced endothelium-dependent dilatation.[29,30] However, in the canine coronary artery, *in vivo*, increasing pulsatile flow induces a dilatation, which involves a NO- and prostacyclin-independent component requiring the activation of IK_{Ca} and SK_{Ca}, but not BK_{Ca}.[36] These data suggest that pulsatile flow evokes an EDHF-mediated vasodilatation that does not rely on the production of cytochrome P450 metabolites.

4.4 CONDUCTED VASODILATATION

The transmission of vasomotor signals along the axis of an arteriole is one of the pathways by which a stimulus, sensed by a localized portion of the artery, is communicated to remote upstream and downstream regions. Local stimulation of endothelial cells with acetylcholine evokes dilatation that can spread along the length of the arteriole.[37,38] This phenomenon is endothelium-dependent, but if NO is involved in some arteries it alone cannot explain the whole phenomenon[39] as in some blood vessels, such as arterioles of murine skeletal muscle, NO actually inhibits conducted vasodilatations.[40] Thus it seems that EDHF-mediated responses are involved strongly in conducted vasodilatation.

Conducted vasodilatation in response to acetylcholine involves a rise in endothelial intracellular calcium, hyperpolarization of the endothelial cells, and endothelium-dependent hyperpolarization of the underlying smooth muscle cells at the site of stimulation. The hyperpolarization of the endothelial cells and the endothelium-dependent hyperpolarization of the smooth muscle cells are transmitted, but not the rise in intracellular calcium.[41-44] In the mesenteric arterioles of the rat, the K_{ATP} opener cromakalim evokes a conducted hyperpolarization and a spreading dilatation without changes in endothelial $[Ca^{2+}]_i$,[44] indicating that the function of the rise in $[Ca^{2+}]_i$ evoked by endothelium-dependent agonists is to activate endothelial K_{Ca}. This explains why the conducted vasodilatation evoked by these endothelium-dependent agonists is blocked by the K_{Ca} blockers charybdotoxin and apamin.[18,45,46]

Intercellular communication is essential in order to observe conducted vasodilatation. Mice deficient for connexin 40 have an impaired conduction of vasodilatation (Figure 74).[47] In vascular beds such as that of the retractor muscle of the hamster or the intestine of the guinea pig, endothelium and smooth muscle are coupled electrically, forming a syncytium. In these arteries, myoendothelial junctions play an essential role and hyperpolarization originating in a single endothelial cell can drive the relaxation of smooth muscle cells throughout an entire arteriolar segment.[48-54] In the feeding artery of the hamster retractor muscle, the cremaster of the mouse, or the mesentery of the rat, the endothelial cells are a privileged pathway for the conduction of hyperpolarization and the resulting vasodilatation.[44,55-57] Cell activation by acetylcholine facilitates communication through myoendothelial gap junctions, as a more pronounced myoendothelial coupling is observed in arteries exposed to the muscarinic agonist than in those in which either the smooth muscle or the endothelial cells are stimulated by injection of negative currents.[51,52] This

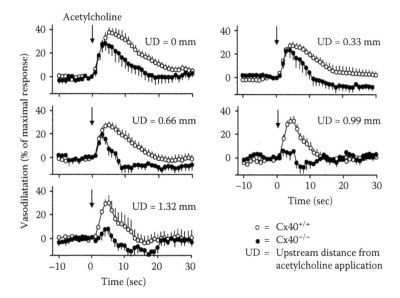

FIGURE 74 Conducted vasodilatation in cremaster arterioles of wild-type (Cx40$^{+/+}$) and connexin-40 knockout mice (Cx40$^{-/-}$). Local pulse pressure application of acetylcholine via a micropipette induces vasodilatation at the site of injection (upstream distance from acetylcholine-application, UD = 0 mm; top left). This local endothelium-dependent vasodilatation is rapidly conducted upstream and is observed almost unaffected up to 1.32 mm from the site of acetylcholine-application (bottom left). In Cx40$^{-/-}$ the vasodilatation to acetylcholine at the site of application is inhibited partially and the conducted vasodilatation is virtually abolished 1 mm upstream of the site of injection. These data indicate that an EDHF-mediated mechanism, which relies on gap-junction communication, is involved in conducted vasodilatation. (By permission of the American Heart Association and Lippincott Williams & Wilkins; modified from de Wit et al.[47])

suggests the involvement of an undefined active membrane process in conducted hyperpolarization.[58] One possibility is that in certain blood vessels, interstitial Cajal cells contribute. Such cells are present in the wall of the spontaneously active rabbit portal vein (where they could act as pacemaker cells),[59,60] but also in that of the mesenteric arteries of the guinea pig (where their function is unknown).[61] In the gut, Cajal cells pace gastrointestinal smooth muscle by initiating slow waves transmitted to the adjacent smooth muscle cells via gap junctions.[62] The presence of these cells may contribute to the active membrane process in conducted hyperpolarization and in the propagation of EDHF-mediated responses in large arteries such as the coronaries.

In other vascular beds, such as those of the cheek pouch and the cremaster of the hamster, the release of a cytochrome P450 metabolite contributes substantially to conducted vasodilatation.[18,63] In the hamster cheek pouch, whereas both myo-myo and endothelial-endothelial junctions are essential pathways, the myoendothelial gap junctions play a limited role.[63] In porcine coronary arterioles, the requirement of K$_{IR}$

activation in conducted vasodilatation[64] strongly suggests that the release and accumulation of potassium ions in the intercellular myoendothelial space plays a critical role in the phenomenon.

4.5 VASOMOTION

Vasomotion is the rhythmic change in arterial or venous diameter occurring spontaneously or upon stimulation with an agonist. The physiological role of this universal phenomenon is not completely understood, although it is likely to contribute to the harmonious and efficient repartition of blood flow.

Vasomotion is generated by cyclical contractions and relaxations of the vascular smooth muscle cells directly associated with oscillatory changes in $[Ca^{2+}]_i$ and in membrane potential. This phenomenon in many arteries relies on the presence of endothelial cells where synchronous calcium waves can also be observed.[65] Nitric oxide release and the subsequent elevation of cyclic-GMP in the smooth muscle cells contribute to the mechanism of underlying vasomotion.[66] In rat mesenteric arteries, the vasomotion evoked by the stimulation of adrenergic receptors is critically dependent on the activation of endothelial IK_{Ca} and SK_{Ca}, indicating that the EDHF-mediated response is essential for the initiation and maintenance of the phenomenon.[67,68] Furthermore, mice deficient for connexin 40 are hypertensive and show an irregular vasomotion.[47,69,70] These findings support the hypothesis that myoendothelial and/or endothelial-to-endothelial cell coupling contributes to synchronous vasomotion, conducted vasodilatation, and to the regulation of peripheral resistance.[57,71]

4.6 HYPOXIA

Acute and chronic hypoxia affect the caliber of arteries and veins by altering the release of endothelial relaxing and contracting factors, and by acting directly on the vascular smooth muscle cells.[72,73] Endothelium-dependent relaxations are inhibited by hypoxia.[74] This is understandable since molecular oxygen, along with L-arginine, is a requested substrate for NO-synthases.[75] However, in blood vessels that exhibit EDHF-mediated responses (e.g., pulmonary, cerebral, and coronary arteries) the EDHF but not the NO component of endothelium-dependent relaxations, is preserved under severe hypoxia.[76–78] In rats subjected to subchronic hypoxia (48 h), the sensitivity to vasoconstrictors and the myogenic responsiveness of the mesenteric artery are reduced. These phenomena persist upon restoration of normoxia. Although an increased NO production explains in part the attenuated systemic vasoconstriction,[79] the endothelium-dependent hyperpolarization of the vascular smooth muscle cells resulting from the augmented production of cytochrome P450 epoxygenase is a major contributor.[80–82] Whether or not EDHF-mediated responses blunt acute hypoxic pulmonary vasoconstriction remains controversial.[83–85]

Thus, EDHF-mediated responses can prevent vasospasm and in some instances compensate for the disappearance of NO-mediated relaxation under hypoxic conditions. This, along with other mechanisms such as the release of prostacyclin and adenosine, as well as the direct activation of potassium channels and the inhibition of Ca_V in the smooth muscle,[86] contributes to the maintenance of a normal caliber of the blood vessel and the preservation of blood flow.

4.7 EXERCISE

Contractions of skeletal muscle increase metabolic demand. In order to provide sufficient oxygen to sustain the augmented activity, the blood flow in exercising muscles should increase rapidly. This is even truer for cardiac muscle, which has a very limited anaerobic capacity. Muscle blood flow is increased within the first second after a single contraction and stabilizes within 30 s during dynamic exercise. This vital physiological adaptation cannot be attributed to a single mediator (or mechanism) and a high level of redundancy allows the suppression of one or more pathways with the preservation of the needed exercise hyperemia. The release of numerous vasodilators in the extracellular compartment helps to explain exercise hyperemia. These include adenosine, ATP, K^+, H^+, NO, prostacyclin, and EDHF.[87,88] The contraction of skeletal muscle evokes conducted vasodilatations allowing the feeding artery supplying the exercising muscle to dilate even without being exposed directly to the vasodilators released in the muscle itself. These conducted vasodilatations have similar characteristics to those produced by either acetylcholine or the injection of currents.[89] Inhibition of the conducted vasodilatation, and therefore the prevention of dilatation of the feeding artery, decreases blood flow to exercising muscle[90] and reduces the availability of oxygen. The link between contraction and ascending vasodilatation is not clear, but EDHF-mediated responses are likely to contribute to skeletal muscle hyperemia.[89] Although, a role for EDHF-mediated responses has been suggested in both flow-induced vasodilatation and in ascending vasodilatation in coronary arteries,[29,30,36,64] the exact contribution of this mechanism during an increase in myocardial oxygen consumption is unknown.

4.8 EDHF AND GENDER

Men and postmenopausal women have a greater incidence of hypertension and coronary artery disease when compared to premenopausal women. The vascular protection observed in the latter has been attributed to a beneficial effect of the female sex hormones, especially estrogens, and justified the now contested hormone replacement therapy. Sex hormones interact with both cytosolic, nuclear, and plasmalemnal receptors in the vasculature, producing both genomic and nongenomic effects. The acute administration of sex hormones, including testosterone, produces rapid vascular relaxation or vasodilation, which can be either endothelium-dependent or –independent.[91,92] However, the physiological relevance of these acute effects is uncertain. The genomic effects of estrogen influence the expression of proteins involved in both the NOS and the EDHF pathways.[91,93]

In isolated mesenteric and tail arteries, as well as in the perfused renal vascular bed taken from sexually mature rats, the contribution of the EDHF-mediated responses is greater in females than in males.[94–97] In resistance arteries of NOS-3/COX-1 double knock-out female mice, an EDHF-mediated mechanism maintains normal endothelium-dependent relaxations, while these relaxations are markedly reduced in the same arteries taken from the corresponding males.[8]

In the mesenteric artery of the female rat, the greater EDHF-mediated response can be attributed to estrogen since in ovariectomized animals this component is markedly reduced, while estrogen supplementation restores it (Figure 75).[98–100] In

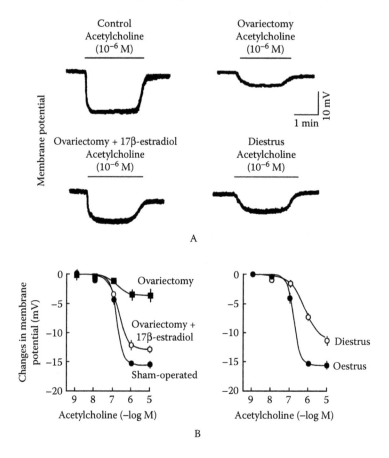

FIGURE 75 Estrogens and endothelium-dependent hyperpolarization in rat isolated mesenteric arteries (in the presence of inhibitors of cyclooxygenases and NO-synthases). (A) Original traces showing the amplitude of the endothelium-dependent hyperpolarization to acetylcholine in isolated mesenteric arteries from female rats. When compared to mesenteric arteries taken from control rat (sham operated), the amplitude of the hyperpolarization is decreased in that from ovariectomized female rats (ovariectomy), while in arteries taken from 17β-estradiol treated ovariectomized animals (ovariectomy + 17β-estradiol) the amplitude of the hyperpolarization is normal. In female rats in the diestrus phase of the estrus cycle, the endothelium-dependent hyperpolarization was suppressed when compared to female rats in the estrus phase. (B) Summary concentration-response curves. These results indicate that estrogen levels potentiate the endothelium-dependent hyperpolarizations. (By permission from the Nature Publishing Group; modified from Liu et al.[98])

this artery, ovariectomy induces reciprocal changes in the respective contribution of EDHF- and NO-mediated responses in the overall endothelium-dependent relaxations so that the disappearance of the EDHF-mediated response is fully compensated by the increase in NO production, a phenomenon associated with a decrease in the expression of the negative regulator of NOS-3 function, caveolin-1.[101] Even short-term alteration in blood levels of estrogens affects the amplitude of the endothelium-dependent hyperpolarizations since, in young female rats at the diestrus phase of the estrus cycle, the amplitude of the hyperpolarizations is reduced significantly (Figure 75).[98] In the gracilis muscle of ovariectomized rats exposed to inhibitors of NO-synthase, flow-induced dilatation depends on the activation of cyclooxygenase, but after estrogen supplementation becomes fully mediated by EDHF.[4] The variation in the amplitude of the EDHF-mediated response produced by ovariectomy and estrogen supplementation is paralleled by variations in the expression of connexin-40 and -43 in the endothelial layer of the mesenteric arterial wall, suggesting that estrogens affect EDHF-mediated responses by controlling the level of cell coupling by gap junctions.[99,101] In addition, estrogen can enhance endothelium-dependent hyperpolarizations by increasing the expression of SK3, the subtype of the endothelial SK_{Ca} channel most likely to be involved in endothelium-dependent hyperpolarizations.[102,103]

By contrast, in the middle cerebral artery of the rat, the endothelium-dependent relaxation is fully NO-mediated in the females, but involves both NO- and an EDHF-mediated component in males.[104] In the middle cerebral artery and pial arterioles of ovariectomized females, the NO-mediated relaxation disappears and an EDHF-mediated component is revealed. Estrogen supplementation reverses the balance between NO and EDHF toward that observed in control non-ovariectomized females.[104,105] These changes are not associated with modifications in calcium homeostasis in the endothelial cells,[106] but rather with changes in the expression of connexin-43.[107] They are not likely explained by NO affecting EDHF function, since manipulations of NO do not alter agonist-induced EDHF-mediated vasodilatations.[108] The inverse modulation by estrogen of the contribution of NO and EDHF/connexin-43 in endothelium-dependent relaxations of cerebral and peripheral blood vessels is puzzling. The observation that in the middle cerebral artery of orchiectomized male rats treatment with testosterone suppresses EDHF-mediated relaxation[109] does not help to clarify the role of sex hormone in the regulation of cerebral vascular tone.

Chronic administration to ovariectomized rats of progesterone, but not medroxyprogesterone, restores the formation of NO in the mesenteric arterial endothelial cells, indicating that progesterone may have a beneficial endothelial effect in some preparations.[110]

In the forearm of postmenopausal women, short-term estrogen treatment improves both the NO-dependent and the EDHF-mediated vasodilatations induced by acetylcholine and substance P, respectively.[111]

Taken in conjunction, these data show that sex hormones exert a prominent role in modulating vascular tone and that the targets affected by sex hormones include EDHF.

4.9 PREGNANCY

The cardiovascular system of females has to adapt during pregnancy. This involves dramatic changes in vascular function. In pregnant rats, the endothelium-dependent relaxations of uterine, but also of other peripheral arteries such as the mesenteric artery, are enhanced. This involves an increased production of NO but also an augmented EDHF-mediated response.[112–115] In pregnant women, similar conclusions have been reached in omental and subcutaneous blood vessels.[116,117] In the myometrium and in subcutaneous arterioles from healthy pregnant women, these EDHF-mediated responses involve gap-junctional communications.[118,119]

5 EDHF and Endothelial Dysfunction

Endothelial dysfunction can lead to vasoconstriction, proliferation of the smooth muscle cells, activation and aggregation of platelets, adhesion and infiltration of white cells, and impaired angiogenesis. The studies focusing on the L-arginine-NO-pathway have demonstrated that a decrease in synthesis of NO and/or a loss of its biological activities is associated with many of the endothelial dysfunctions observed in various cardiovascular diseases.[1] However, alteration of the EDHF pathway can also contribute to endothelial dysfunction.[2] Whether endothelial dysfunction in cardiovascular disease is a risk factor, a risk marker, or a surrogate end point is still a matter of debate.[3]

5.1 HYPERTENSION

The formation of EETs, the endothelium-derived mediator that activates BK_{Ca} on vascular smooth muscle cell, and the production of 20-HETE, the potent vasoconstrictor that conversely inhibits BK_{Ca}, and the expression of cytochrome P450 in the kidney, are altered in several genetic and experimental animal models of hypertension, suggesting that the alteration of one of the EDHF pathways may contribute to the pathogenesis of hypertension.[4] However, hypertension *per se* may also affect the endothelial responses indirectly by accelerating the aging of the vascular wall.

5.1.1 SPONTANEOUSLY HYPERTENSIVE RAT (SHR)

Most of the studies on EDHF in hypertension have been performed in the mesenteric artery of the SHR. They show an impairment of the endothelium-dependent relaxation to acetylcholine, which can be attributed to a marked attenuation of the EDHF component and a concomitant production of cyclooxygenase-derived contractile prostanoids (EDCF) with no or little alteration in the production of NO (Figure 76).[5–10]

This decrease in EDHF-mediated response has been associated with, but not yet causally linked to, a change in the expression profile of gap junctions in endothelial cells. Indeed, Cx37 and Cx40 are lower in artery of the SHR than in that of the WKY, while the reverse is observed for Cx43.[11,12] In addition, the conducted vasodilatation, evoked by iontophoretically applied acetylcholine, is reduced in the mesenteric artery of the SHR, a phenomenon also observed in arteries of the retractor muscle of hypertensive hamster.[13,14] In the WKY, such conducted vasodilatation is facilitated by the activation of K_{IR}, a phenomenon that disappears in the SHR.[13]

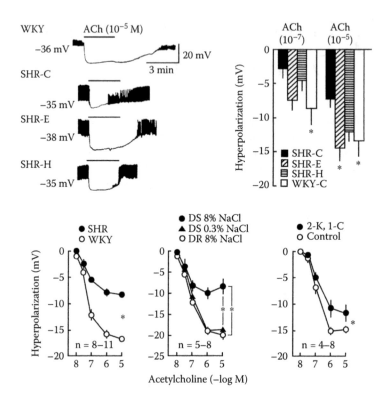

FIGURE 76 Endothelium-dependent hyperpolarization in various rat models of hypertension (in the presence of inhibitors of cyclooxygenases and NO-synthases). Top left: Original traces showing the acetylcholine (ACh) -induced endothelium-dependent hyperpolarization in normotensive Wistar-Kyoto rats (WKY) and the spontaneously hypertensive strain (SHR) either under control conditions or after treatment with antihypertensive agents, either angiotensin-converting enzyme inhibitor (enalapril: SHR-E) or a vasodilator (hydralazine: SHR-H) at equi-effective doses. Top right: Summary of the impairment of the endothelium-dependent hyperpolarization in the SHR and of the restoration of the response by enalapril and to a lesser extent by hydralazine. Bottom: Acetylcholine-induced endothelium-dependent hyperpolarizations are impaired in various model of hypertension SHR, Dahl-salt sensitive rat (DS), and two-kidney one-clip models. However, the degree of impairment is different depending on the hypertension model studied, suggesting that elevated blood pressure *per se* is not the major determinant of the impaired endothelium-dependent hyperpolarization. (By permission of the American Heart Association and Lippincott Williams & Wilkins; modified from Onaka et al.[193])

A loss of EDHF-mediated responses with no or little alteration in the production of NO is also observed in the renal vascular bed of the SHR.[15,16] In the isolated renal artery of the WKY, the EDHF-mediated response relies on the release of potassium ions. In this artery, a total and selective disappearance of this response occurs in aged SHRs, while in young ones, a compensatory mechanism involving a cyto-chrome P450-dependent pathway preserves the endothelium-dependent hyperpolar-ization.[17] Similarly, in the tail artery of the young SHR, endothelium-dependent relaxations are maintained. This has been attributed to a preserved functional role for EDHF resulting from an increase in the heterocellular myoendothelial coupling, which compensates for other structural changes in the arterial media.[18]

In the coronary vascular bed, results are controversial with sometimes a decrease,[19,20] no change,[21,22] or even an increase in endothelium-dependent relax-ations.[23,24] Likewise, in the studies that have addressed the role of EDHF-mediated responses in these coronary arteries, the contribution of EDHF was either reduced[20] or increased.[22,24] These discrepancies do not seem to be explained by differences in age, sex, or arterial blood pressure of the animals, but may rather be due to differences in methodology (isometric contractions in isolated rings for the former study[20] vs. changes in diameter in pressurized preparations).[22,24]

In female SHRs, endothelial dysfunction is less pronounced than in males. Thus, the contraction of the vascular smooth muscle to prostanoids is smaller and the NO-mediated component of the endothelium-dependent relaxation is larger, possibly due to a greater sensitivity of the smooth muscle cells to NO. The EDHF component appears to be inhibited to the same extent in both genders.[25]

Altering the diet modifies endothelium-dependent responses. Thus, in the mesen-teric artery of the SHR, a high-salt diet increases the release of EDCF and NO and decreases EDHF-mediated responses, while a high-cholesterol diet augments EDHF-mediated responses and reduces the responses attributed to NO.[26]

5.1.2 OTHER GENETIC MODELS OF HYPERTENSION

In the SHRSP, a more severe model of genetic hypertension than the SHR, the endothelial dysfunction in the mesenteric artery is more pronounced than in the SHR, and both the EDHF-mediated responses and the NO activity are decreased.[11,27]

In the Munich Wistar Fromter, a genetically hypertensive rat with spontaneous albuminuria, both EDHF- and NO-mediated responses are reduced in the coronary artery, while the endothelium-dependent relaxation of the mesenteric artery is unal-tered.[24] In the mesenteric artery of the Lyon hypertensive rat strain, both the NO- and EDHF-mediated responses are virtually unaltered when compared to those observed in the corresponding normotensive controls.[28]

In the renal circulation of the salt-sensitive Dahl rat, fed with a high-salt diet, the NO component is altered with no modification of the EDHF-mediated responses[15] while in the afferent arterioles, the two endothelial mechanisms are impaired.[29] In the isolated mesenteric artery of that strain, the endothelium-dependent hyperpolar-ization to acetylcholine is reduced when compared to control animals or to salt-sensitive animals fed a low-salt diet (Figure 76).[30] In nongenetically selected animals, such as the weanling Sprague-Dawley rat, a high-salt diet also induces hypertension but, in contrast to the Dahl salt-sensitive rats, no apparent endothelial dysfunction is observed because an increase in the EDHF-component compensates for the decrease in the NO-mediated relaxation.[31]

5.1.3 Induced Hypertension

In the mesenteric artery of the DOCA salt-hypertensive rat, the endothelium-dependent hyperpolarization, in response to acetylcholine, is reduced.[32–35] However, in the renal artery of this model, the NO component is decreased with no modification of the EDHF component.[15]

In the mesenteric artery of the rat with two-kidney one-clip hypertension, EDHF-mediated responses are impaired (Figure 76),[30] while in the carotid artery, the contribution of this component is enhanced.[36]

In rats with renal mass reduction, early endothelial dysfunction is observed in the mesenteric vascular bed even with no or minimal elevation of systolic blood pressure. Nitric oxide availability is decreased and this is partially compensated for by an increase in the production of prostacyclin, with no changes in EDHF-mediated responses.[37] However, in animals subjected to a high-salt diet, which develop hypertension, both the NO- and EDHF-mediated responses are reduced.[38] In a more severe model of subtotal nephrectomy leading to chronic renal failure, the NO-dependent component of the endothelium-dependent relaxation is not altered, while the EDHF-mediated response is abolished and the relaxation to K-ATP openers inhibited.[39]

5.1.4 NO-Synthase Inhibition or Deletion

Rats treated chronically with a NOS inhibitor develop hypertension. In the renal circulation, the normal endothelium-dependent responses, despite a decrease in the availability of NO, suggest an increase in EDHF-mediated response, although a contribution of prostacyclin has not been excluded.[40] Studies of the perfused mesentery indeed suggest that the decrease in the availability of NO is compensated for by an increased EDHF component.[41,42] However, in isolated rings of mesenteric artery, no increase in the contribution of the EDHF-pathway is obvious.[43–45] The discrepancies between these studies are not related to the dose used or the duration of the treatment with the NOS inhibitor and probably reflect the different methodologies (perfused vascular bed vs. isolated rings). They could be explained by the large variability observed in the contribution of NO and EDHF along different sections of the mesenteric artery of rat and mouse.[46–48]

In the guinea pig, NOS inhibitors are also potent pressor agents,[49] but chronic treatment with L-nitro-arginine-methyl ester (L-NAME) does not significantly affect endothelium-dependent hyperpolarizations, at least in the isolated carotid artery (Figure 77).[50] By contrast, in dogs chronic NOS inhibition provokes a transient increase in arterial blood pressure followed by a return to control value. In that species, although systemic vascular resistances are increased, the ability to decrease heart rate and cardiac output prevents the occurrence of systemic hypertension.[51] In the coronary arteries of these dogs, the endothelium-dependent hyperpolarizations are not or are only minimally affected by the chronic L-nitro-arginine (L-NA) treatment (Figure 77).[50]

NOS-3 knock-out mice are moderately hypertensive[52] and, as mentioned earlier (see section 4.1), both prostacyclin and, in particular, EDHF-mediated responses play a compensatory role (Figure 71).[47,48,53–57]

(ex-vivo experiments; in the presence of NO-synthase and cyclooxgenase inhibitors)

A Canine coronary artery

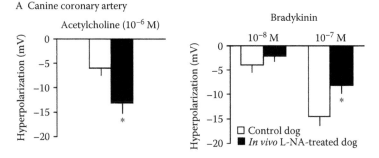

B Guinea-pig carotid artery

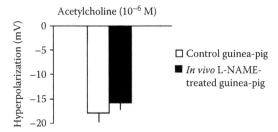

FIGURE 77 Endothelium-dependent hyperpolarization and chronic treatment with an inhibitor of NO-synthase (in the presence of inhibitors of cyclooxygenases and NO-synthases). (A) Coronary arteries harvested from dog chronically treated with an intravenous infusion of L-nitroarginine (L-NA, 20 mg/kg, intravenously for seven days). The endothelium-dependent hyperpolarization to acetylcholine (left) is enhanced significantly, while that to bradykinin (right) is impaired. (B) In carotid arteries from guinea pig treated with L-nitroarginine-methyl-ester (L-NAME, 60 mg/kg in drinking water during 7 to 15 days), the endothelium-dependent hyperpolarization to acetylcholine is not affected. These results confirm that hypertension *per se* does not impair EDHF-mediated responses and suggest that these responses are not necessarily a compensatory mechanism that preserves endothelial function when the NO-synthase pathway is altered. (By permission from Lippincott Williams & Wilkins; modified from Corriu et al.[50])

5.1.5 PULMONARY HYPERTENSION

Acute and chronic hypoxia induces pulmonary hypertension. Whether or not EDHF-mediated responses blunt hypoxic pulmonary vasoconstriction remains controversial.[58–61] Relaxations attributed to endothelium-dependent hyperpolarizations have been observed in pulmonary arteries from animals with both normal and increased pulmonary pressure. However, the mechanisms underlying these responses differ, since in pulmonary arteries of control rats, EDHF-mediated responses depend on activation of cytochrome P450 and gap junction communication, while the mechanism in arteries from rats with pulmonary hypertension remains undefined.[61]

5.1.6 HUMAN HYPERTENSION

Few studies combined inhibitors of both cyclooxygenases and the NO-synthases to assess the role of EDHF-mediated responses in the endothelial dysfunction characteristic of human hypertension. The mechanism underlying the endothelial dysfunction in essential hypertension most likely is linked to oxidative stress and the activation of a cyclooxygenase pathway, which reduces availability of NO. The presence of an EDHF-like mechanism compensates for this decreased bioavailability in order to sustain endothelium-dependent vasodilatation (Figure 78).[62–64] This compensatory pathway can be depressed by additional aggravating factors such as hyperhomocysteinemia.[64] Indeed, in rats with chronic hyperhomocysteinemia, induced by a high-methionine/low-vitamin B diet, a severely reduced EDHF-mediated vasodilatation is observed in the renal vasculature.[65,66] The effects of other risk factors, such as smoking, on EDHF-mediated responses in humans (and animals) are unknown.

Taken in conjunction, these results show that hypertension *per se* does not produce a consistent alteration of the EDHF-mediated responses. The question remains, in models where such an alteration is observed, whether this dysfunction could contribute to the genesis of the syndrome or is a consequence of the hypertensive process.

5.2 AGING

The SHR, at least in regard to the endothelial dysfunction phenotype, can be considered as a prematurely aging rat. Indeed, in the aging normotensive rat, a progressive and slow decrease in the release of NO is observed together with a reduction in EDHF-mediated responses, which occurs earlier, and an increase in the production of EDCF.[7,8,67,68]

There is a significant inhibition of endothelium-dependent hyperpolarizations with aging in human coronary and peripheral arteries, although in the human it is difficult to separate what can be ascribed to aging *per se* rather than to other risk factors such as atherosclerosis, hyperlipidemia, or smoking (Figure 69).[69,70] The measurement of forearm blood flow *in vivo* suggests that aging and hypertension produce the same endothelial dysfunction; in other words, that hypertension causes premature aging of the endothelium. With aging, oxidative stress and the activation of the metabolism of arachidonic acid by cyclooxygenase reduce the availability of NO, which is, at least initially, compensated for by the presence of an EDHF-like mechanism.[71,72]

5.3 ECLAMPSIA

Eclampsia is a pregnancy-specific disorder characterized by hypertension, proteinuria, and alterations in endothelial cell function. Human omental arterioles taken from pre-eclamptic women exhibit a specific deficit in the endothelial production of prostacyclin.[73] In myometrial arterioles from pre-eclamptic mothers, the upregulation of the EDHF-mediated responses observed in normal pregnancy does not occur.[74,75] These endothelial dysfunctions may contribute to the clinical features of the disease.

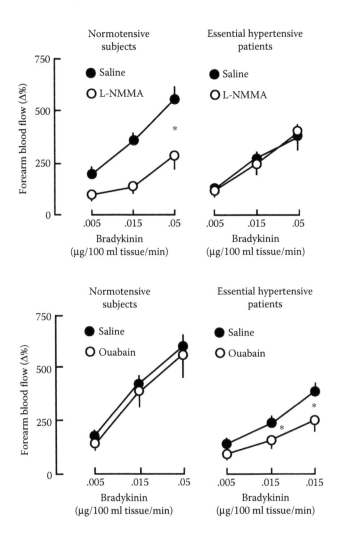

FIGURE 78 EDHF-mediated responses in hypertensive patients. Forearm blood flow was measured by strain-gauge plethysmography in normotensive human and essential hypertensive patients. Top: In normotensive subjects (left), the endothelium-dependent vasodilator bradykinin causes an increase in forearm blood flow, which is inhibited by the administration of the NO-synthase inhibitor N-methyl-L-arginine (L-NMMA). In essential hypertensive patients (right) the vasodilatation in response to bradykinin is impaired and is no longer affected by the administration of L-NMMA. Bottom: In normotensive subjects (left), the blocker of Na+/K+-ATPase ouabain does not affect the increase in forearm blood flow evoked by bradykinin, while in essential hypertensive patients (right), the cardiac glycoside inhibits the vasodilator response. These results show that the contribution of NO-mediated vasodilatation is impaired in essential hypertensive patients and suggest a compensatory role for an EDHF-mediated mechanism. These results need to be confirmed with a more specific blocker of EDHF-mediated responses, in the presence or not of inhibitors of cyclooxygenases and NO-synthases inhibitor. (By permission of the American Heart Association and Lippincott Williams & Wilkins; modified from Taddei et al.[63])

5.4   ATHEROSCLEROSIS AND HYPERCHOLESTEROLEMIA

One of the first studies designed to investigate whether or not EDHF-mediated responses are altered in atherosclerosis was performed in isolated carotid arteries of hypercholesterolemic rabbits.[76] They led to the conclusion that an enhanced contribution of EDHF compensated for the decrease in NO-mediated relaxation. However, this interpretation has been revisited since, in this artery, the production of NO appears to be solely responsible for endothelium-dependent relaxations, implying that the EDHF pathway is not involved (Figure 37).[77] Indeed, in the carotid artery of the rabbit NO produces a cyclic-GMP-independent activation of BK_{Ca} of the smooth muscle cells and the contribution of this NO-dependent hyperpolarizing pathway is increased during hypercholesterolemia. The resistance of this endothelium-dependent hyperpolarization to hypercholesterolemia has been confirmed in the same artery[78,79] and demonstrated in the renal artery.[80–82] Similarly, in the renal artery of SHR subjected to high cholesterol intake, the contribution of the EDHF pathway is enhanced, while that of NO is decreased.[26] By contrast, in the rabbit aorta, hypercholesterolemia impairs the global endothelium-dependent relaxations and both the hyperpolarization-dependent and -independent components are affected.[84,85] None of the experiments involving rabbit arteries allow the sure determination whether the EDHF pathway does play a role or whether the observed responses can be attributed solely to the release of NO.

However, in pressurized segments of murine gracilis artery (free of atherosclerotic lesions), the contribution of EDHF-mediated responses is increased in dyslipidemic mice expressing the human apolipoprotein B-100 [hApoB(+/+)] when compared to that of wild-type mice. This augmented contribution is attributed to the generation of an additional EDHF mechanism in the hApoB(+/+). In arteries from the wild-type mice, the EDHF-mediated response relies on the release of K^+ ions, while in those of the dyslipidemic mice the additional involvement of EETs is observed.[86]

In isolated gastroepiploic arteries from atherosclerotic patients, endothelium-dependent hyperpolarizations are inhibited.[70] The prolonged duration of hypercholesterolemia and the severity of the atherosclerosic process in the human may contribute to the degree of dysfunction of the EDHF pathway.

Hypercholesterolemia is associated with an increase in oxidized LDL, which plays an important role in the initiation and progression of atherosclerosis. Controversial results have been published concerning the effects of oxidized LDL on EDHF-mediated response. In porcine coronary arteries, the endothelial dysfunction induced by acute incubation with oxidized LDL involves only the NO-dependent pathway. The EDHF-mediated responses due to either endothelial hyperpolarization or production of cytochrome P450 metabolites remain unaffected.[87] By contrast, in human coronary arteries, oxidized LDL decreases the expression of cytochrome P450 and the associated EDHF-mediated responses.[88] However, EDHF-mediated responses, in general, appear less sensitive to oxidized lipid and oxidative stress than NO-dependent relaxations.[89–92] In oxidized LDL, the elevated level of lysophospholipids (e.g., lisophosphatidylcholine) can generate endothelial dysfunction. The presence of lisophosphatidylcholine inhibits NO-mediated relaxations but the lysophospholipids are even more potent in inhibiting EDHF-dependent responses as shown in rat mesenteric and porcine coronary arteries and possibly (see the limitation mentioned above) in the rabbit aorta.[93–96] This inhibitory effect of lisophosphatidyl-choline cannot be attributed to cellular toxicity as it is fully reversible.

Whether or not, as has been hypothesized, EDHF-mediated responses compensate for the impairment of NO-dependent relaxations in the early phase of atherosclerosis remains to be demonstrated.[90]

5.5   HEART FAILURE

Congestive heart failure, which results in reduced peripheral blood flows, is mainly attributed to dysfunction of the cardiac myocytes. However, an endothelial dysfunction with a decreased bioavailability of NO contributes to the development of heart failure.[97,98]

The endothelium-dependent relaxations of isolated mesenteric arteries of rats with coronary artery ligation resulting in myocardial infarction and subsequent heart failure is affected minimally. However, the NO-mediated component is reduced, while that due to EDHF is augmented, compensating for the decrease in the availability of NO.[99] Opposite results have been described in the aorta of the same model.[100] However, in the latter blood vessel, the endothelium-dependent relaxations are due solely to the release of NO.[101,102] A decrease of the NO-dependent component associated with a compensatory role of EDHF-mediated relaxations has also been described in the forearm vasculature of patients with congestive heart failure in response to the administration of acetylcholine.[103]

5.6   ISCHEMIA-REPERFUSION

Ischemia-reperfusion is associated with endothelial dysfunction and a decreased availability of NO.[104,105] Little information seems available on the potential role EDHF-mediated responses in ischemia-reperfusion.

In anesthetized dogs, one hour of occlusion of the left circumflex coronary artery followed by two hours of reperfusion does not induce major changes in NO-mediated relaxation of the ischemic coronary artery rings but potentiates the contribution of EDHF-mediated dilatations.[106] In isolated perfused rat hearts, ischemia-reperfusion does not reduce the EDHF-mediated responses, while the availability of NO is decreased.[107] Several pathways could be involved in the preserved EDHF-mediated responses. These include the endothelial release of CNP,[108] the production of CO[109] and the formation of kinins, and the subsequent involvement of EDHF-mediated responses following the activation of endothelial B_2 bradykinin receptor subtypes.[110] In severe cases of ischemia-reperfusion, the production of peroxynitrite may impair EDHF-mediated responses by directly decreasing K_{Ca} activity.[111]

Similar results have been obtained in the cerebral circulation. A decrease in the NO component with an associated increase in EDHF-mediated responses has been observed in the middle cerebral artery of the rat.[112] Similarly, in a model of head injury, the endothelium-dependent NO-mediated relaxation of the middle cerebral artery is abolished, while an increase in the EDHF-mediated component maintains endothelium-dependent relaxations.[113] These effects cannot be explained by the disappearance of a tonic repression of the EDHF component exerted by NO under control conditions,[114] as has been suggested for coronary arteries and peripheral blood vessels.[115,116] The increased contribution of the EDHF component is due to the augmented intracellular endothelial calcium concentration observed after stimulation of endothelial receptors.[117]

5.7 ANGIOPLASTY

Restenosis and altered vasomation frequently occur after coronary angioplasty. Changes in morphology and functionality are observed in the regenerated endothelial cells, which recolonize the site of angioplasty.[118] Segments of porcine coronary arteries with regenerated endothelium show a selective impairment of Gi-mediated endothelium-dependent relaxations.[118,119] In this artery, the endothelium-dependent relaxation to mediators for which signal transduction does not involve Gi, such as bradykinin, is not impaired because the reduced production of NO is compensated for by an increased contribution of endothelium-dependent hyperpolarization. For mediators such as serotonin, for which signal transduction involves mainly Gi, both NO production and endothelium-dependent hyperpolarization are impaired (Figure 79).[120,121] In the same

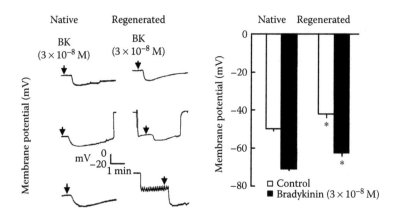

FIGURE 79 Endothelium-dependent hyperpolarization in porcine coronary arteries covered with regenerated endothelium (in the presence of inhibitors of cyclooxygenases and NO-synthases). Left: Original traces showing that the amplitude of the endothelium-dependent hyperpolarization in response to bradykinin (BK) is more variable in arteries with regenerated endothelium (Regenerated) than in control arteries with native endothelium (Native). As the EDHF-mediated response tends to drive the membrane potential of the smooth muscle cells toward E_K (see also Figure 40), the amplitude of the hyperpolarization is larger in arteries with depolarized smooth muscle cells. Right: Summary bar graph showing that the resting membrane potential is altered in smooth muscle cells from arteries with regenerated endothelium, but that the absolute amplitude of the endothelium-dependent hyperpolarization remains unaffected. The preservation of the endothelium-dependent hyperpolarization compensates for the loss of the NO-mediated responses. (By permission from the Nature Publishing Group; modified from Thollon et al.[120])

preparation, both the NO- and the EDHF-mediated responses elicited by substance P were inhibited six weeks after angioplasty.[122] In addition, intracoronary beta-radiation therapy, a clinical procedure used to limit restenosis after angioplasty, produces further inhibition of endothelium-dependent and -independent relaxations. In arteries subjected to angioplasty and radiation, the NO- and EDHF-mediated responses are completely prevented up to six weeks after the surgical procedure.[122]

In the carotid artery of the rat, after endothelial denudation by balloon angioplasty, the endothelium-dependent relaxations are also impaired and both the NO- and EDHF-mediated responses are reduced. However, the alteration in NO-dependent responses is transient and is restored 28 days after the procedure, while the reduction of the EDHF-mediated responses is sustained.[123,124] The cell membrane of the smooth muscle is more depolarized in carotid arteries subjected to angioplasty and the endothelium-dependent hyperpolarization is reduced.[125] This dysfunction is associated with a decreased endothelial expression of both SK_{Ca} and IK_{Ca}.[123]

These results indicate that regeneration of the endothelial cells is likely to alter the functionality and/or expression of K_{Ca} and therefore EDHF-mediated responses. In the porcine coronary artery, the different effects of angioplasty reported could be explained by the difference in the agonists used. In this artery, the endothelium-dependent hyperpolarization evoked by substance P relies exclusively on the activation of endothelial K_{Ca}, while that produced by bradykinin involves both the activation of endothelial K_{Ca} and the release of cytochrome P450 metabolites.[126,127] Thus, it seems logical that the EDHF-mediated response to substance P is impaired, while it is plausible that the hyperpolarization to bradykinin is preserved and plays a compensatory role because of an increased involvement of EETs.

5.8 TRANSPLANTATION

Coronary graft vasculopathy develops in the majority of heart transplant recipients. The accelerated atherosclerosis of the coronary arteries is preceded by reduced endothelium-dependent relaxations. In a porcine model of heterotopic heart transplantation, without immunosuppression but with immunological typing in order to mimic the kinetics of slow low-grade rejection observed in human transplantation, obvious signs of rejection are present in the grafted heart six weeks after transplantation. In the isolated coronary arteries of those hearts, an impairment of NO- and the EDHF-mediated relaxations is obvious. The amplitude of the endothelium-dependent hyperpolarization, *per se*, was not affected, suggesting an uncoupling between the hyperpolarization of the smooth muscle cells and their ability to relax.[128,129]

Cardioplegic hyperkalemic solution used to produce heart arrest and organ preservation solutions, such as the University of Wisconsin or the Euro-Collins solution, produce an alteration in the EDHF-mediated responses in porcine coronary and pulmonary arteries.[130–134] Preservation of the organ in St Thomas's Hospital or in Histidine-Tryptophan-Ketoglutarate solution appears to be less damaging for EDHF-mediated responses.[132,135] The addition in the preservation solution of hyperpolarizing substances such as 11,12 EETs or the K-ATP opener, nicorandil, attenuates the deleterious effect of these solutions.[136–138]

5.9 DIABETES

Macro- and microvascular complications are the principal causes of morbidity and mortality in patients with insulin-dependent (type I) and non-insulin-dependent (type II) diabetes mellitus. The macrovascular complication is diabetes-accelerated athero-sclerosis, while diabetes microvascular diseases include retinopathy, nephropathy, and neuropathy. These are leading causes of blindness, end-stage renal diseases, myocardial infarction, stroke, and peripheral lesions leading to limb amputation. Animal and clinical studies show a strong relationship between plasma glucose levels and microvascular diseases, while both hyperglycemia and insulin resistance, with the associated dyslipidemia, play an important role in the pathogenesis of macrovas-cular diseases.

5.9.1 INSULIN-DEPENDENT DIABETES

Most of the studies designed to assess EDHF-mediated responses in diabetes have been performed in streptozotocin-treated rats, a model of hyperglycemia and insulinopenia that mimics type I diabetes in human. In this model, a decrease in EDHF-mediated response has been reported in mesenteric and carotid arteries, in arterioles supplying the sciatic nerves, and in the renal circulation.[139–144] These data suggest that the impaired EDHF-mediated responses are involved in diabetic microangiopathy and especially in diabetic neuropathy and nephropathy. These alterations in EDHF-mediated responses have been observed sometimes with, but most often without, a change in the NO-mediated responses. For instance, in rats treated with steptozotocin, a comparison was performed between two arteries that do or do not exhibit EDHF-mediated responses, the mesenteric and femoral artery, respectively. The endothelium-dependent response to acetylcho-line was impaired only in the mesenteric artery of the diabetic rat and, in this blood vessel, the NO-mediated component was maintained fully when compared to that of control animals (Figure 80).[145] The endothelium-dependent hyperpo-larization in response to 1-EBIO, an opener of intermediate conductance calcium-activated potassium channel (IK_{Ca}), which acts at the endothelial cell level, was also inhibited, indicating that the transduction of the signal from the endothelial muscarinic receptor was not involved.[145] In this model, the alteration in the EDHF-mediated responses can be attributed to a reduction in the action of cyclic-AMP with an increase in phosphodiesterase-3 activity and a decreased expression and activity of protein kinase A.[146,147] Indeed, cyclic-AMP facilitates EDHF-mediated responses by enhancing electrotonic conduction via gap junctions (see section 3.8.1.6).[148] This interpretation provides a rational explanation for the specific alteration in the EDHF-mediated responses in the mesenteric artery of streptozotocin-treated rats. However, when the endothelium-dependent vasodi-latation induced by histamine was studied in the perfused mesenteric vascular bed of streptozotocin-treated rats, an opposite conclusion was reached, with an increased vasodilator response attributed to a potentiation of the EDHF-mediated component and a decreased contribution of NO.[149] The reasons for these discrep-ancies are unknown.

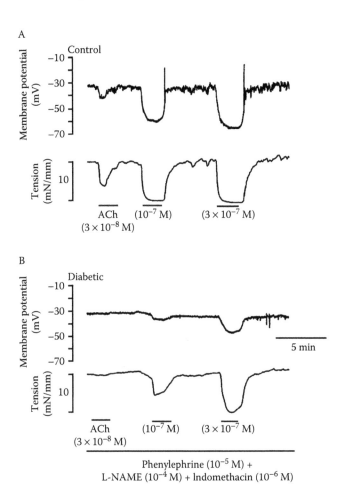

FIGURE 80 Endothelium-dependent hyperpolarization and streptozotocin-induced diabetes (in the presence of inhibitors of cyclooxygenases and NO-synthases). Changes in tension and membrane potential were measured in the smooth muscle cells of the rat mesenteric artery taken from (A) control and (B) hyperglycemic, insulinopenic streptozotocin-treated rats. Acetylcholine (ACh)-induced EDHF-mediated responses (hyperpolarization and relaxation) are markedly impaired in the mesenteric artery of streptozotocin-treated rats. (By permission of the American Physiological Society; modified from Wigg et al.[145])

In patients with type I diabetes under good glycemic control and without albuminuria, endothelial function appears normal. This has been demonstrated by various clinical investigations and confirmed in isolated subcutaneous arteries where both the NO- and EDHF-mediated responses were preserved.[141,150] However, in patients with microalbuminuria, impairment of the endothelium-dependent vasodilatation is observed.[141] In these patients, it is not known whether the various components of the endothelium-dependent vasodilatation are differentially affected by the disease.

5.9.2 INSULIN-INDEPENDENT DIABETES

Type II diabetes, in its early phase before pancreatic beta cell failure, is characterized by insulin-resistance, hyperinsulinemia, moderate hyperglycemia, and dyslipidemia and is often associated with hypertension (syndrome X, metabolic syndrome). This cluster of abnormalities contributes to the development of atherosclerosis.[151,152] In these diabetic patients, endothelial dysfunction has also been reported.[141]

Various genetic and nongenetic animal models have been created to study type II diabetes. In the fructose-fed rat, the EDHF-mediated responses of the isolated mesenteric and coronary arteries are impaired markedly while, at least in the mesenteric artery, the NO-dependent response is not affected by the diet.[153–156] Similarly, in the leptin-deficient, genetically obese and mildly hypertensive Zucker rat, EDHF-mediated responses are reduced in the mesenteric vascular bed without alteration in the NO-mediated dilatation.[34,157] In the Otsuka Long-Evans Tokushima fatty rat, a strain that spontaneously develops hyperglycemia, hyperinsulinemia, insulin resistance, and mild obesity, similar observations have been made.[158,159] Finally, in the nicotinamide- and streptozotocin-treated rats, a model of partial pancreatic beta cell destruction that appears closer to type II than to type I diabetes, both the NO- and the EDHF-mediated responses are attenuated.[160]

In humans, the impairment of endothelium-dependent relaxations observed in the penile artery involves both the NO- and EDHF-mediated components.[161] Type II diabetes is often associated with a procoagulant state linked, in part, to the presence of microparticles, shed membrane vesicles released during cell activation, which possess procoagulant and inflammatory properties. Microparticles derived from T lymphocytes impair the NO and prostacyclin components of the endothelium-dependent vasodilatation of murine mesenteric arteries without impeding the EDHF-mediated responses.[162] In the arteries of the various animal models of type II diabetes, the alteration of endothelium-dependent relaxations involves mainly the EDHF-mediated component, indicating that microparticles play no, or a minor role, in the endothelial dysfunction. However, in humans, type II diabetes is a long-lasting chronic disease and the procoagulant state is likely to exacerbate endothelial dysfunction.

In the db/db–/– mouse, a genetic model of type II diabetes with a mutation on the leptin receptor, the NO-mediated relaxation of the mesenteric artery is reduced, while the EDHF component is preserved, though the mechanism involved in these responses differs in the diabetic and the db/db+/? mice. In control mice, the EDHF-mediated response involves gap junctional communication and K^+ ions, while in the diabetic mice an additional contribution of EETs is observed.[163,164]

In the mesenteric artery of streptozotocin-treated rat, K_{ATP} channels of the smooth muscle are not affected by diabetes.[139] However, in other arteries of this and other models, as well as in human coronary arterioles taken from type I and type II diabetic patients, dysfunctions of K_{ATP} and other potassium channels (K_V, K_{IR}) exist,[144,155,165,166] indicating that diabetes can also alter the intrinsic ability of vascular smooth muscle cells to hyperpolarize.

5.9.3 Hyperglycemia

Acute hyperglycemia *per se* affects endothelial function by facilitating the production of vasoconstrictor prostanoid.[167,168] It also directly affects potassium channel activity. Thus, in the isolated rat coronary artery, the activity of K_V is inhibited by acute hyperglycemia, as is that of K_{ATP} in human omental artery. In both blood vessels, this is due to an increased production of superoxide anions.[169,170] Human umbilical endothelial cells isolated after gestational diabetic pregnancies are hyperpolarized. In these cells, the transport of L-arginine and the production of NO are increased, while the synthesis of prostacyclin is diminished.[171,172] These phenomena are mimicked by acute short-term exposure of umbilical endothelial cells isolated from normal pregnancies to elevated glucose.[173] In endothelial cells from other blood vessels, hyperglycemia is associated with the production of reactive oxygen species and a decrease in the bioavailability of NO.[174] It is not known whether prolonged exposure to elevated glucose *per se* alters endothelial calcium-activated potassium channels and thus EDHF-mediated responses.

5.9.4 Hyperinsulinemia

Insulin is a physiological vasodilator and in the guinea pig submucosal arterioles, EDHF-mediated responses are one of the mechanisms involved in insulin-induced endothelium-dependent vasodilatation.[175] Similarly, in humans, the acute infusion of insulin enhances the acetylcholine-induced increases in forearm blood flow. This response has been attributed to the L-arginine-nitric oxide pathway in normotensive subjects, and possibly to the EDHF pathway in hypertensive patients.[62]

However, insulin can also exert deleterious effect on endothelium-dependent vasodilatation. Thus, acute incubation of the isolated mesenteric artery of the rat with insulin selectively depresses the acetylcholine-induced, EDHF-mediated relaxation, but only when NO-synthase is blocked.[176] These deleterious effects of insulin may contribute to the endothelial dysfunction observed in hyperinsulinic and glucose-intolerant patients.

In conclusion, in most of the animal models of type I or type II diabetes as well as in human diabetic patients, EDHF-mediated responses are altered. However, a direct link between hyperglycemia and impairment of the endothelium-dependent hyperpolarizations has not been established.

5.10 SEPSIS

The release of bacterial lipopolysaccharides (LPS) is in part responsible for vasoplegia and the hyperdynamic state of the circulatory system observed during sepsis. LPS produces vasodilatation by endothelium-dependent and -independent relaxations of the vascular smooth muscle[177,178] and by promoting the expression of NOS-2.[179] However, LPS also attenuates endothelium-dependent relaxations, an effect that can be linked to NOS-2 expression.

In porcine coronary and mesenteric arteries of the rat, LPS inhibits endothelium-dependent hyperpolarizations. This reduction is prevented by inhibiting NOS-2 expression and reversed by the specific blockade of the enzyme.[116,180,181] However, in another study in the mesenteric artery of the rat, LPS did not affect the EDHF-mediated responses.[182] These discrepancies can be explained by the differences in the duration of exposure to and the concentration of LPS tested. Indeed, in the rat, repeated *in vivo* administration of stepwise increasing doses of LPS over four days impairs the EDHF-mediated relaxation of the subsequently isolated renal artery, an effect prevented by treatment with a NO-scavenger. By contrast, a single adminis-tration of LPS, although at a higher dose than the cumulative doses given to the preceding group, did not produce an inhibitory effect.[83] The bacterial release of LPS during sepsis is associated with the generation of cytokines, in particular interleukin-1 (IL-1). In the rat mesenteric artery, IL-1 does not affect endothelium-dependent relaxations, while tumor-necrosis factor-α (TNFα) specifically inhibits the NO-component without affecting the EDHF-mediated responses. The combination of TNFα plus IL-1 is not more effective than TNFα alone.[183]

NO produced by NOS-2 can inhibit EDHF-mediated responses by reducing the permeability of connexin 37-containing gap junctions[184] or by interacting with the heme moeity of cytochrome P450 and inhibiting enzyme activity.[116] However, LPS could also inhibit EDHF-mediated vasodilatation by other means, for example, by depressing the expression of connexin 43[185] or by activating potassium channels in the vascular smooth muscle,[186,187] which would prevent any further endothelium-dependent hyperpolarization.[188]

5.11 CANCER

In the mesenteric artery of patients with colonic adenocarcinoma, endothelial cells express BK_{Ca} and the expression of endothelial IK_{Ca} is increased, resulting in enhanced endothelial hyperpolarization in response to bradykinin.[189]

Vascular abnormalities such as stenosis and poor surgical wound healing are a major cause of postoperative complications of irradiated tissues. In the rabbit aorta and human cervical arteries from patients receiving radiation therapy, the observed endothelial dysfunction exclusively involves the NO component, with no alteration in EDHF-mediated responses.[190,191]

5.12 LEAD INTOXICATION

Environmental and occupational hazards can cause a variety of health problems including cardiovascular disease. The EDHF-mediated hyperpolarizations and relax-ations of isolated mesenteric arteries from rats chronically exposed to lead are inhibited, while the NO-dependent relaxations are preserved.[192]

6 EDHF and Therapeutic Interventions

No drug is available that has been designed to target EDHF-mediated responses. Although endothelial dysfunction is a common denominator of various cardiovascular diseases, whether or not the improvement of endothelial function will have a positive impact on the morbidity/mortality rate is a matter of debate. Large-scale clinical trials are required to demonstrate that treatment of endothelial dysfunction can lead to a better prognosis in patients with essential hypertension or other cardiovascular diseases.[1] Nevertheless, several drugs with beneficial effects on the cardiovascular system improve endothelial function and some augment EDHF-mediated responses, suggesting that these actions contribute to their overall beneficial effects.

6.1 ANGIOTENSIN-CONVERTING ENZYME INHIBITORS AND ANTAGONISTS OF THE AT₁-ANGIOTENSIN RECEPTOR (AT1)

Angiotensin-converting enzyme inhibitors potentiate the endothelium-dependent relaxations and vasodilatations in response to bradykinin *in vitro* in a variety of isolated arteries, including the human coronary arteries, and also *in vivo*.[1-6] The potentiation of endothelium-dependent responses during acute exposure to angiotensin-converting enzyme inhibitors (cilazaprilat, perindoprilat) involves not only NO but also EDHF (Figure 69).[1,4,5] This is explained not only by the angiotensin-converting enzyme inhibitors preventing the degradation of bradykinin,[7] but also by direct interactions of these inhibitors with the sequestration and the sensitivity of the bradykinin B_2 receptors.[8,9] Angiotensin-converting enzyme inhibitors also cause vasodilatation in some arterial beds by potentiating the effect of locally produced bradykinin or inhibiting the local production of angiotensin II. Cilazaprilat *in vivo* produces a bradykinin B_2 receptor-dependent vasodilatation of the canine juxtamedullary afferent artery via NO- and EDHF-mediated responses.[10] In addition, chronic treatment with angiotensin-converting enzyme inhibitors in animals as well as in humans improves endothelial function by mechanisms that do not appear to be directly related to bradykinin.[1,6] In murine mesenteric arteries, the EDHF-mediated responses, attributed to the NOS-3-dependent production of H_2O_2 are enhanced by a chronic treatment with temocapril because of an upregulation of the expression of NOS-3.[11]

In the SHR, chronic treatment with angiotensin-converting enzyme inhibitors (ramipril, enalapril) normalizes blood pressure and enhances arterial dilatation by improving the EDHF-mediated component of the response to acetylcholine (Figure 76).[12,13] In a very similar manner, treatment with AT_1 receptor antagonists (losartan, candesartan) improves EDHF-mediated responses. No additional effects are obtained by treating the animals with the combination of an angiotensin-converting enzyme inhibitor and an AT_1 receptor antagonist.[14,15] The beneficial effect of AT_1 receptor antagonists can be explained by the preferential activation of AT_2 receptor subtype during AT_1 receptor blockade. The activation of AT_2 receptor, like angiotensin-converting enzyme inhibition, leads to kinin-mediated effects.[16] The improvement in EDHF-mediated responses caused by candesartan in the SHR may involve normalization of the expression of gap junctions in the endothelial cells. Indeed, the decreased expression of Cx37 and Cx40, and the increased expression of Cx43 observed in the SHR as compared to the WKY, are corrected by candesartan.[17]

In aging Wistar rats, enalapril or candesartan also prevents the age-related decline in EDHF-mediated relaxation and hyperpolarization.[18,19]

In rats chronically treated with a NOS inhibitor, the controversial changes in the contribution of EDHF to the overall endothelium-dependent relaxation, are likely due to differences in the experimental procedures used (see section 5.1.4). Likewise, quinapril does not affect EDHF-mediated response in animals where L-NAME causes an increased contribution of endothelium-dependent hyperpolarizations,[20] while losartan increases EDHF-mediated response when L-NAME attenuates the relaxation attributed to this pathway.[21]

In the Doca-salt hypertensive rat, a model of hypertension resistant to angiotensin-converting enzyme inhibitors, the diminished EDHF-mediated response is not restored by chronic treatment with quinapril.[22]

6.2 DIURETICS

Loop and non-loop diuretics have direct vasodilator properties in blood vessels from various species including the human.[23–25] Torasemide and furosemide release prostacyclin from human endothelial cells.[26] In the guinea pig mesenteric artery, piretanide and furosemide produce EDHF-like responses.[27] In the SHR, chronic treatment with hydrochlorothiazide, in combination with hydralazine, normalizes arterial blood pressure and improves EDHF-mediated response to acetylcholine, but to a lesser extent than treatment with the angiotensin-converting enzyme inhibitor, enalapril.[13] Hydrochlorothiazide plus hydralazine does not mimic the effects of candesartan in restoring the altered expression of connexins (Cx37 and Cx40, Cx43) in the endothelial cells.[17] In aged WKY, the same combination does not prevent the age-related decline in either EDHF-mediated relaxation or hyperpolarization.[18,19]

To judge from clinical studies, the effects of diuretics on endothelial function are less clear-cut that those involving angiotensin-converting enzyme inhibitors or AT_1 receptor antagonists. Aldosterone antagonists (spironolactone and second-generation antagonists) improve endothelial function in patients with heart failure.[28,29] Furosemide, a loop diuretic, augments endothelium-dependent dilatations in normotensive subjects.[30] However, clinical studies on patients with essential hypertension and heart failure and treated with diuretics are mostly negative.[31–35]

6.3 CALCIUM CHANNEL BLOCKERS

NO- and EDHF-mediated responses require an increase in $[Ca^{2+}]_i$ in the endothelial cells.[36-39] Conversely, a decrease in the extracellular concentration of calcium attenuates EDHF-mediated responses.[40,41] Endothelial cells do not express L-type calcium channels (CaV).[42,43] Therefore the various calcium channel blockers do not prevent EDHF-mediated responses.[44] In isolated blood vessels, certain dihidropyridines can evoke endothelium-dependent, NO-dependent relaxations, a property that is not linked to the calcium blocking properties of these compounds.[45-47] In the SHR, chronic treatment with benidipine decreases blood pressure and improves endothelial function in the renal arteries, but exclusively by the NO-mediated pathway.[48] In L-NAME-treated rats, diltiazem also reduces arterial blood pressure and partially restores NO-mediated relaxation.[20] In addition, in the porcine coronary artery, nifedipine increases the expression of endothelial cytochrome P450, the production of EETs, and the amplitude of endothelium-dependent hyperpolarizations.[49]

Calcium antagonists, including dihydropyridines, are used in the treatment of angina pectoris and hypertension. Their mechanism of action is based on the inhibition of the L-type calcium current of the vascular smooth muscle ($Ca_V1.2$). However, some compounds in this therapeutic class, such as nifedipine or lacidipine, can restore the impaired endothelial function in patients.[1,50] These endothelial effects have been attributed to either an additional property of these dihydropyridines to directly stimulate the release of NO or to increase its bioavailability by virtue of their antioxidant properties. Treatment of hypertensive patients with lercanidipine improves NO availability without beneficial effects on the EDHF-mediated responses.[51,52]

These studies suggest that EDHF-mediated responses are not or are only minimally affected by calcium channel blockers.

6.4 β-ADRENOCEPTOR BLOCKERS

In SHRSP chronically treated with carvedilol, the arterial blood pressure is decreased and the endothelium-dependent relaxation of the mesenteric artery improved when compared to non-treated SHRSP. However, it is not completely restored when compared to WKY. Although the absolute value of the endothelium-dependent hyperpolarization in response to acetylcholine is not enhanced by carvedilol, the resting membrane potential of the smooth muscle cells of the mesenteric artery is more negative than in that of arteries from untreated animals, which may contribute to the improved endothelial function and the antihypertensive effect of the drug.[53] In the SHR treated with celiprolol, an apparent improvement in the EDHF-mediated relaxation of the mesenteric artery is observed with both hypotensive and non-hypotensive doses of the drug.[54] Whether this improvement is linked to a real increase in endothelium-dependent hyperpolarization or to a shift in the resting membrane potential is unknown.

As a rule, β-adrenoceptor blockers do not improve endothelial function in the human and atenolol, for instance, may even have negative effects on the peripheral microcirculation.[1,55] Human coronary artery endothelial cells express β_3-adrenoceptor and their stimulation evokes NO- and EDHF-mediated responses,[56] suggesting that treatment with a nonselective β-adrenoceptor or a selective β_3-adrenoceptor blocker should block this response. However, besides reducing cardiac output, some β_1-adrenoceptor blockers induce vasodilatation and decrease peripheral arterial resistance independently of α-adrenoceptor blockade or compensatory β_2-adrenoceptor stimulation. Nebivolol stimulates the release of NO, and carvedilol, because of antioxidant properties, increases its bioavailability.[1,57,58] Whether or not these or other β-adrenoceptor blockers affect EDHF-mediated responses in patients or healthy volunteers is unknown.

6.5 STATINS

The majority of the clinical benefits obtained with inhibitors of HMG-CoA reductase are a direct result of their lipid-lowering properties. However, statins display additional cholesterol-independent ("pleiotropic") effects that include improvement of endothelial function and augmented synthesis of NO.[59] In the endothelial cells of the human umbilical vein, cerivastatin activates BK_{Ca}, a contributing factor in the statin-induced NO release.[60] In rat arteries, the drug directly elicits hyperpolarization of the smooth muscle by opening K_V.[61] The activation of endothelial K_{Ca} by statins may imply that these drugs should improve EDHF-mediated responses but the available data are controversial. In isolated carotid arteries from hypercholesterolemic rabbits, the impaired endothelium-dependent relaxation is restored by a chronic treatment with lovastatin. However, the improvement observed in the endothelial function following statin administration appears to involve the NO pathway without any modification of EDHF-mediated responses.[62] Again, the involvement of EDHF-mediated response in the rabbit carotid artery being minimal (see section 3.3.5 and Figure 37),[63] this study cannot be considered as conclusive. In the mesenteric artery of the SHRSP, a more appropriate vascular bed for studying EDHF-mediated responses, fluvastatin improves endothelium-dependent relaxations to acetylcholine by a mechanism that exclusively depends on the NO component, while the impaired EDHF-mediated responses are not affected by chronic treatment with the drug.[64] By contrast, in rats with streptozotocin-induced diabetes, rosuvastatin reverses the deficit in nitrergic nerve-mediated vasodilatation of the corpora cavernosa and corrects the decrease in the EDHF-mediated component of the acetylcholine-induced dilatation in the mesenteric vascular bed.[65]

6.6 ET-A RECEPTOR ANTAGONISTS

In the mesenteric artery of the WKY, endothelin-1 and 3 evoke endothelium-dependent hyperpolarizations through the activation of the endothelial ET-B receptor subtype, a response that is reduced with aging.[66,67] In the isolated aorta of the WKY and the SHR, chronic treatment with an antagonist of the endothelin ETA receptor, LU 135252, does not affect blood pressure but increases the contribution of EDHF-mediated responses to the endothelium-dependent relaxations evoked by acetylcholine.[68]

Since the involvement of EDHF in the aorta of the rat is minimal, these results need to be confirmed in other vascular preparations.

6.7 ANTI-DIABETIC DRUGS

Endothelial dysfunction is associated with both diabetic micro- and macroangiopathy, but good metabolic control restores virtually normal endothelial function.[69,70] For instance, appropriate insulin therapy in poorly controlled type II diabetic patients improves endothelium-dependent and independent vasodilatations.[71,72] However, a short-term reduction in HbA1c levels is not sufficient to improve endothelial function.[73] In addition, since the metabolic syndrome associates multiple risk factors such as hypertension, obesity, dyslipidemia, and procoagulant state, cardiovascular drugs such as angiotensin-converting enzyme antagonist, angiotensin receptor antagonist, and statins have beneficial effects on the endothelial function of diabetic and/or obese patients.[74,75]

6.7.1 ORAL HYPOGLYCEMIC AGENTS

Oral antihyperglycemic agents are currently subdivided into five major classes: biguanides (metformin), sulfonylureas, glinides, glitazones, and α-glucosidase inhibitors. Metformin improves both insulin resistance and endothelial function in type II diabetic patients without the confounding collection of risk factors seen in the metabolic syndrome, but does not affect arterial blood pressure.[76,77] In type II diabetics, sulfonylureas do not have beneficial or deleterious effects on endothelium-dependent responses.[78] Some beneficial effects of sulfonylureas, such as gliclazide, are related to antiatherosclerotic actions, e.g., a decreased adhesion of monocytes to the endothelial cells.[79] Treatment with peroxisome proliferator-activated receptor-gamma (PPAR-γ) agonist improves endothelial function in coronary patients and nondiabetic patients with metabolic syndrome, as well as in type II diabetic patients.[80–82] Rosiglitazone increases NO production and blood flow in the skin of type II diabetics[83] and decreases the levels of an endogenous inhibitor of NO-synthase, asymmetric dimethylarginine.[84] However, overall, the effects of oral anti-hyperglycemic agents on cardiovascular diseases and endothelial function remain controversial.[85]

PPAR-α agonists (fibrates) are a widely used class of lipid-regulating agents prescribed in dyslipidemic patients of various etiologies, including type II diabetics. In these patients, ciprofibrate reduces postprandial lipemia and oxidative stress and improves both fasting and postprandial endothelial function.[86]

In the mesenteric artery from insulin-resistant rats, both chronic treatment and acute *in vitro* treament with metformin improve the impaired endothelium-dependent relaxation to acetylcholine.[87] This effect is mediated by an improvement in the NO-dependent relaxation with no changes in the EDHF-dependent component.

Rosiglitazone lowers blood pressure in the mildly hypertensive, insulin-resistant Zucker fatty rat. It improves the impaired endothelium-dependent relaxation of isolated mesenteric arteries of that strain.[88] Similarly, in a transgenic murine model expressing both human renin and angiotensinogen transgens, the thiazolidinedione decreases arterial blood pressure. It improves the endothelium-dependent relaxation of the carotid arteries, but also causes direct relaxation of the vascular smooth muscle of the isolated arteries.[89]

The data available from both clinical and experimental studies on the effects of oral diabetic agents on vascular and endothelial function are scarce and the mechanism of the effect of these compounds on endothelial function is mostly unknown. Whether they affect EDHF-mediated responses is essentially unknown.

6.7.2 DRUGS TARGETING THE VASCULAR COMPLICATIONS OF DIABETES

The accumulation of glucose in the endothelial cells and the resulting overproduction of superoxide anions by the mitochondrial electron-transport chain appears to be the unifying mechanism at the origin of the four pathogenic pathways described in hyperglycemia-induced damage: (1) increased polyol flux, (2) increased hexosamine pathway flux, (3) activation of protein kinase C β and/or δ isoforms, and (4) increased advanced glycation end-products (AGE) formation. The mitochondrial production of superoxide anions induced by hyperglycemia increases DNA strand breaks and the subsequent activation of poly(ADP-ribose) polymerase (PARP) leading to NAD^+ depletion. The latter event and the direct interaction with superoxide anions decrease GAPDH activity, leading to the accumulation of glycolytic metabolites upstream of GAPDH, as well as glucose itself.[90]

In streptozotocin-treated rats, chronic treatment with an aldose reductase inhibitor, WAY 121509, does not affect the elevated plasma blood glucose. However the inhibitor partially restores the impaired endothelium-dependent vasodilatation of the mesenteric vasculature and this involves a significant improvement in the EDHF-mediated response.[91] Elevated protein kinase C (PKC) activity has been associated with diabetic nephropathy, neuropathy, retinopathy, and in general with the vascular complications of diabetes. In streptozotocin-treated rats, chronic administration with the PKC inhibitor, LY 333531, again partially restores the impaired EDHF-mediated response evoked by acetylcholine in the mesenteric vascular bed.[92] Chronic administration of a PARP inhibitor, PJ34, restores the impaired endothelium-dependent relaxation of the isolated aorta of streptozotocin-treated mice. Isolated rings from PARP-deficient mice are resistant to loss of endothelial function induced by high glucose.[93] Whether or not EDHF-mediated responses will also be improved by drugs targeting this enzyme remains to be determined.

Calcium dobesilate is an angioprotective agent with a poorly understood mechanism of action, which has been extensively used in the treatment of diabetic retinopathy. This compound potentiates endothelium-dependent relaxations of the rabbit aorta.[94] It also potentiates the EDHF-mediated contribution to endothelium-dependent relaxations of human penile arteries.[95]

Clinical trials with inhibitors of aldose reductase and PKC have been deceptive, possibly because these compounds were targeting only a single pathway. Upstream intervention, such as catalytic antioxidant (SOD or catalase mimetics), PARP inhibitors (PJ34, INO-1001), and supplementation with benfotiamine (a lipid-soluble derivative of thiamine that increases transketolase activity), may be more beneficial.[90,93]

In conclusion, the impairment of endothelial function observed in diabetes, at least in animal models of the disease, consistently involves the EDHF-mediated component. However, very little is known about the effects of drugs designed to treat the vascular complications of diabetes on these EDHF-mediated responses.

6.8 ANESTHETICS

Anesthetic agents encompass a wide variety of structurally different molecules. They have multiple effects on ionic channels and therefore affect many targets other than the nervous system.[96] They have direct and indirect actions on the vasculature and can modulate endothelium-dependent responses including endothelium-dependent hyperpolarizations.

6.8.1 GENERAL ANESTHETICS

The volatile anesthetic agents, isoflurane, enflurane, sevoflurane, or halothane, produce arterial relaxation both *in vivo* and *in vitro*.[97,98] For instance, isoflurane dilates porcine coronary resistance arterioles. This dilatation is partially mediated by K_{ATP} and is highly dependent on the presence of a functioning endothelium.[99,100] However, these compounds can also inhibit endothelium-dependent relaxations, but when this is the case the direct inhibitory effect on the smooth muscle cells often overcomes the reduction in endothelium-dependent responses. Both NO- and EDHF-mediated responses are inhibited by these volatile anesthetics.[101,102] The inhibition of the latter has been attributed to the inhibition of cytochrome P450 and thus to the prevention of the release of EETs.[103] However, the prevention of EDHF-mediated responses by volatile anesthetics also can be explained by their modulating role on ionic channels. They activate not only K_{2P} channels, TREK-1 and TASK, which may partially explain their anesthetic properties,[104,105] but also, in various arteries, K_{ATP} and BK_{Ca}.[106–109] They block GIRK, K_V, K_VLQT1, ClCa, and Ca_V,[110–114] More importantly, they also inhibit IK_{Ca} and, at high concentrations, SK_{Ca},[115–117] the two endothelial potassium channels involved in the initiation of EDHF-mediated responses. In addition, theoretically, halothane also uncouples gap junctions and may therefore disrupt myoendothelial gap junction communication.[118] However, the elevated concentrations required to inhibit gap junction communications are unlikely to be clinically relevant.

Intravenous anesthetics such as etomidate and thiopental inhibit the activity of cytochrome P450, while phenobarbital does not. In rat coronary and human renal arteries, etomidate and thiopental inhibit EDHF-mediated responses, while phenobarbital does not.[119,120] Furthermore, etomidate and ketamine inhibit K_{ATP}-mediated relaxation of vascular smooth muscle.[121] Propofol produces a vasodilatation involving both direct relaxation of vascular smooth muscle by the activation of K_{ATP} and BK_{Ca}[122,123] and an endothelium-dependent component. These vasodilator effects can cause undesirable hypotension. Propofol affects EDHF-mediated responses differently depending on the concentration tested or the vascular bed studied. Thus, in the human omental artery, at low and clinically relevant concentrations, the drug enhances EDHF-mediated responses, while at higher concentrations it inhibits them.[124] In the canine pulmonary artery, this anesthetic agent inhibits the EDHF-mediated component.[125] The differences between the two arteries can be linked to a differential involvement of cytochrome P450 in their EDHF-mediated responses.

Halothane, sevoflurane, etomidate, and ketamine inhibit the increase in endothelial-free calcium concentration evoked by acetylcholine or bradykinin.[126–128] Cytochrome P450 modulates the endothelial calcium influx in response to calcium store depletion.[129] Whether this effect of anesthetics on the intracellular calcium concentration of the endothelial cells is linked to their inhibitory properties on cytochrome P450 is unknown.

6.8.2 LOCAL ANESTHETICS

Local anesthetic drugs (e.g., lidocaine, tetracaine, bupivacaine, ropivacaine) inhibit endothelium-dependent vasodilatations by reducing the production of NO.[130–132] In addition, these compounds inhibit K_{ATP}, K_V, TASK-2, TREK-1, and HERG.[133–138] By contrast, procaine stimulates the release of NO in the rat aorta and, despite its intrinsic depolarizing effect on vascular smooth muscle cells, enhances EDHF-mediated responses in porcine coronary arteries.[139,140]

Most of the effects of these anesthetics are observed at clinically relevant concentrations. Therefore, in the design of *in vivo* experiments aimed at studying endothelial function, the choice of the anesthetic agent should be considered carefully. For instance, in anesthetized rats and hamsters, isoflurane and pentobarbital inhibit EDHF-mediated vasodilatation while halothane, ketamine, and urethane do not.[141,142]

6.9 DIETARY SUPPLEMENTATION

Various dietary measures have been proposed to alleviate the symptoms of cardiovascular diseases. Some of these dietary supplements significantly improve endothelial function.

In the pig, a diet rich in omega-3 polyunsaturated fatty acids increases endothelium-dependent relaxations and prevents the endothelial dysfunction provoked by balloon denudation of the coronary endothelium and/or hypercholesterolemia.[143,144] The improvement in EDHF-mediated response contributes to this beneficial effect of omega-3 polyunsaturated fatty acids (Figure 81).[145] Some omega-6 fatty acids such as linolenic acid, which is contained in evening primrose oil, also improve endothelial function. In the mesentery of controls as well as of rats with streptozotocin-induced diabetes, chronic treatment with evening primrose oil increases EDHF-mediated responses.[146]

In diabetes, endothelial dysfunction has been linked to the generation of reactive oxygen species, which neutralize NO and interfere with vasodilator function. Dietary supplements aimed at reducing the production of reactive oxygen species either by scavenging superoxide anions and/or by chelating transition metal (folate, α-lipoic acid, trientine) restore EDHF-mediated responses in the mesenteric and renal vascular beds of diabetic animals.[147–149] An antioxidant treatment can also improve altered endothelium-dependent relaxations in hypertension. Thus, chronic administration of tempol decreases arterial blood pressure and restores the EDHF-mediated responses in the mesenteric artery of DOCA-salt hypertensive rats.[150]

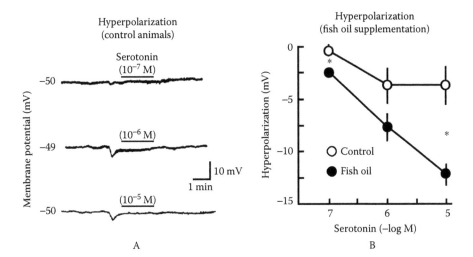

FIGURE 81 Endothelium-dependent hyperpolarizations and a diet rich in omega-3 polyunsaturated fatty acid. The membrane potential was recorded in the smooth muscle cells of coronary arteries taken from pigs fed with a standard chow or a standard chow supplemented with fish oil (in the presence of inhibitor of cyclooxygenase and an antagonist of the 5-HT$_2$ serotoninergic receptor subtype). (A) In coronary arteries from control animals, serotonin evokes minimal endothelium-dependent hyperpolarization. (B) Summary graph showing that a diet rich in omega-3 polyunsaturated fatty acid increases the endothelium-dependent hyperpolarization to serotonin. These results suggest that the improvement of EDHF-mediated responses may contribute to the beneficial cardiovascular therapeutic effects of fish oil. (By permission of Lippincott Williams & Wilkins; modified from Nagao et al.[145])

Various edible fruits, roots, and vegetables, as well as plant extracts used in traditional medicines, evoke vascular relaxations that involve the release of NO, EDHF-mediated responses, and, in some instances, the formation of prostacyclin.[151–155] Polyphenolic derivatives, contained in fruits and vegetables, contribute to these endothelial effects.[156] In particular, in isolated porcine coronary arteries[157–159] and in rat mesenteric arteries,[160] polyphenols derived from red wine induce not only the release of NO[151,161,162] but also EDHF-mediated hyperpolarizations and relaxations. The endothelium-dependent effects of polyphenols and other natural products have often been attributed to their antioxidant properties (reducing or preventing oxidative stress).[52,155] However, in the porcine coronary artery, the endothelial effects of red wine-derived polyphenols involve the endothelial formation of superoxide anions by a flavin-dependent enzyme.[159,163] Thus, in this case, a pro-oxidant effect of polyphenols in the endothelial cells is required to elicit the endothelium-dependent response.[155] Subsequently, an increase in endothelial $[Ca^{2+}]_i$ and the activation of the PI3-kinase/Akt pathway is observed, leading to both the polyphenol-induced NO- and EDHF-dependent responses.[158,159,163] The similar transduction pathway required for polyphenol-induced NO release and EDHF-mediated responses is disturbing. Additional experiments designed to unequivocally rule out the involvement of residual NO formation should be performed. The hypothesis that receptors for polyphenols could be present in the vascular wall deserves attention. The first receptor for a dietary polyphenol, the 67 KDa laminin receptor, which binds epigallocatechin-3-gallate, a polyphenol derived from green tea, has been identified.[164] This may allow a better understanding of the mechanism underlying the vasodilator effect of polyphenols and possibly the design of selective and potent ligands of these receptors with potential therapeutic applications. Red wine-derived polyphenol-induced EDHF-mediated responses may contribute to the so-called "French paradox" in the epidemiology of cardiovascular diseases.

Phytoestrogens can be used as a substitutive treatment for hormone replacement therapy in post-menopausal women. Genistein, a phytoestrogen, found in soybeans, produces acute NO-dependent dilatation in human forearm, potentiates endothelium-dependent vasodilatation to acetylcholine, and, when given chronically to postmeno-pausal women, improves endothelial function to a similar extent as does an estrogen/progesterone regimen.[165,166] The multiple vascular effects of these phytoestrogens, including an enhanced NOS-3 expression and endothelium-dependent relaxations, as well as direct stimulation of smooth muscle BK_{Ca}, is not necessarily linked to the activation of estrogen receptors.[165,167–169] In male rats, seven days of treatment with daidzein mimics the effect of 17 beta-estradiol in stimulating the contribution of an EDHF-mediated response to the endothelium-dependent relaxations of isolated aortic rings.[170] Additional work is required to assess the potentially beneficial role of phytoestrogen on endothelium-dependent relaxations and hyperpolarizations, and to determine whether they are a safe alternative to hormone replacement therapy.

Alteration of mineral intake also influences endothelial function and EDHF-mediated responses. Hence, a diet rich in milk minerals (Whey diet) improves endothelial function in the mesenteric artery of DOCA-salt hypertensive rats without reducing arterial blood pressure. An EDHF-mediated component contributes to the positive effect of the Whey diet.[171] Similar results have been obtained in the SHR, though in this strain the Whey diet also significantly reduces arterial blood pressure.[172] The effects of the Whey diet can be attributed to the increase in potassium intake since, in the SHR and in the Dahl salt-sensitive rat, a diet enriched with potassium has similar positive effects.[172,173]

Dietary calcium intake correlates inversely with the value of arterial blood pressure both in the human and in animal models of hypertension. In the mesenteric vascular bed of rats wih various types of hypertension (including DOCA-salt hypertensive rats, rats chronically treated with a NO-synthase inhibitor, and SHR), an increase in calcium intake decreases arterial blood pressure and improves EDHF-mediated responses.[174–178] In the SHR, the combination of a supplementation in dietary calcium and potassium produces additional effects, virtually normalizing arterial blood pressure and endothelium-dependent relaxations. An improvement in NO-mediated and EDHF-mediated responses and a decrease in the production of endothelial constricting prostanoids concur to improve the endothelium-dependent responses.[176] A diet rich in calcium, which does not cause hypercalcemia, normalizes blood pressure and endothelium-dependent relaxations, including EDHF-mediated responses, in NaCl-hypertensive rats. However, in the same hypertensive model, chronic hypercalcemia induced by oral 1α-hydroxy-vitamin D3 administration is not associated with an improvement in endothelial function.[178] By contrast, in the

Relaxation in the presence of indomethacin (10^{-5} M)

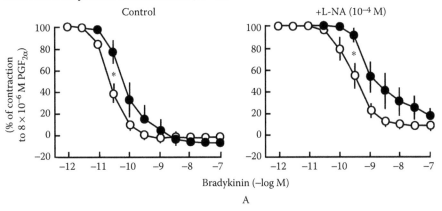

Hyperpolarization in the presence of indomethacin + L-NA

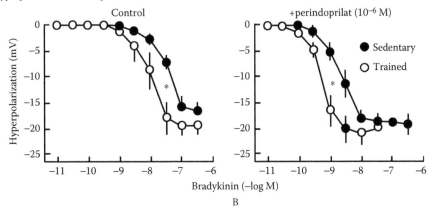

FIGURE 82 Endothelium-dependent hyperpolarizations and exercise training. Changes in tension and membrane potential were measured in isolated coronary arteries taken from sedentary or exercise-trained dogs (in the presence of an inhibitor of cyclooxygenases). (A) Changes in tension. The endothelium- and concentration-dependent relaxations evoked by bradykinin were significantly improved in exercise-trained dogs either under control conditions (left) or after treatment with an inhibitor of NO synthase, L-nitro-arginine (L-NA; right). (B) Changes in membrane potential of canine coronary smooth muscle cells. The endothelium dependent hyperpolarization in response to bradykinin (in the presence of inhibitors of cyclooxygenases and NO-synthases) is improved by exercise training either under control conditions (left) or after treatment with an inhibitor of the angiotensin-converting enzyme, perindoprilat (right). These results indicate that exercise improves endothelial function by increasing both the NO- and the EDHF-mediated responses. This improvement is not linked to a change in angiotensin-converting enzyme activity. The improvement of the EDHF-mediated response may contribute to the overall beneficial effect of exercise on cardiovascular function. (By permission from the Nature Publishing Group; modified from Mombouli et al.[183])

mesenteric vascular bed of SHR, oral supplementation with cholecalciferol, a pre-cursor of vitamin D3, improves the EDHF-mediated responses.[179] In the latter study the precise role of calcium ions was not analyzed.

In conclusion, various so-called nonpharmacological therapeutic strategies help to reverse endothelial dysfunction, including blunted EDHF-mediated responses. Whether the improvement of these EDHF-mediated responses contributes to the beneficial effects of these dietary maneuvers can be suspected but is far from being demonstrated.

6.10 EXERCISE

Regular exercise is recommended as a nonpharmacological measure for the treatment of the metabolic syndrome (insulin resistance, obesity, hypertension) and various other cardiovascular diseases. Aerobic physical activity reduces oxidative stress and can prevent endothelial dysfunction.[52]

Exercise *per se* improves endothelial function in lean and healthy animals and in humans. This phenomenon has been attributed to enhanced eNOS expression and NO production, possibly because of a chronic increase in the shear stress exerted by the flowing blood.[180–182] EDHF-mediated responses are also increased in healthy exercise-trained animals (Figure 82).[183] A positive effect of exercise training on EDHF-mediated responses, as well as on NO-dependent relaxations, is also obtained in various animal models of diseases such as the SHR,[184] pigs submitted to progres-sive coronary occlusion,[185] and the Otsuka Long Evans Tokushima fatty rat (a model of type II diabetes).[186]

The beneficial effect of exercise on endothelial function has also been demon-strated in the human during the aging process or in the course of essential hyper-tension.[52,187,188] In the human, the potential role of EDHF-mediated responses in this beneficial effect of exercise remains to be determined.

7 Conclusions

The available data show that endothelium-dependent relaxations, independent of the production of NO and prostacyclin, are likely to play an important role in cardiovascular physiology in numerous animal species and in the human. The number of published papers exploring the role of EDHF-mediated responses under physiological or pathological conditions has grown steadily over the years. However, a lot of confusion surrounds this field of research because of the use of unspecific pharmacological tools and the frequent absence of electrophysiological measurements. The scientists involved in the field, as well as journal referees and editors, could markedly augment clarity by applying or suggesting simple rules:

1. The existence of an EDHF-mediated response should be demonstrated unequivocally.

An EDHF-mediated response should be reported only when the evidence for a third pathway, besides the L-arginine-NO-synthase and the arachidonic acid-cyclooxygenase pathways, is demonstrated beyond any doubt. A relaxation and/or hyperpolarization resistant to the combined presence of inhibitors of NO-synthases and cycloxygenases does not necessarily imply the existence of an EDHF-mediated response. An EDHF-mediated response, i.e., an endothelium-dependent relaxation and/or a hyperpolarization observed in the combined presence of inhibitors of NO-synthases and cycloxygenases should still be observed after the subsequent addition of a NO scavenger (e.g., oxyhemoglobin, carboxy-PTIO). These experiments are needed to rule out the possible involvement of residual NO production, either from the NO-synthase itself or from NO-synthase-independent sources (e.g., NO stores). Theoretically, any paper mentioning an EDHF-mediated response in which this type of experiment has not been performed is subject to caution. However, performing experiments in the presence of multiple pharmacological agents can create problems. Thus, once unequivocally documented in a given blood vessel, the EDHF-mediated response can be studied with the conventional combination of inhibitors of NO-synthases and cycloxygenases.

2. The mechanism of an EDHF-mediated response should be determined whenever possible.

The so-called EDHF-mediated response is a mixed bag that encompasses different endothelial mediators or pathways that occur independently or in some cases co-exist. The production of H_2O_2, the production of EETs by cytochrome P450, or that of CNP does not require hyperpolarization of endothelial cells, although in a similar manner to the release of NO itself, the membrane potential of the endothelial cells can be a regulatory factor. H_2O_2, EETs, and CNP, again like NO, are released by the endothelial cells and activate the potassium channels situated on the smooth muscle cells. Appropriate tools are now available to properly identity these mediators.

Once their involvement is confirmed in a given vascular bed, they should be referred to by their proper name, i.e., endothelium-derived NO, prostacyclin, H_2O_2, EETs, or CNP, and no longer termed "EDHF."

We propose that only the pathway requiring the activation of endothelial SK_{Ca} and IK_{Ca} and hyperpolarization of the endothelial cells should be referred to as "endothelium-dependent hyperpolarizations" or "EDHF-mediated" responses, as there is, at present, no better name to characterize it. To implement this simple nomenclature will dramatically simplify the interpretation of incoming studies.

3. Use the most specific pharmacological tools available.

The use of nonselective drugs, in particular inhibitors of cytochrome P450, which also inhibit calcium-activated potassium channels, and charybdotoxin, which inhibits both IK_{Ca} and BK_{Ca}, has been responsible for major confusion in the EDHF field. Henceforth, authors performing studies in which so-called EDHF-mediated responses are examined should select more specific inhibitors of cytochrome P450 and of antagonists of the EETs, such as the EEZE compounds, iberiotoxin, to specifically inhibit BK_{Ca}, and TRAM 34 or TRAM 39 in order to specifically block IK_{Ca}. These measures, if applied, will in most of the cases allow the NO-, prostacyclin-resistant responses to be identified properly and, therefore, to be named properly.

By implementing these three simple rules, the clarification of EDHF-mediated responses will progress dramatically. The discovery of more specific tools interfering with the EDHF pathways, particularly if they can be used *in vivo,* will allow a better understanding of the physiological and pathological role of EDHF-mediated responses, if any. In this respect, specific and potent activators of SK_{Ca} and IK_{Ca} are lacking.

A question remains whether EDHF is a back-up, an alternative mechanism when NO-dependent responses have disappeared, or is contributing to endothelial function in parallel with the other pathways. Once again the answer may depend on the mechanism involved in the EDHF-mediated response studied. For instance, the production of H_2O_2 by NO-synthase occurs mainly when the production of NO is jeopardized. Similarly, NO may inhibit cytochrome P450 activity and prevent the release of EETs, suggesting that inhibition of the production of NO will enhance that of EETs. However, there is no evidence that NO alters the activation of endothelial IK_{Ca} and SK_{Ca}, suggesting that EDHF-mediated responses relying on the transmission of the endothelial hyperpolarization via myoendothelial gap junctions and/or K^+ accumulation in the intercellular space should not be markedly affected by changes in the production or availability of NO.

The endothelium-dependent hyperpolarizations are independent of NO and prostacyclin under various pathological conditions. Several pharmacological and "nonpharmacological" therapeutic strategies can restore these responses. However, whether the improvement in EDHF-mediated responses contributes to the improvement in endothelial function and can be translated into a therapeutic benefit is still speculative. The answer to this question is of utmost importance for determining whether this pathway can constitute a new therapeutic target(s) and thus generate new therapeutic drugs. Increasing the production of H_2O_2 is unlikely to be suitable in any given pathology and the activation of cytochrome P450 in order to increase EET production is questionable since this enzyme also produces reactive oxygen species. However, activating endothelial IK_{Ca} and/or SK_{Ca} or increasing their expression, as well as improving myoendothelial communication, for instance, by increasing the expression of connexin(s), may be valuable therapeutic targets. The studies describing the effect of estrogens on the expression of SK3 and connexin 43, as well as the experiments in mice over-expressing SK3, are in favor of such a strategy in cardiovascular drug discovery.

References

CHAPTER 1

1. Choi, K., Kennedy, M., Kazarov, A., Papadimitriou, J.C., Keller, G. (1998) A common precursor for hematopoietic and endothelial cells. *Development* **125**: 725–732.
2. Huttner, I.G., Gabbiani, G. (1983) Vascular endothelium in hypertension. In *Hypertension*, J. Genest et al., eds., McGraw-Hill, New York, pp. 473–488.
3. Jaffe, E.A. (1987) Cell biology of endothelial cells. *Hum. Pathol.* **18**: 234–239.
4. Pucovsky, V., Moss, R.F., Bolton, T.B. (2003) Non-contractile cells with thin processes resembling cells of Cajal found in the wall of guinea-pig mesenteric arteries. *J. Physiol.* **552**: 119–133.
5. Rhodin, J.A.G. (1980) Architecture of the vessel wall. In *Handbook of Physiology, The Cardiovascular System (2),* D.F. Bohr et al., eds., Am. Phys. Soc., Bethesda, MD, pp. 1–13.
6. Kalebic, T., Garbisa, S., Glaser, B. (1983) Basement membrane collagen degradation by migrating endothelial cells. *Science* **221**: 281–284.
7. Kalluri, R. (2003) Basement membranes: structure, assembly and role in tumour angiogenesis. *Nat. Rev. Cancer* **3**: 422–433.
8. Michiels, C. (2003) Endothelial cells functions. *J. Cell. Physiol.* **193**: 430–443.
9. Van Hinsbergh, V.W.M. (1997) Endothelial permeability for macromolecules: mechanistic aspects of pathophysiological modulation. *Arterioscler. Thromb. Vasc. Biol.* **17**: 1018–1023.
10. Michel, C.C., Curry, F.E. (1999) Microvascular permeability. *Physiol. Rev.* **79**: 703–761.
11. Félétou, M., Bonnardel, E., Canet, E. (1996) Bradykinin and changes in microvascular permeability in the hamster cheek pouch: Role of NO. *Br. J. Pharmacol.* **118**: 1371–1376.
12. Félétou, M., Staczek, J., Duhault, J. (2001) VEGF and in vivo plasma extravasation in the hamster cheek pouch. *Br. J. Pharmacol.* **132**: 1342–1348.
13. Boeynaems, J-M., Pirotton, S. (1994) Regulation of the vascular endothelium: signals and transduction mechanisms. CRC Press, Boca Raton, FL, pp. 1–117.
14. Aurrand-Lions, M., Johnson-Léger, C., Imhof, B.A. (2003) Role of interendothelial adhesion molecules in the control of vascular functions. *Vascular Pharmacol.* **39**: 239–246.
15. Carmeliet, P. (2003) Angiogenesis in health and disease. *Nat. Med.* **9**: 653–660.
16. Folkman, J., D'Amore, P.A. (1996) Blood vessel formation: what is its molecular basis? *Cell* **87**: 1153–1155.
17. Egeblad, M., Werb, Z. (2002) New functions for the matrix metalloproteinases in cancer progression. *Nature Rev. Cancer* **2**: 161–174.
18. Sylvestre, J-S., Levy, B. (2002) Vasculogenèse et angiogenèse. In *Biologie et Pathologie du Coeur et des Vaisseaux,* F. Pinet et al., eds., Medecine-Sciences, Flammarion, Paris, pp. 463–468.
19. Gillis, C.N., Pitt, B.R. (1982) The fate of circulating amines within the pulmonary circulation. *Annu. Rev. Physiol.* **44**: 269–281.

20. Corvol, P., Michaud, A., Soubrier, F., Williams, T.A. (1995) Recent advances in the knowledge of the structure and function of the angiotensin I converting enzyme. *J. Hypertens.* **13**: S3–S10.

21. Turner, A.J., Tanzawa, K. (1997) Mammalian membrane metallopeptidases: NEP, ECE, KELL, and PEX. *FASEB J.* **11**: 355–364.

22. Lambeir, A.M., Durinx, C., Scharpe, S., De Meester, I. (2003) Dipeptidyl-peptidase IV from bench to bedside: an update on structural properties, functions, and clinical aspects of the enzyme DPP IV. *Crit. Rev. Clin. Lab. Sci.* **40**: 209–294.

23. Worthley, M.I., Corti, R., Worthley, S.G. (2004) Vasopeptidase inhibitors: will they have a role in clinical practice? *Br. J. Clin. Pharmacol.* **57**: 27–36.

24. Majesky, M.W. (2003) Vascular smooth muscle diversity: insight from developmental biology. *Curr. Atherosclerosis Reports* **5**: 208–213.

25. Berk, B.C. (2001) Vascular smooth muscle growth: autocrine growth mechanisms. *Pharmacol. Rev.* **81**: 999–1030.

26. Hao, H., Gabbiani, G., Bochaton-Piallat, M-L. (2003) Arterial smooth muscle heterogeneity. Implications for atherosclerosis and restenosis development. *Arterioscler. Thromb. Vasc. Biol.* **23**: 1510–1520.

27. Mochizuki, S., Brassard, B., Hinek, A. (2002) Signaling pathways transduced through the elastin receptor facilitate proliferation of arterial smooth muscle cells. *J. Biol. Chem.* **277**: 44854–44863.

28. Brooke, B.S., Bayes-Genis, A., Li, D.Y. (2003) New insight into elastin and vascular disease. *Trends Cardiovasc. Med.* **13**: 176–181.

29. Folkow, B. (1989) The Benjamin W. Zweifach award lecture. Functional and structural "autoregulation" — some personal considerations concerning the century-old development of these microvascular concepts. *Microvasc. Res.* **37**: 242–255.

30. Intengan, H.D., Schiffrin, E.L. (2000) Structure and mechanical properties of resistance arteries in hypertension; Role of adhesion molecules and extracellular matrix determinants. *Hypertension* **36**: 312–318.

31. Mallat, Z., Tedgui, A. (2000) Apoptosis in the vasculature: mechanisms and functional importance. *Br. J. Pharmacol.* **130**: 947–962.

32. Mulvany, M.J. (2002) Small artery remodeling and significance in the development of hypertension. *News Physiol. Sci.* **17**: 105–109.

33. Shepherd, J.T., Vanhoutte, P.M. (1979) The *Human Cardiovascular System. Facts and Concepts.* Raven Press, New York, pp. 1–351.

34. Bolton, T.B. (1979) Mechanism of action of transmitters and other substances on smooth muscle. *Physiol. Rev.* **59**: 606–718.

35. Félétou, M., Alya, G., Tricoche, R., Walden, M. (1986) Sources of calcium and cholinergic contraction of the rat portal vein and the sheep coronary artery. *Arch. Int. Pharmacodyn. Therapeut.* **283**: 254–271.

36. Somlyo, A.V., Somlyo, A.P. (1994) Signal transduction and regulation in smooth muscle. *Nature* **372**: 231–236.

37. Somlyo, A.V., Somlyo, A.P. (2000) Signal transduction by G-proteins rho-kinase and protein phosphatase to smooth muscle and non-muscle myosin II. *J. Physiol.* **522**: 177–185.

38. Ganitkevich, V., Hasse, V., Pfitzer, G. (2002) Ca^{2+}-dependent and Ca^{2+}-independent regulation of smooth muscle contraction. *J. Muscle Res. Cell Motility* **23**: 47–52.

39. Jones, A.W. (1980) Content and fluxes of electrolytes. In *Handbook of Physiology, The Cardiovascular System (2)*, D.F. Bohr et al., eds., American Physical Society, Bethesda, MD, pp. 253–299.

40. Goldman, D.E. (1943) Potential, impedance and rectification in the membrane. *J. Gen. Physiol.* **27**: 37–60.
41. Hodgkin, A.L., Katz, B. (1949) The effects of sodium ions on the electrical activity of the giant axon of the squid. *J. Physiol.* **108**: 37–77.
42. Large, W.A., Wang, Q. (1996) Characteristics and physiological role of the Ca^{2+}-activated Cl^- conductance in smooth muscle. *Am. J. Physiol.* **271**: C435–C454.
43. Hatem, S., Bril, A. (2002) Bases biophysiques et moléculaires de l'électrogénèse. In *Biologie et Pathologie du Coeur et des Vaisseaux*, F. Pinet et al., eds., Medecine-Sciences, Flammarion, Paris, pp. 3–12.
44. Beaugé, L.A., Gadsby, D.C., Garrahan, P.J. (1997) *Na/K-ATPase & related transport ATPases.* Annals of the New York Academy of Sciences, New York, Vol. 834, pp. 1–694.
45. Fleming, W.W. (1980) The electrogenic Na^+, K^+ pump in smooth muscle. Physiologic and pharmacologic significance. *Ann. Rev. Pharmacol. Toxicol.* **20**: 129–149.
46. Lansbery, K., Mendenhall, M.L., Vehige, L.C., Taylor, J.A., Sanchez, G., Blanco, G., Mercer, R.W. (2003) Isoforms of the Na,K-ATPase. In *EDHF 2002*, P.M. Vanhoutte, ed., Taylor and Francis, London, pp. 27–34.
47. Arystarkhova, E., Wetzel, R.K., Asinovski, N.K., Sweadner, K.J. (1999) The gamma subunit modulates Na^+ and K^+ affinity of the renal Na, K-ATPase. *J. Biol. Chem.* **274**: 33183–33185.
48. Beguin, P., Wang, X., Firsov, D., Puoti, A., Claeys, D., Horisberger, J.D., Geering, K. (1997) The γ subunit is a specific component of the Na, K-ATPase and modulates its transport function. *EMBO J.* **16**: 4250–4260.
49. Therein, A.G., Karlish, S.J.D., Blostein, R. (1999) Expression and functional role of the γ subunit of the Na, K-ATPase in mammalian cells. *J. Biol. Chem.* **274**: 12252–12256.
50. Juhaszova, M., Blaustein, M.P. (1997) Distinct distribution of different Na^+ pump α subunit isoforms in plasmalemna. Physiological implications. In *Na/K-ATPase and Related Transport ATPases.* L.A. Beaugé et al., eds., Annals of the New York Academy of Sciences, New York, Vol. 834, pp. 524–536.
51. Weston, A.H., Richards, G.R., Burnham, M.P., Félétou, M., Vanhoutte, P.M., Edwards, G. (2002) K^+-induced hyperpolarization in rat mesenteric artery: identification, localization and role for Na^+/K^+-ATPases. *Br. J. Pharmacol.* **136**: 918–926.
52. Shieh, C-C., Coghlan, M., Sullivan, J.P., Gopalakrishnan, M. (2000) Potassium channels: molecular defects, diseases and therapeutic opportunities. *Pharmacol. Rev.* **52**: 557–593.
53. Lesage, F., Lazdunski, M. (2000) Molecular and functional properties of two-pore-domain potassium channels. *Am. J. Physiol.* **279**: F793–F801.
54. Ford, J.W., Stevens, E.B., Treheme, J.M., Packer, J., Bushfield, M. (2002) Potassium channels: gene therapeutic relevance, high throughput screening technologies and drug discovery. *Prog. Drug Res.* **58**: 133–168.
55. IUPHAR Ion Channel Compendium (2002) Potassium channels. International Union of Pharmacology. www.iuphar.org.
56. Cole, W.C., Walsh, M.P. (2001) Delayed rectifier K+ channels of vascular smooth muscle cells: characterization, function, and regulation by phosphorylation. In *Potassium Channels in Cardiovascular Biology,* S. Archer, N. Rusch, eds., Kluwer Academic/Plenium Publishers, New York, pp. 485–503.
57. Archer, S. (2001) The role of potassium channels in the control of the pulmonary circulation. In *Potassium Channels in Cardiovascular Biology,* S. Archer, N. Rusch, eds., Kluwer Academic/Plenium Publishers, New York, pp. 543–570.

58. Quignard, J-F., Chataigneau, T., Corriu, C., Duhault, J., Félétou, M., Vanhoutte, P.M. (1999) Potassium channels involved in EDHF-induced hyperpolarization of the smooth muscle cells of the isolated guinea-pig carotid artery. In *Endothelium-Dependent Hyperpolarizations*, P.M. Vanhoutte, ed., Harwood Academic Publishers, Amsterdam, pp. 201–208.

59. Quignard, J-F., Harley, E.A., Duhault, J., Vanhoutte, P.M., Félétou, M. (2003) K$^+$ channels in bovine retinal pericytes: effects of β-adrenergic stimulation and elevated glucose. *J. Cardiovasc. Pharmacol.* **42**: 379–388.

60. Chandy, K., Gutman, G.A. (1993) Nomenclature for mammalian potassium channel genes. *Trends Pharmacol. Sci.* **14**: 344–440.

61. Clement-Chomienne, O., Ishii, K., Walsh, M.P., Cole, W.C. (1999). Identification, cloning and expression of rabbit vascular smooth muscle Kv1.5 and comparison with native delayed rectifier K$^+$ current. *J. Physiol.* **515**: 653–667.

62. Pongs, O. (1995) Regulation of the activity of voltage-gated potassium channels by β-subunits. *Semin. Neurosc.* **7**: 137–146.

63. Rhodes, K.J., Monaghan, M.M., Barrezueta, N.X., Nawoshick, S., Bekele-Arcuri, Z., Matos, M.F., Nakahira, K., Schechter, L.E., Trimmer, J.S. (1996) Voltage-gated K$^+$ channels β-subunits: expression and distribution of Kvβ1 and Kvβ2 in adult rat brain. *J. Neurosci.* **16**: 4846–4860.

64. Trimmer, J.S., Rhodes, K.J. (2001) Heteromultimer formation in native K+ channels. In *Potassium Channels in Cardiovascular Biology*, S. Archer and N. Rusch, eds., Kluwer Academic/Plenium Publishers, New York, pp. 163–175.

65. Quignard, J-F., Félétou, M., Edwards, G., Duhault, J., Weston, A.H., Vanhoutte, P.M. (2000) Role of endothelial cells hyperpolarization in EDHF-mediated responses in the guinea-pig carotid artery. *Br. J. Pharmacol.* **129**: 1103–1112.

66. Sadanaga, T., Ohya, Y., Ohtsubo, T., Goto, K., Fujii, K., Abe, I. (2002) Decreased 4-aminpyridine sensitive K$^+$ currents in endothelial cells from hypertensivre rats. *Hypertens. Res.* **25**: 589–596.

67. Jentsch, T.J. (2000) Neuronal KCNQ potassium channels: physiology and role in diseases. *Nat. Rev. Neurosci.* **1**: 21–30.

68. Nerbonne, J.M. (2001) Molecular mechanisms controlling functional voltage-gated K+ channel diversity and expression in the mammalian heart. In *Potassium Channels in Cardiovascular Biology*, S. Archer and N. Rusch, eds., Kluwer Academic/Plenium Publishers, New York, pp. 297–335.

69. Ohya, S., Horowitz, N., Greenwood, I.A. (2002) Functional and molecular identification of ERG channels in murine portal vein myocytes. *Am. J. Physiol.* **283**: C866–C877.

70. Ohya, S., Sergeant, G.P., Greenwood, I.A., Horowitz, N. (2003) Molecular variants of KCNQ channels expressed in murine portal vein myocytes. A role in delayed rectifier current. *Circ. Res.* **92**: 1016–1023.

71. Köhler, M., Hirschberg, B., Bond, C.T., Kinzie, J.M., Marrion, N.V., Maylie, J., Adelman, J.P. (1996) Small-conductance calcium-activated K$^+$ channels from mammalian brain. *Science* **273**: 1709–1714.

72. Zhang, B.M., Kohli, V., Adachi, R., Lopez, J.A., Udden, M.M., Sullivan, R. (2001a) Calmodulin binding to the C-terminus of the small conductance Ca^{2+}-activated K$^+$ channel HSK1 is affected by alternative splicing. *Biochemistry* **40**: 3189–3195.

73. Wittekindt, O.H., Visan, V., Tomita, H., Imtiaz, F., Gargus, J.J., Lehman-Horn, F., Grissmer, S., Morris-Rosendahl, D.J. (2004) An apamin- and scyllatoxin-insensitive isoform of the human SK3 channel. *Mol. Pharmacol.* **65**: 788–801.

74. Schumacher, M.A., Rivard, A.F., Bachinger, H.P., Adelman, J.P. (2001) Structure of the gating domain of a Ca^{2+}-activated K^+ channel complexed with Ca^{2+}/calmodulin. *Nature* **410**: 1120–1124.

75. Xia, X.M., Fakler, B., Rivard, A., Wayman, G., Johnson-Pais, T., Keen, J.E., Ishii, T., Hirschberg, B., Bond, C.T., Lutsenko, S., Maylie, J., Adelman, J.P. (1998) Mechanism of calcium-gating in small-conductance calcium-activated potassium channels. *Nature* **395**: 503–507.

76. Hirschberg, B., Maylie, J., Adelman, J.P., Marrion, N.V. (1998) Gating of recombinant small-conductance Ca-activated K^+ channels by calcium. *J. Gen. Physiol.* **111**: 565–581.

77. Barfod, E.T., Moore, A.L., Lidofsky, S.D. (2001) Cloning and functional expression of a liver isoform of the small conductance Ca^{2+}-activated K^+ channel SK3. *Am. J. Physiol.* **280**: C836–C842.

78. Campos Rosa, J., Galanakis, D., Piergentili, A., Bhandari, K., Ganellin, C.R., Dunn, P.M., Jenkinson, D.H. (2000) Synthesis, molecular modeling, and pharmacological testing of bis-quinolinium cyclophanes: potent, non-peptidic blockers of the apamin-sensitive Ca(2+)-activated K(+) channel. *J. Med. Chem.* **43**: 420–431.

79. Castle, N.A. (1999) Recent advances in the biology of small conductance calcium-activated potassium channels. *Perspect. Drug Discov. Design* **15/16**: 131–154.

80. Dunn, P.M. (1999) UCL 1684: a potent blocker of Ca^{2+}-activated K^+ channel in rat adrenal chromaffin cells in culture. *Eur. J. Pharmacol.* **368**: 119–123.

81. Strobaek, D., Joergensen, T.D., Christophersen, P., Ahring, P.K., Olesen, S.P. (2000) Pharmacological characterization of small-conductance Ca^{2+}-activated K^+ channels stably expressed in HEK 293 cells. *Br. J. Pharmacol.* **129**: 627–630.

82. Liegeois, J.F., Mercier, F., Graulich, A., Graulich-Lorge, F., Scuvee-Moreau, J., Setin, V. (2003) Modulation of small conductance calcium-activated potassium (SK) channels: a new challenge in medicinal chemistry. *Curr. Med. Chem.* **10**: 625–647.

83. Grunnet, M., Jensen, B.S., Olesen, S-P., Klaerke, D.A. (2001) Apamin interacts with all subtypes of cloned small conductance Ca^{2+}-activated K^+ channels. *Pflügers Atrch. Eur. J. Physiol.* **441**: 544–550.

84. Shah, M., Haylett, D.G. (2000). The pharmacology of hSK1 Ca^{2+}-activated K^+ channels expressed in mammalian cell lines. *Br. J. Pharmacol.* **129**: 627–630.

85. Pedarzini, P., D'hoedt, D., Doorty, K.B., Wadsworth, J.D.F., Joseph, J.S., Jeyaseelan, K., Manjunatha Kini, R., Gadre, S.V., Sapanekar, S.M., Stocker, M., Strong, P.N. (2002) Tapamin, a venom peptide from the Indian red scorpion (*Mesobuthus tamulus*) that targets small conductance Ca^{2+}-activated K^+ channels and afterhyperpolarization currents in central neurons. *J. Biol. Chem.* **277**: 46101–46109.

86. Cao, Y., Dreixler, J.C., Roizen, J.D., Roberts, M.T., Houamed, K.M. (2001) Modulation of recombinant small-conductance Ca(2+)-activated K(+) channels by the muscle relaxant chlorzoxazone and structurally related compounds. *J. Pharacol. Exp. Ther.* **296**: 683–689.

87. Ishii, T.M., Maylie, J., Adelman, J.P. (1997a) Determinants of apamin and d-tubocurarine block in SK potassium channels. *J. Biol. Chem.* **272**: 23195–23200.

88. Wadsworth, J.D., Torelli, S., Doorty, K.B., Strong, P.N. (1997) Structural diversity among subtypes of small-conductance Ca^{2+}-activated-K^+ channels. *Arch. Biochem. Biophys.* **346**: 151–160.

89. Joiner, W.J., Wang, L.Y., Tang, M.D., Kaczmarek, L.K. (1997) hSK4, a member of a novel subfamily of calcium-activated potassium channels. *Proc. Natl. Acad. Sci. USA* **94**: 11013–11018.

90. Ishii, T.M., Silvia, C., Hirschberg, B., Bond, C.T., Adelman, J.P., Maylie, J. (1997) A human intermediate conductance calcium-activated potassium channel. *Proc. Natl. Acad. Sci. USA* **94**: 11651–11656.

91. Logsdon, N.J., Kang, J., Togo, J.A., Christian, E.P., Aiyar, J. (1997) A novel gene, hKCa4, encodes the calcium-activated potassium channel in human lymphocytes. *J. Biol. Chem.* **272**: 32723–32726.

92. Vergara, C., Latorre, R., Marrion, N.V., Adelman, J.P. (1998) Calcium-activated potassium channels. *Curr. Opin. Neurobiol.* **8**: 321–329.

93. Gardos, G. (1958) The function of calcium in the potassium permeability of human erythrocytes. *Biochem. Biophys. Acta* **30**: 653–654.

94. Wulff, H., Miller, M.J., Haensel, W., Grissner, S., Cahalan, M.D., Chandy, K.G. (2000) Design of potent and selective inhibitor of the intermediate-conductance Ca^{2+}-activated K^+ channel, IKCa1: a potential immunosuppressant. *Proc. Natl. Acad. Sci. USA* **97**: 8151–8156.

95. Edwards, G., Gardener, M.J., Félétou, M., Brady, G., Vanhoutte, P.M., Weston, A.H. (1999) Further investigation of endothelium-derived hyperpolarizing factor (EDHF) in rat hepatic artery: studies using 1-EBIO and ouabain. *Br. J. Pharmacol.* **128**: 1064–1070.

96. Begenisich, Y.T., Nakamoto, T., Ovitt, C., Nehrke, K., Brugnara, C., Alper, S.L., Melvin, J.E. (2004) Physiological roles of the intermediate conductance, Ca^{2+}-activated K channel, Kcnn4. *J. Biol. Chem.* **279**: 47681–47687.

97. Marchenko, S.M., Sage, S.O. (1996) Calcium-activated potassium channels in the endothelium of intact aorta. *J. Physiol.* **492**: 53–60.

98. Frieden, M., Sollini, M., Beny, J-L. (1999) Substance P and bradykinin activate different types of Kca currents to hyperpolarize cultured porcine coronary artery endothelial cells. *J. Physiol.* **519**: 361–371.

99. Burnham, M.P., Bychkov, R., Félétou, M., Richards, G.R., Vanhoutte, P.M., Weston, A.H., Edwards, G. (2002) Characterization of an apamin-sensitive small conductance Ca^{2+}-activated K^+ channel in porcine coronary artery endothelium: relevance to EDHF. *Br. J. Pharmacol.* **135**: 1133–1143.

100. Ding, H., Jiang, Y., Triggle, C.R. (2003) The contribution of D-tubocurarine and apamin-sensitive potassium channels to endothelium-derived hyperpolarizing factor-mediated relaxation of small arteries from e-NOS-/- mice. In *EDHF 2002*, Vanhoutte P.M., ed., Taylor and Francis, London, pp. 283–296.

101. Eichler, I., Wibawa, J., Grgic, I., Knorr, A., Brakemier, S., Pries, A.R., Hoyer, J., Köhler, R. (2003) Selective blockade of endothelial Ca^{2+}-activated small- and intermediate-conductance K^+ channels suppresses EDHF-mediated vasodilation. *Br. J. Pharmacol.* **138**: 594–601.

102. Taylor, M.S., Bonev, A.D., Gross, T.P., Eckman, D.M., Brayden, J.E., Bond, C.T., Adelman, J.P., Nelson, M.T. (2003) Altered expression of small-conductance Ca^{2+}-activated K^+ (SK3) channels modulate arterial tone and blood pressure. *Circ. Res.* **93**: 124–131.

103. Vogalis, F., Goyal, R.K. (1997) Activation of small conductance Ca(2+)-dependent K^+ channels by purinergic agonists in smooth muscle of the mouse ileum. *J. Physiol.* **502**: 497–508.

104. Gebremedhin, D., Kaldunski, M., Jacobs, E.R., Harder, D.R., Roman, R.J. (1996) Coexistence of two types of calcium activated potassium channels in rat renal arterioles. *Am. J. Physiol.* **270**: F69–F81.

105. Marchenko, S.M., Sage, S.O. (2001) Single channel properties of Ca^{2+}-activated K^+ channels in the vascular endothelium. In *Potassium Channels in Cardiovascular Biology*, S. Archer, N. Rusch, eds., Kluwer Academic/Plenium Publishers, New York, pp. 651–666.

106. Bychkov, R., Burnham, M.P., Richards, G.R., Edwards, G., Weston, A.H., Félétou, M., Vanhoutte, P.M. (2002) Characterization of a charybdotoxin-sensitive intermediate conductance Ca^{2+}-activated K^+ channel in porcine coronary endothelium: relevance to EDHF. *Br. J. Pharmacol.* **138**: 1346–1354.

107. Ohya, S., Kimura, S., Kitsukawa, M., Muraki, K., Watanabe, M., Imaizumi, Y. (2000) SK4 encodes intermediate conductance Ca^{2+}-activated K^+ channels in mouse urinary bladder smooth muscle cells. *Jpn. J. Pharmacol.* **84**: 97–100.

108. Neylon, C.B., Lang, R.J., Fu, Y., Bobik, A., Reinhart, P.H. (1999) Molecular cloning and characterization of the intermediate-conductance Ca(2+)-activated K(+) channel in vascular smooth muscle: relationship between K(Ca) channel diversity and smooth muscle function. *Circ. Res.* **85**: 33–43.

109. Kohler, R., Wulff, H., Eichler, I., Kneifel, M., Neumann, D., Knorr, A., Grgi, I., Kampfe, D., Si, H., Wibawa, J., Real, R., Borner, K., Brakemeier, S., Orzechowski, H.D., Reusch, H.P., Paul, M., Chandy, K.G., Hoyer, J. (2003) Blockade of intermediate-conductance calcium-activated potassium channel as a new therapeutic strategy for restonosis. *Circulation* **108**: 1119–1125.

110. Kestler, H.A., Janko, S., Haussler, U., Muche, R., Hombach, V., Hoher, M., Wiecha, J. (1998) A remark on the high-conductance calcium-activated potassium channel in human endothelial cells. *Res. Exp. Med.* **198**: 133–143.

111. Jow, F., Sullivan, K., Sokol, P., Numann, R. (1999) Induction of Ca^{2+}-activated K^+ current and transient outward currents in human capillary endothelial cells. *J. Membrane Biol.* **167**: 53–64.

112. Sollini, M., Frieden, M., Bény, J-L. (2002). Charybdotoxin-sensitive small conductance KCa channel activated by bradykinin and substance P in endothelial cells. *Br. J. Pharmacol.* **136**: 1201–1209.

113. Meera, P., Wallner, M., Toro, L. (2001) Molecular biology of high-conductance, Ca^{2+}-activated potassium channels. In *Potassium Channels in Cardiovascular Biology,* S. Archer, N. Rusch, eds., Kluwer Academic/Plenium Publishers, New York, pp. 49–70.

114. Gribkoff, V.K., Starett, J.E. Jr., Dworetzky, S.I. (2001) Maxi-K potassium channels: form, function and modulation of a class of endogenous regulators of intracellular calcium. *Neuroscientist* **7**: 166–177.

115. Wallner, W., Meera, P., Toro, L. (1996) Determinant for β subunit regulation in high-conductance voltage-activated and Ca^{2+}-sensitive K^+ channels: an additional transmembrane region in the N-terminus. *Proc. Natl. Acad. Sci.* **93**: 14922–14927.

116. Joiner, W.J., Tang, M., Wang, L., Dworetzky, S., Boissard, C., Gan, L., Gribkoff, V.K., Kaczmarek, L.K. (1998) Formation of intermediate-conductance calcium-activated potassium channels by interaction of Slack and Slo subunits. *Nat. Neurosci.* **1**: 462–469.

117. Quignard, J-F., Chataigneau, T., Corriu, C., Edwards, G., Weston, A.H., Félétou, M., Vanhoutte, P.M. (2002). Endothelium-dependent hyperpolarization and lipoxygenase derived metabolites of arachidonic acid in the carotid artery of the guinea pig. *J. Cardiovasc. Pharmacol.* **40**: 467–477.

118. Nelson, M.T., Cheng, H., Rubart, M., Santana, L.F., Bonev, A.D., Knot, H.J., Lederer, W.J. (1995) Relaxation of arterial smooth muscle by calcium sparks. *Science* **270**: 633–637.

119. Bychkov, R., Gollasch, M., Ried, C., Luft, F.C., Haller, H. (1997) Regulation of spontaneous transient outward potassium currents in human coronary arteries. *Circulation* **95**: 503–510.

120. Perez, G.J., Bonev, A.D., Patlak, J.B., Nelson, M.T. (1999) Functional coupling of ryanodine receptors to KCa channels in smooth muscle cells from rat cerebral arteries. *J. Gen. Physiol.* **113**: 229–238.

121. Pluger, S., Faulhaber, J., Furstenau, M., Lohn, M., Waldschutz, R., Gollasch, M., Haller, H., Luft, F.C., Ehmke, H., Pongs, O. (2000) Mice with disrupted BK channel beta1 subunit gene feature abnormal Ca(2+) sparks/STOC coupling and elevated blood pressure. *Circ. Res.* **87**: E53–E60.

122. Baron, A., Frieden, M., Chabaud, F., Beny, J-L. (1996) Ca^{2+}-dependent non-selective cation and potassium channels activated by bradykinin in pig coronary endothelial cells. *J. Physiol.* **49**: 699–706.

123. Rusko, J., Tanzi, F., van Breemen, C., Adams, D.J. (1992) Calcium-activated potassium channels in native endothelial cells from rabbit aorta. Conductance, Ca^{2+}-sensitivity and block. *J. Physiol.* **455**: 601–621.

124. Brakemier, S., Eichler, I., Knorr, A., Fassheber, T., Köhler, R., Hoyer, J. (2003a) Modulation of Ca^{2+}-activated K^+ channel in renal artery endothelium in situ by nitric oxide and reactive oxygen species. *Kidney Int.* **64**: 199–207.

125. Köhler, R., Degenhardt, C., Kühn, M., Runkel, N., Paul, M., Hoyer, J. (2000) Expression and function of endothelial Ca^{2+}-activated K^+ channels in human mesenteric artery—a single cell reverse transcriptase-polymerase chain reaction and electrophysiological study in situ. *Circ. Res.* **87**: 496–503.

126. Papassotiriou, J., Köhler, R., Prenen, J., Krause, H., Akbar, M., Eggermont, J., Paul, M., Distler, A., Nilius, B., Hoyer, J. (2000) Endothelial K^+ channel lacks the Ca^{2+} sensitivity-regulating β subunit. *FASEB J.* **14**: 885–894.

127. Nelson, M.T., Quayle, J.M. (1995) Physiological roles and properties of potassium channels in arterial smooth muscle. *Am. J. Physiol.* **268**: C799–C822.

128. Quayle, J.M., Nelson, M.T., Standen, N.B. (1997) ATP-sensitive potassium channels and inwardly rectifying potassium channels in smooth muscle. *Physiol. Rev.* **77**: 1165–1232.

129. Bradley, K.K., Jaggar, J.H., Bonev, A.D., Heppner, T.J., Flynn, E.R., Nelson, M.T., Horowitz, B. (1999) Kir2.1 encodes the inward rectifier potassium channel in rat arterial smooth muscle cells. *J. Physiol.* **515**: 639–651.

130. Edwards, G., Richards, G.R., Gardener, M.J., Félétou, M., Vanhoutte, P.M., Weston, A.H. (2003) Role of the inward-rectifier K^+ channel and Na^+/K^+-ATPase in the hyperpolarization to K^+ in rat mesenteric arteries. In *EDHF 2002*, P.M. Vanhoutte, ed., Taylor and Francis, London, pp. 309–317.

131. Hashitani, H., Suzuki, H. (1997) K^+ channels which contribute to the acetylcholine-induced hyperpolarization in smooth muscle of the guinea-pig submucosal arterioles. *J. Physiol.* **501**: 319–29.

132. Coleman, H.A., Tare, M., Parkington, H.C. (2001a) K^+ currents underlying the action of endothelium-derived hyperpolarizing factor in guinea-pig, rat and human blood vessels. *J. Physiol.* **531**: 359–373.

133. Coleman, H.A., Tare, M., Parkington, H.C., (2001b) EDHF is not K^+ but may be due to spread of current from the endothelium in guinea pig arterioles. *Am. J. Physiol.* **280**: H2478–H2483.

134. Quayle, J.M., McCarron, J.G., Brayden, J.E., Nelson, M.T. (1993) Inward rectifier K^+ currents in smooth muscle from rat resistance-sized cerebral arteries. *Am. J. Physiol.* **265**: C1363–C1370.

135. Robertson, B.E., Bonev, A.D., Nelson, M.T. (1996) Inward rectifier K+ currents in smooth muscle cells from rat coronary arteries: block by Mg^{2+}, Ca^{2+}, and Ba^{2+}. *Am. J. Physiol.* **271**: H696–H705.

136. Knot, H.J., Zimmermann, P.A., Nelson, M.T. (1996) Extracellular potassium-induced hyperpolarization and dilatations of rat coronary and cerebral arteries involve inward rectifier potassium channels. *J. Physiol.* **492**: 419–430.

137. Zaritsky, J.J., Eckman, D.M., Wellman, G.C., Nelson, M.T., Schwarz, T.L. (2000) Targeted disruption of Kir2.1 and Kir2.2 genes reveal the essential role of the inwardly rectifying K^+ current in K^+-mediated vasodilation. *Circ. Res.* **87**: 160–166.

138. Droogmans, G., Nilius, B. (2001) Overview: potassium channels in vascular endothelial cells. In *Potassium Channels in Cardiovascular Biology,* S. Archer, N. Rusch, eds., Kluwer Academic/Plenium Publishers, New York, pp. 639–650.

139. Nilius, B., Droogmans, G. (2001) Ion channels and their functional role in vascular endothelium. *Physiol. Rev.* **81**: 1416–1459.

140. Forsyth, S.E., Hoger, A., Hoger, J.H. (1997) Molecular cloning and expression of a bovine endothelial rectifier potassium channel. *FEBS Lett.* **409**: 277–282.

141. Kamouchi, M., Van Den Brent, K., Eggermont, J., Droogmans, G., Nilius, B. (1997) Modulation of inwardly rectifying potassium channels in cultured bovine pulmonary artery endothelial cells. *J. Physiol.* **504**: 545–556.

142. Olesen, S.P., Bundgaard, M. (1993) ATP-dependent closure and reactivation of inward rectifier K^+ channels in endothelial cells. *Circ. Res.* **73**: 492–495.

143. Sadja, R., Alagem, N., Reuveny, E. (2003) Gating of GIRK channels: details of an intricate, membrane-delimited complex. *Neuron* **39**: 9–12.

144. Krapivinsky, G., Gordon, E., Wickman, K., Velimirovic, B., Krapivinsky, L., Clapham, D.E. (1995) The G-protein-gated atrial K^+ channel, IKACh, is a heteromultimer of two inwardly rectifying K^+ channel proteins. *Nature* **374**: 135–141.

145. Ren, Y.J., Xu, X.H., Zhong, C.B., Feng, N., Wang, X.L. (2001) Hypercholesterolemia alters vascular functions and gene expression of potassium channels in rat aortic smooth muscle cells. *Acta Pharmacol. Sin.* **22**: 274–278.

146. Ahluwalia, A., MacAllister, R.J., Hobbs, A.J. (2004) Vascular actions of natriuretic peptides. CyclicGMP-dependent and -independent mechanisms. *Basic Res. Cardiol.* **99**: 83–89.

147. Terzic, A., Vivaudou, M. (2001) Molecular pharmacology of ATP-sensitive K^+ channels: how and why. In *Potassium Channels in Cardiovascular Biology,* S. Archer, N. Rusch, eds., Kluwer Academic/Plenium Publishers, New York, pp. 257–277.

148. Gross, G.J., Peart, J.N. (2003) K_{ATP} channels and myocardial preconditioning: an update. *Am. J. Physiol.* **285**: H921–H930.

149. Cole, W.C., Clement-Chomienne, O. (2003) ATP-sensitive K^+ channels of vascular smooth muscle cells. *J. Cardiovasc. Electrophysiol.* **14**: 94–103.

150. Cui, Y., Gibblin, J.P., Clapp, L.H., Tinker, A. (2001) A mechanism for ATP-sensitive potassium channel diversity: functional assembly of two pore-forming subunits. *Proc. Natl. Acad. Sci. USA.* **98**: 729–734.

151. Cui, Y., Tran, S., Tinker, A., Clapp, L.H. (2002) The molecular composition of K(ATP) channels in human pulmonary artery smooth muscle cells and their modulation by growth. *Am. J. Respir. Cell. Mol. Biol.* **26**: 135–143.

152. Katnick, C., Adams, D.J. (1997) Characterization of ATP-sensitive potassium channels in freshly dissociated rabbit aortic endothelial cells. *Am. J. Physiol.* **272**: H2507–H2511.

153. Schnitzler, M.M., Derst, C., Daut, J., Preisig-Muller, R. (2000) ATP-sensitive potassium channels in capillaries isolated from guinea-pig heart. *J. Physiol.* **525**: 307–317.

154. Murai, T., Muraki, K., Imaizumi, Y., Watanabe, M. (1999) Levcromakalim causes indirect endothelial hyperpolarization via myoendothelia pathway. *Br. J. Pharmacol.* **128**: 1491–1496.

155. White, R., Hiley, C.R. (2000) Hyperpolarization of rat mesenteric endothelial cells by ATP-sensitive K(+) channel openers. *Eur. J. Pharmacol.* **397**: 279–290.

156. Edwards, G., Félétou, M., Gardener, M.J., Glen, C.D., Richards, G.R., Vanhoutte, P.M., Weston, A.H. (2001) Further investigations into the endothelium-dependent hyperpolarizing effects of bradykinin and substance P in porcine coronary artery. *Br. J. Pharmacol.* **133**: 1145–1153.

157. Ketchum, K.A., Joiner, W.J., Sellers, A.J., Kaczmarek, L.K., Goldstein, S.A.N. (1995) A new family of outwardly rectifying potassium channel proteins with two pore domains in tandem. *Nature* **376**: 690–695.

158. Lesage, F., Guillemare, E., Fink, M., Duprat, F., Lazdunski, M., Romey, G., Barhanin, J. (1996) TWIK-1, a ubiquitous human weakly inward rectifying K+ channel with a novel structure. *EMBO J.* **15**: 1004–1011.

159. Yost, C.S. (2003) Update on tandem pore (2P) domain K+ channels. *Curr. Drugs Targets* **4**: 347–351.

160. Lesage, F., Reyes, R., Fink, M., Duprat, F., Guillemare, E., Lazdunski, M. (1996) Dimerization of TWIK-1 K+ channel subunits via a disulfide bridge. *EMBO J.* **15**: 6400–6407.

161. Czirjak, G., Enyedi, P. (2002) Formation of functional heterodimers between the TASK-1 and TASK-3 two-pore domain potassium channel subunits. *J. Biol. Chem.* **277**: 5426–5432.

162. Gurney, A.M., Osipenko, O.N., MacMillan, D., McFarlane, K.M., Tate, R.J., Kempsill, F.E. (2003) Two-pore domain K channel, TASK-1, in pulmonary artery smooth muscle cells. *Circ. Res.* **93**: 957–964.

163. Duprat, F., Lesage, F., Fink, M., Rehyes, R., Heurteaux, C., Lazdunski, M. (1997) TASK, a human background K+ channel to sense external pH variations near physiological pH. *EMBO J.* **16**: 5464–5471.

164. Kim, Y., Bang, H., Kim, D. (1999) TBAK-1 and TASK-1, two pores K+ channel subunits: kinetic properties and expression in rat heart. *Am. J. Physiol.* **277**: H1669–H1678.

165. Patel, A.J., Honore, E., Lesage, F., Fink, M., Romey, G., Lazdunski, M. (1999) Inhalational anaesthetics activate two-pore domain background K+ channels. *Nat. Neurosci.* **2**: 422–426.

166. Fink, M., Duprat, F., Lesage, F., Reyes, R., Romey, G., Heurteaux, C., Lazdunski, M. (1996) Cloning, functional expression and brain localization of a novel unconventional outward rectifier K+ channel. *EMBO J.* **15**: 6854–6862.

167. Fink, M., Lesage, F., Duprat, F., Heurteaux, C., Reyes, R., Fosset, M., Lazdunski, M. (1998) A neuronal two P domain K+ channel stimulated by arachidonic acid and polyunsaturated fatty acids. *EMBO J.* **17**: 3297–3308.

168. Lesage, F., Maingret, F., Lazdunski, M. (2000) Cloning and expression of human TRAAK, a polyunsaturated fatty acids-activated and mechano-sensitive K+ channels. *FEBS Lett.* **471**: 137–140.

169. Maingret, F., Fosset, M., Lesage, F., Lazdunski, M., Honore, E. (1999) TRAAK is a mammalian neuronal mechano-gated K+ channel. *J. Biol. Chem.* **274**: 1381–1387.

170. Maingret, F., Patel, A.J., Lesage, F., Lazdunski, M., Honore, E. (1999) Mechano- or acid stimulation, two interactive modes of activation of the TREK-1 potassium channel. *J. Biol. Chem.* **274**: 26691–26696.

171. Koh, S.D., Monaghan, K., Sergeant, G.P., Ro, S., Walker, R.L., Sanders, K.M., Horowitz, B. (2001) TREK-1 regulation by nitric oxide and cGMP-dependent protein kinase. An essential role in smooth muscle inhibitory neurotransmission. *J. Biol. Chem.* **276**: 44338–44346.

172. Jentsch, T.J., Stein, V., Weinreich, F., Zdebik, A.A. (2002) Molecular structure and physiological function of chloride channels. *Physiol. Rev.* **82**: 503–568.

173. Chipperfield, A.R., Harper, A.A. (2000) Chloride in smooth muscle. *Prog. Biophys. Mol. Biol.* **74**: 175–221.

174. Kitamura, K., Yamazaki, J. (2001) Chloride channels and their functional roles in smooth muscle tone in the vasculature. *Jpn. J. Pharmacol.* **85**: 351–357.

175. Lamb, F.S., Barna, T.J. (1998) The endothelium modulates the contribution of chloride current to norepinephrine-induced vascular contraction. *Am. J. Physiol.* **275**: H161–H168.

176. Graves, J.E., Greenwood, I.A., Large, W.A. (2000) Tonic regulation of vascular tone by nitric oxide and chloride ions in rat isolated small coronary arteries. *Am. J. Physiol.* **279**: H2604–H2611.

177. Ellershaw, D.C., Greenwood, I.A., Large, W.A. (2000) Dual modulation of swelling-activated chloride current by NO and NO donors in rabbit portal vein myocytes. *J. Physiol.* **528**: 15–24.

178. Nelson, M.T., Conway, M.A., Knot, H.J., Brayden, J.E. (1997) Chloride channel blockers inhibit myogenic tone in rat cerebral arteries. *J. Physiol.* **502**: 259–264.

179. Welsh, D.G., Nelson, M.T., Eckman, D.M., Brayden, J.E. (2000) Swelling-activated cation channels mediate depolarization on rat cerebro-vascular smooth muscle by hypoosmolarity and intravascular pressure. *J. Physiol.* **527**: 139–148.

180. Hogg, R.C., Wang, Q., Large, W.A. (1994) Action of niflumic acid on evoked and spontaneous calcium-activated chloride and potassium currents in smooth muscle cells from rabbit portal vein. *Br. J. Pharmacol.* **112**: 977–984.

181. Criddle, D.N., Soares de Moura, R., Greenwood, I.A., Large, W.A. (1996) Effect of niflumic acid on noradrenaline-induced contractions of the rat aorta. *Br. J. Pharmacol.* **118**: 1065–1071.

182. Nilius, B., Voets, T., Eggermont, J., Droogmans, G. (1999) Vrac: a multifunctional volume-regulated anion channel in vascular endothelium. In *Chloride Channels*, R. Kozlowski, ed., Isis Medical Media, Oxford, UK, pp. 47–63.

183. Groschner, K., Graier, W.F., Kukovetz, W.R. (1994) Histamine-induces K^+, Ca^{2+} and Cl^- currents in human vascular endothelial cells: role of ionic currents in stimulation of nitric oxide biosynthesis. *Circ. Res.* **75**: 304–314.

184. Nilius, B., Prenen, J., Szucs, G., Wei, L., Tanzi, F., Voets, T., Droogmans, G. (1997) Calcium-activated chloride channels in bovine pulmonary artery endothelial cells. *J. Physiol.* **498**: 381–396.

185. Li, Z., Niwa, Y., Sakamoto, S., Chen, X., Nakaya, Y. (2000) Estrogens modulate a large conductance chloride channel in cultured porcine aortic endothelial cells. *J. Cardiovasc. Pharmacol.* **35**: 506–510.

186. Ertel, E.A., Campbell, K.P., Harpold, M.M., Hofmann, F., Mori, Y., Perez-Reyes, E., Schwartz, A., Snutch, T.P., Tanabe, T., Birnbaumer, L., Tsien, R.W., Catterall, W.A. (2000) Nomenclature of voltage-gated calcium channels. *Neuron* **25**: 533–535.

187. Triggle, D.J. (1997) Cardiovascular T-type calcium channels: physiological and pharmacological significance. *J. Hypertens.* **15**: S9–S15.

188. Morita, H., Cousins, H., Onoue, H., Ito, Y., Inoue, R. (1999) Predominant distribution of nifedipine-insensitive, high voltage-activated Ca^{2+} channels in the terminal mesenteric artery of guinea pig. *Circ. Res.* **85**: 596–605.

189. Keef, K.D., Hume, J.R., Zhong, J. (2001) Regulation of cardiac and smooth muscle Ca^{2+} channels ($Ca_V1.2a,b$) by protein kinases. *Am. J. Cell. Physiol.* **281**: C1743–C1756.

190. Ito, Y., Kuriyama, H. (1971) Membrane properties of the smooth muscle fibres of the guinea-pig portal vein. *J. Physiol.* **214**: 427–441.

191. Félétou, M., Hoeffner, U., Vanhoutte, P.M. (1989) Endothelium-dependent relaxing factors do not affect the smooth muscle of portal mesenteric veins. *Blood Vessels* **26**: 21–32.

192. Sanders, K.M. (2001) Signal transduction in smooth muscle: Mechanism of calcium handling in smooth muscles. *J. Appl. Physiol.* **91**: 1438–1449.

193. Harder, D.R., Bellardinelli, C., Sperelakis, N., Rubio, R., Berne, R.M. (1979) Differential effects of adenosine and nitroglycerin on the action potentials of large and small coronary arteries. *Circ. Res.* **44**: 176–182.

194. Félétou, M., Vanhoutte, P.M. (1988) Endothelium-dependent hyperpolarisation of canine coronary smooth muscle. *Br. J. Pharmacol.* **93**: 515–524.

195. McFadzean, I., Gibson, A. (2002) The developing relationship between receptor-operated and store-operated calcium channels in smooth muscle. *Br. J. Pharmacol.* **135**: 1–13.

196. Nilius, B. (2003) From TRPs to SOCs, CCEs and CRACs: consensus and controversies. *Cell Calcium* **33**: 293–298.

197. Voets, T., Nilius, B. (2003) The pore of TRP channels: trivial or neglected? *Cell Calcium* **33**: 299–302.

198. Inoue, R., Okada, T., Onoue, H., Hara, Y., Shimizu, S., Naitoh, S., Ito, Y., Mori, Y. (2001) The transient receptor protein homologue TRP6 is the essential component of vascular α1-adrenoceptor-activated Ca^{2+}-permeable cation channel. *Circ. Res.* **88**: 325–332.

199. Xu, S.Z., Beech, D.J. (2001) TrpC1 is a membrane spanning subunit of store-operated Ca^{2+} channels in native vascular smooth muscle cells. *Circ. Res.* **88**: 84–87.

200. Walker, R.L., Hume, J.R., Horrowitz, B. (2001a) Differential expression and alternative splicing of TRP channel genes in smooth muscles. *Am. J. Physiol.* **280**: C1184–C1192.

201. Albert, A.P., Large, W.A. (2003) Store-operated Ca^{2+}-permeable non-selective cation channels in smooth muscle cells. *Cell Calcium* **33**: 345–356.

202. Landsberg, J.W., Yuan, J.X.J. (2004) Calcium and TRP channels in pulmonary vascular smooth muscle proliferation. *News Physiol. Sci.* **19**: 44–50.

203. Kamishima, T., Davies, N.W., Standen, N.B. (2000) Mechanisms that regulate $[Ca^{2+}]_i$ following depolarization in rat systemic arterial smooth muscle cells. *J. Physiol.* **522**: 285–295.

204. Van Breemen, C., Chen, Q., Laher, I. (1995) Superficial buffer barrier function of smooth muscle sarcoplasmic reticulum. *Trends Pharmacol. Sci.* **16**: 98–105.

205. Horowitz, A., Menice, C.B., Laporte, R., Morgan, K.G. (1996) Mechanisms of smooth muscle contraction. *Physiol. Rev.* **76**: 967–1003.

206. Taggart, M.J. (2001) Smooth muscle excitation-contraction coupling: a role for caveolae and caveolins? *News Physiol. Sci.* **16**: 61–65.

207. Blaunstein, M.P., Golovina, V.A., Song, H., Choate, J., Lencesova, L., Robinson, S.W., Wier, W.G. (2002) Organization of Ca^{2+} stores in vascular smooth muscle: functional implications. *Novartis Found. Symp.* **246**: 125–141.

208. Wu, K.D., Lee, W.S., Wey, J., Bungard, D., Lytton, J. (1995) Localization and quantification of endoplasmic reticulum Ca^{2+}-ATPase isoform transcripts. *Am. J. Physiol.* **269**: C775–C784.

209. Kuriyama, H., Kitamura, K., Nabata, H. (1995) Pharmacological and physiological significance of ion channels and factors that modulate them in vascular tissues. *Pharmacol. Rev.* **47**: 387–573.

210. Patel, S., Joseph, S.K., Thomas, A.P. (1999) Molecular properties of inositol 1,4,5-trisphosphate receptors. *Cell Calcium* **25**: 247–264.

211. Otsu, K., Willard, H.F., Khanna, V.K., Zorzato, F., Green, N.M., MacLennan, D.H. (1990) Molecular cloning of cDNA encoding the Ca2+ release channel (ryanodine receptor) of rabbit cardiac muscle sarcoplasmic reticulum. *J. Biol. Chem.* **265**: 13472–13483.

212. del Valle-Rodriguez, A., Lopez-Barneo, J., Urena, J. (2003) Ca^{2+} channel-sarcoplasmic reticulum coupling: a mechanism of arterial myocyte contraction without Ca^{2+} influx. *EMBO J.* **22**: 4337–4345.

213. Carafoli, E., Garcia-Martin, E., Guerini, D. (1996) The plasma membrane calcium pump: recent development and future perspectives. *Experientia* **52**: 1091–1100.

214. Jaggar, J.H., Porter, V.A., Lederer, W.J., Nelson, M.T. (2000) Calcium sparks in smooth muscle. *Am. J. Physiol.* **278**: C235–C256.

215. Boittin, F.X., Dipp, M., Kinnear, N.P., Galione, A., Evans, A.M. (2003) Vasodilation by the calcium-mobilizing messenger ADP-ribose. *J. Biol. Chem.* **278**: 9602–9608.

216. Jackson, W.F. (2000) Ion channels and vascular tone. *Hypertension* **35**: 173–178.

217. Gauthier, K.M., Rusch, N.J. (2003) Potassium channels and membrane potential in vascular endothelial and smooth muscle cells. In *EDHF 2002,* P.M. Vanhoutte, ed., Taylor and Francis, London, pp. 1–12.

218. Terasawa, K., Nakajima, T., Iida, H., Iwasawa, K., Oonuma, H., Jo, T., Morita, T., Nakamura, F., Fujimori, Y., Toyo-oka, T., Nagai, R. (2002) Non-selective cation currents regulate membrane potential of rabbit coronary arterial cell: modulation by lysophosphatidylcholine. *Circulation* **106**: 3111–3119.

219. Sturek, M., Hermsmeyer, K. (1986) Calcium and sodium channels in spontaneously contracting vascular muscle cells. *Science* **233**: 475–478.

220. Quignard, J-F., Ryckwaert, F., Albat, B., Nargeot, J., Richard, S. (1997) A novel tetrodotoxin-sensitive Na^+ current in culture human coronary myocytes. *Circ. Res.* **80**: 377–382.

221. Chobi, C., Mangoni, M.E., Boccara, G., Nargeot, J., Richard, S. (2000) Evidence for tetrodoxin-sensitive sodium currents in primary cultured myocytes from human, pig and rabbit arteries. *Pflugers Arch.* **440**: 149–152.

222. Nelson, M.T., Huang, Y., Brayden, J.E., Hescheler, J., Standen, N.B. (1990) Arterial dilations in response to calcitonin gene related peptide involve activation of K^+ channels. *Nature* **344**: 770–773.

223. Itoh, T., Seki, N., Suzuki, S., Ito, S., Kajikuri, J., Kuriyama, H. (1992) Membrane hyperpolarisation inhibits agonist-induced synthesis of inositol 1,4,5-trisphosphate in rabbit mesenteric artery. *J. Physiol.* **451**: 307–328.

224. Busse, R., Edwards, G., Félétou, M., Fleming, I., Vanhoutte, P.M., Weston, A.H. (2002) Endothelium-dependent hyperpolarization, bringing the concepts together. *Trends Pharmacol. Sci.* **23**: 374–380.

225. Busse, R., Fichtner, H., Luckhoff, A., Kohlhardt, M. (1988) Hyperpolarisation and increased free calcium in acetylcholine-stimulated endothelial cells. *Am. J. Physiol.* **255**: H965–H969.

226. Wang, X., Lau, F., Li, L., Yoshikawa, A., Van Breemen, C. (1995) The actylcholine-sensitive intracellular Ca^{2+}-store in fresh endothelial cells and evidence for ryanodine receptors. *Circ. Res.* **77**: 37–42.

227. Sakai, T. (1990) Acetylcholine-induces Ca-dependent K currents in rabbit endothelial cells. *Jpn. J. Pharmacol.* **53**: 235–246.

228. Marchenko, S.M., Sage, S.O. (1993) Electrical properties of resting and acetylcholine-stimulated endothelium in intact rat aorta. *J. Physiol.* **462**: 735–751.

229. Wu, S., Moore, T.M., Brough, G.H., Whitt, S.R., Chinkers, M., Li, M., Stevens, T. (2000) Cyclic-nucleotide-gated channels mediate membrane depolarization following activation of store-operated calcium entry in endothelial cells. *J. Biol. Chem.* **275**: 18887–18896.

230. Cioffi, D.L., Wu, S., Stevens, T. (2003) On the endothelial cell I_{SOC}. *Cell Calcium* **33**: 323–336.

231. Nilius, B., Droogmans, G., Wondergen, R. (2003) Transient receptor potential channels in endothelium: solving the calcium entry puzzle? *Endothelium* **10**: 5–15.
232. Plant, T.D., Schaefer, M. (2003) TRPC4 and TRPC5: receptor-operated Ca^{2+}-permeable non selective cation channels. *Cell Calcium* **33**: 441–450.
233. Freichel, M., Suh, S.H., Pfeifer, A., Schweig, U., Trost, C., Weissgerber, P., Biel, M., Philipp, S., Freise, D., Droogmans, G., Hofman, F., Flockerzi, V., Nilius, B. (2001) Lack of endothelial store-operated Ca^{2+} current impairs agonist-dependent vasorelaxation in TRP4-/- mice. *Nat. Cell Biol.* **3**: 121–127.
234. Tiruppathi, C., Freichel, M., Vogel, S.M., Paria, B.C., Mehta, D., Flockerzi, V., Malik, A.B. (2002) Impairment of store-operated Ca^{2+} entry in TRPC4-/- mice interferes with increase in lung microvascular permeability. *Circ. Res.* **91**: 70–76.
235. Brough, G.H., Wu, S., Cioffi, D., Moore, T.M., Li, M., Dean, N., Stevens, T. (2001) Contribution of endogenously expressed Trp1 to a Ca^{2+}-selective store-operated Ca^{2+} entry pathway. *FASEB J.* **15**: 1727–1738.
236. Wissenbach, U., Boding, M., Freichel, M., Flockerzi, V. (2000) Trp12, a novel Trp related protein from kidney. *FEBS Lett.* **485**: 127–134.
237. Watanabe, H., Vriens, J., Prenen, J., Droogmans, G., Voets, T., Nilius, B. (2003) Anandamide and arachidonic acid use epoxyeicosatrienoic acids to activate TRPV4 channels. *Nature* **424**: 434–438.
238. Liedtke, W., Choe, Y., Marti-Renom, M.A., Bell, A.M., Denis, C.S., Sali, A., Hudspeth, A.J., Friedman, J.M., Heller, S. (2000) Vanilloid receptor-related osmotically activated channel (VR-OAC), a candidate vertebrate osmoreceptor. *Cell* **103**: 525–535.
239. Launay, P., Fleig, A., Perraud, A.L., Scharenbereg, A.M., Penner, R., Kinet, J.P. (2002) TRPM4 is a Ca^{2+}-activated nonselective cation channel mediating cell membrane depolarization. *Cell* **109**: 397–407.
240. Suh, S.H., Watanabe, H., Droogmans, G., Nilius, B. (2002) ATP and nitric oxide modulate a Ca^{2+}-activated non-selective cation current in macrovascular endothelial cells. *Pflügers Arch. Eur. J. Physiol.* **444**: 438–445.
241. deMello, D.E., Sawyer, D., Galvin, N., Reid, L.M. (1997) Early fetal development of lung vasculature. *Am. J. Respir. Cell Mol. Biol.* **16**: 568–581.
242. Nagao, T., Illiano, S.C., Vanhoutte, P.M. (1992) Heterogeneous distribution of endothelium-dependent relaxations resistant to N^G-nitro-L-arginine in rats. *Am. J. Physiol.* **263**: H1090–H1094.
243. Shimokawa, H., Yasutake, H., Fujii, K., Owada, M.K., Nakaike, R., Fukumoto, Y., Takayanagani, T., Nagao, T., Egashira, K., Fujishima, M., Takeshita, A. (1996) The importance of the hyperpolarizing mechanism increases as the vessel size decrease in endothelium-dependent relaxations in rat mesenteric circulation. *J. Cardiovasc. Pharmacol.* **28**: 703–711.
244. Quilley, J., Fulton, D., McGiff, J.C. (1997) Hyperpolarizing factors. *Biochem. Pharmacol.* **54**: 1059–1070.
245. Christ, G.J., Spray, D.C., el Sabban, M., Moore, L.K., Brink, P.R. (1996) Gap junctions in vascular tissues. Evaluating the role of intercellular communication in the modulation of vasomotor tone. *Circ. Res.* **79**: 631–646.
246. Dhein, S. (1998) Gap junction channels in the cardiovascular system: pharmacological and physiological modulation. *Trends Pharmacol. Sci.* 19: 229–241.
247. Gros, D., Abran, P., Alcolea, S., Dupays, L., Mialhe, A., Hervé, J-C., Miquerol, L., Ruissy-Theveniau, M. (2002) Connexines et canaux jonctionnels. Leur rôle dans la propagation de l'influx dans le cœur. In *Biologie et Pathologie du Cœur et des Vaisseaux*, F. Pinet et al., eds., Medecine-Sciences, Flammarion, Paris, pp. 31–43.

248. Willecke, K., Eiberger, J., Degen, J., Eckardt, D., Romualdi, A., Guldenagel, M., Deutsch, U., Sohl, G. (2002) Structural and functional diversity of connexin genes in the mouse and human genomes. *Biol. Chem.* **383**: 725–737.

249. Edwards, G., Félétou, M., Gardener, M.J., Thollon, C., Vanhoutte, P.M., Weston, A.H. (1999) Role of gap junctions in the responses to EDHF in rat and guinea-pig small arteries. *Br. J. Pharmacol.* **128**: 1788–1794.

250. Willecke, K. (2001) Cardiovascular gap-junctions: functional diversity, complementation and specialization of connexins. In *EDHF 2000*, P.M. Vanhoutte, ed., Taylor and Francis, New York, pp. 14–21.

251. Gardener, M.J., Burnham, M., Weston, A.H., Edwards, G. (2001) Role of gap-junctions in endothelium-dependent hyperpolarizations. In *EDHF 2000*, P.M. Vanhoutte, ed., Taylor and Francis, New York, pp. 53–61.

252. Li, X., Simard, J.M. (2001) Connexin45 gap junction channels in rat cerebral vascular smooth muscle cells. *Am. J. Physiol.* **281**: H1890–H1898.

253. Yeh, H.I., Rothery, S., Dupont, E., Coppen, S.R., Severs, N.J. (1998) Individual gap junction plaques contain multiple connexins in arterial endothelium. *Circ. Res.* **83**: 1248–1263.

254. Li, X., Simard, J.M. (1999) Multiple connexins form gap junction channel in rat basilar artery smooth muscle cells. *Circ. Res.* **84**: 1277–1284.

255. Brink, P.R., Beyer, E.C., Christ, G.J. (2001) What do gap junctions do anyway? In *EDHF 2000*, P.M. Vanhoutte, ed. Taylor and Francis, New York, pp. 1–13.

256. Neljssen, J., Herberts, C., Drijfhout, J.W., Relts, E., Janssen, L., Neefjes, J. (2005) Cross-presentation by intercellular peptide transfer through gap junctions. *Nature* **434**: 83–88.

257. Griffith, T.M. (2004) Endothelium-dependent smooth muscle hyperpolarization: do gap junctions provide a unifying hypothesis? *Br. J. Pharmacol.* **141**: 881–903.

258. Furchgott, R.F., Zawadzki, J.V. (1980) The obligatory role of the endothelial cells in the relaxation of arterial smooth muscle by acetylcholine. *Nature* **288**: 373–376.

CHAPTER 2

1. Furchgott, R.F., Zawadzki, J.V. (1980) The obligatory role of the endothelial cells in the relaxation of arterial smooth muscle by acetylcholine. *Nature* **288**: 373–376.

2. Smith, W.L., Marnett, L.J. (1991) Prostaglandin endoperoxide synthase: structure and catalysis. *Biochem. Biophys. Acta* **1083**: 1–17.

3. Smani, T., Zakharov, S.I., Leno, E., Csutora, P., Trepakova, E.S., Bolotina, V.M. (2003) Ca^{2+}-independent phosphpolipase A2 is a novel determinant of store operated Ca^{2+} entry. *J. Biol. Chem.* **278**: 11909–11915.

4. Kudo, I., Murakami, M. (2002) Phospholipase A_2 enzymes. *Prostaglandins Other Lipid Mediat.* **68–69**: 3–58.

5. Vane, J., Bakhle, Y.S., Botting, R.M. (1998) Cyclooxygenases 1 and 2. *Annu. Rev. Pharmacol. Toxicol.* **38**: 97–120.

6. Moncada, S., Gryglewski, R.J., Bunting, S., Vane, J.R. (1976) An enzyme isolated from arteries transforms prostaglandin endoperoxides to an unstable substance that inhibits platelet aggregation. *Nature* **263**: 663–665.

7. De Witt, D.L., Smith, W.L. (1988) Primary structure of prostaglandin G/H synthase from sheep vesicular gland determined from the complementary DNA sequence. *Proc. Natl. Acad. Sci. USA* **85**: 1412–1416.

8. Merlie, J.P., Fagan, D., Mudd, J., Needleman, P. (1988) Isolation and characterization of the complementary DNA for sheep seminal prostaglandins endoperoxide synthase (cyclooxygenase). *J. Biol. Chem.* **263**: 3550–3553.
9. Yokohama, C., Takai, T., Tanabe, T. (1988) Primary structure of sheep prostaglandin endoperoxide synthase deduced from cDNA sequence. *FEBS Lett.* **231**: 347–351.
10. O'Banion, M.K., Winn, V.D., Young, D.A. (1992) cDNA cloning and functional activity of a glucocorticoid-regulated inflammatory cyclooxygenase. *Proc. Natl. Acad. Sci. USA* **89**: 4888–4892.
11. Hla, T., Neilson, K. (1992) Human cyclooxygenase-2 cDNA. *Proc. Natl. Acad. Sci. USA* **89**: 7384–7388.
12. Doroudi, R., Gan, L.M., Selin Sjogren, L., Jern, S. (2000) Effects of shear stress on eicosanoid gene expression and metabolites production in vascular endothelial cells as studied in a novel biomechanical perfusion model. *Biochem. Biophys. Res. Comm.* **269**: 257–264.
13. Davidge, S.T. (2001) Prostaglandin H synthase and vascular function. *Circ. Res.* **89**: 650–660.
14. De Witt, D.L., Day, J.S., Sonnenburg, W.K., Smith, W.L. (1983) Concentrations of prostaglandin endoperoxide synthase and prostaglandin I2 synthase in the endothelium and smooth muscle of bovine aorta. *J. Clin. Invest.* **72**: 1882–1888.
15. Wolin, M.S. (2000) Interactions of oxidants with vascular signalling systems. *Arterioscler. Thromb. Vasc. Biol.* **20**: 1430–1442.
16. Morrow, J.D., Hill, K.E., Burk, R.F., Nannour, T.M., Badr, K.F., Roberts, L.J. II, (1990) A series of prostaglandins F2-like compounds are produced in vivo in humans by a non-cyclooxygenase, free radical catalysed mechanism. *Proc. Natl. Acad. Sci. USA* **87**: 9383–9387.
17. Watkins, M.T., Patton, G.M., Soler, H.M., Albadawi, H., Humphries, D.E., Evans, J.E., Kadowaki, K. (1999) Synthesis of 8-epi-prostaglandin F2α by human endothelial cells: role of prostaglandin H2 synthase. *Biochem. J.* **344**: 747–75.
18. Pratico, D., Lawson, J.A., Rokach, J., FitzGerald, G.A. (2001) The isoprostanes in biology and medicine. *Trends Endocrinol. Metabol.* **12**: 243–247.
19. Janssen, L.J. (2002) Are endothelium-derived hyperpolarizing and contracting factors isoprostanes. *Trends Pharmacol. Sci.* **23**: 59–62.
20. Yang, D., Zhang, J-N., Vanhoutte, P.M., Félétou, M. (2004) NO and inactivation of the endothelium-dependent contracting factor released by acetylcholine in SHR. *J. Cardiovasc. Pharmacol.* **43**: 815–820.
21. Taba, Y., Sasaguri, T., Miyagi, M., Abumiya, T., Miwa, Y., Ikeda, T., Mitsumata, M. (2000) Fluid shear stress induces lipocalin-type prostaglandin D2 synthase in vascular endothelial cells. *Circ. Res.* **86**: 967–973.
22. Urade, Y., Eguchi, N. (2002) Lipocalin-type and hematopoietic prostaglandin D synthases as a novel example of functional convergence. *Prostaglandins Other Lipid Mediat.* **68–69**: 375–382.
23. Law, R.E., Goetze, S., Xi, X.P., Jackson, S., Kawano, Y., Demer, L., Fisbein, M.C., Meehan, W.P., Hsueh, W.A. (2000) Expression and function of PPARgamma in rat and human vascular smooth muscle cells. *Circulation* **101**: 1311–1318.
24. Jackson, S.M., Parhami, F., Xi, X.P., Berliner, J.A., Hsueh, W.A., Law, R.E., Demer, L.L. (1999) Peroxisome proliferator-activated receptor activators target human endothelial cells to inhibit leukocyte-endothelial interaction. *Arteriosceler. Thromb. Vasc. Biol.* **19**: 2094–2104.
25. Wakino, S., Law, R.E., Hsueh, W.A. (2002) Vascular protective effects by activation of nuclear receptor PPARγ. *J. Diabetes Complications* **16**: 46–49.

26. Murakami, M., Nakatami, Y., Tanioka, T., Kudo, I. (2002) Prostaglandin E synthase. *Prostaglandins Other Lipid Mediat.* **68–69**: 383–399.
27. Jakobsson, P.J., Thoren, S., Morgenstern, R., Samuelsson, B. (1999) Identification of human prostaglandin E synthase: a microsomal, glutathione-dependent, inducible enzyme, constituting a potential novel drug target. *Proc. Natl. Acad. Sci. USA* **96**: 7220–7225.
28. Qian, Y.M., Jones, R.L., Chan, K.M., Stock, A.I., Ho, J.K. (1994) Potent contractile actions of prostanoid EP3-receptor agonists on human isolated pulmonary artery. *Br. J. Pharmacol.* **113**: 369–374.
29. Baxter, G.S., Clayton, J.K., Coleman, R.A., Marshall, K., Sangha, R., Senior, J. (1995) Characterization of the prostanoid receptors mediating constriction and relaxation of human isolated uterine artery. *Br. J. Pharmacol.* **116**: 1692–1696.
30. Jones, R.L., Qian, Y.M., Chan, K.M., Yim, A.P. (1998) Characterization of a prostanoid EP3-receptor in guinea-pig aorta: partial agonist action of the non-prostanoid ONO-AP-324. *Br. J. Pharmacol.* **125**: 1288–1296.
31. Watanabe, K. (2002) Prostaglandin F synthase. *Prostaglandins Other Lipid Mediat.* **68–69**: 401–407.
32. Moncada, S., Herman, A.G., Higgs, E.A., Vane, J.R. (1977) Differential formation of prostacyclin (PGX or PGI2) by layers of the arterial wall. An explanation for the anti-thrombotic properties of vascular endothelium. *Thromb. Res.* **11**: 323–344.
33. Moncada, S., Vane, J.R. (1979) Pharmacology and endogenous roles of prostaglandin endoperoxides, thromboxane A2 and prostacyclin. *Pharmacol. Rev.* **30**: 293–331.
34. Wise, H., Jones, R.L. (1996) Focus on prostacyclin and its novel mimetics. *Trends Pharmacol. Sci.* **17**: 17–21.
35. Radomski, M.W., Palmer, R.M.J., Moncada, S. (1987) The role of nitric oxide and cGMP in platelet adhesion to vascular endothelium. *Biochem. Biophys. Res. Comm.* **148**: 1482–1489.
36. Radomski, M.W., Palmer, R.M.J., Moncada, S. (1987) Comparative pharmacology of endothelium-derived relaxing factor, nitric oxide and prostacyclin in platelets. *Br. J. Pharmacol.* **92**: 181–187.
37. Radomski, M.W., Palmer, R.M.J., Moncada, S. (1987) The anti-aggregating properties of vascular endothelium: interactions between prostacyclin and nitric oxide. *Br. J. Pharmacol.* **92**: 639–646.
38. Shen, R-F., Tai, H-H. (1998) Thromboxanes: synthase and receptors. *J. Biomed. Sci.* **5**: 153–172.
39. Wang, L-H., Kulmacz, R.J. (2002) Thromboxane synthase: structure and function of protein and gene. *Prostaglandins Other Lipid Mediat.* **68–69**: 409–422.
40. Shiokoshi, T., Ohsaki, Y., Kawabe, J., Fujino, T., Kikuchi, K. (2002) Down-regulation of nitric oxide accumulation by cyclooxygenase induction and thromboxane A2 production in interleukin-1beta stimulated rat aortic smooth muscle cells. *J. Hypertens.* **20**: 455–461.
41. Pfister, S.L., Deinhart, D.D., Campbell, W.B. (1998) Methacholine-induced contraction of rabbit pulmonary artery: role of platelet-endothelial transcellular thromboxane synthesis. *Hypertension* **31**: 206–212.
42. Tsuboi, K., Sugimoto, Y., Ichikawa, A. (2002) Prostanoid receptor subtypes. *Prostaglandins Other Lipid Mediat.* **68–69**: 535–556.
43. Hirata, T., Ushikubi, F., Kakizuka, A., Okuma, M., Narumiya, S. (1996) Two thromboxane A_2 receptor isoforms in human platelets. Opposite coupling to adenylate cyclase with different sensitivity to Arg60 to Leu mutation. *J. Clin. Invest.* **97**: 949–956.

44. Namba, T., Oida, H., Sugimoto, Y., Kakikuza, A., Negeshi, M., Ichikawa, A., Narumiya, S. (1994) cDNA cloning of a mouse prostacyclin receptor. Multiple signalling pathways and expression in thymic medulla. *J. Biol. Chem.* **269**: 9986–9990.

45. Kobayashi, T., Narumiya, S. (2002) Function of prostanoid receptors: studies on knock-out mice. *Prostaglandins Other Lipid Mediat.* **68–69**: 557–573.

46. Hirai, H., Tanaka, K., Yoshie, O., Ogawa, K., Kemmotsu, K., Takamori, Y., Ichimasa, M., Sugamura, K., Nakamura, M., Takano, S., Nagata, K. (2001) Prostaglandin D2 selectively induces chemotaxis in T helper type 2 cells, eosinophils, and basophils via seven-transmembrane receptor CRTH2. *J. Exp. Med.* **193**: 255–261.

47. Funk, C.D., Chen, X-S., Johnson, E.N., Zhao, L. (2002) Lipoxygenase genes and their targeted disruption. *Prostaglandins Other Lipid Mediat.* **68–69**: 303–312.

48. Schneider, C., Brash, A.R. (2002) Lipoxygenase-catalyzed formation of R-configuration hydroperoxides. *Prostaglandins Other Lipid Mediat.* **68–69**: 291–301.

49. Radmark, O. (2002) Arachidonate 5-lipoxygenase. *Prostaglandins Other Lipid Mediat.* **68–69**: 211–234.

50. Voelkel, N.F., Tuder, R.M., Wade, K., Hoper, M., Lepley, R.A., Goulet, J.L., Koller, B.H., Fitzpatrick, F. (1996) Inhibition of 5-lipoxygenase-activating protein (FLAP) reduces pulmonary vascular reactivity and pulmonary hypertension in hypoxic rats. *J. Clin. Invest.* **97**: 2491–2498.

51. Yoshimoto, T., Takahashi, Y. (2002) Arachidonate 12-lipoxygenases. *Prostaglandins Other Lipid Mediat.* **68–69**: 245–262.

52. Gonzales-Nunez, D., Claria, J., Rivera, F., Poch, E. (2001) Increased levels of 12(*S*)-HETE in patients with essential hypertension. *Hypertension* **37**: 334–338.

53. Natarajan, R., Pei, H., Gu, J.L., Sarma, J.M., Nadler, J. (1999) Evidence for 12-lipoxygenase induction in the vessel wall following balloon injury. *Cardiovasc. Res.* **41**: 489–499.

54. Kuhn, H., Walther, M., Kuban, R.J. (2002) Mammalian arachidonate 15-lipoxygenases: structure, function and biological implications. *Prostaglandins Other Lipid Mediat.* **68–69**: 263–290.

55. Vane, J., Gryglewski, R.J., Botting, R.M. (1987) The endothelial cell as a metabolic and endocrine organ. *Trends Pharmacol. Sci.* **8**: 491–496.

56. Kuhn, H., Bellkner, J., Wiesner, R., Brash, A.R. (1990) Oxygenation of biological membrane by the pure reticulocyte lipoxygenase. *J. Biol. Chem.* **265**: 18351–18361.

57. Chaitidis, P., Schewe, T., Sutherland, M., Kuhn, H., Nigam, S. (1998) 15-lipoxygenation of phospholipids may precede the sn-2 cleavage by phospholipase A2: reaction specificities of secretory and cytosolic phospholipases A2 towards native and 15-lipoxygenated arachidonyl phospholipids. *FEBS Lett.* **434**: 437–441.

58. Holtzhutter, H.G., Wiesner, R., Rathman, J., Stober, R., Kuhn, H. (1997) Kinetic studies on the interaction of nitric oxide with a mammalian lipoxygenase. *Eur. J. Biochem.* **245**: 608–616.

59. O'Donnell, V.B., Taylor, K.B., Parthasarathy, S., Khun, H., Koesling, D., Friebe, A., Bloodsworth, A., Darley-Usmar, V.M., Freeman, B.A. (1999) 15-lipoxygenase catalytically consumes nitric oxide and impairs activation of guanylate cyclase. *J. Biol. Chem.* **274**: 2083–2091.

60. Capdevila, J.H., Falck, J.R. (2002) Biochemical and molecular properties of the cytochrome P450 arachidonic acid monooxygenases. *Prostaglandins Other Lipid Mediat.* **68–69**: 325–344.

61. Roman, R.J. (2002) P450 Metabolites of arachidonic acid in the control of cardiovascular function. *Physiol. Rev.* **82**: 131–185.

62. Capdevila, J.H., Falck, J.R., Harris, R.C. (2000) Cytochrome P450 and arachidonic acid bioactivation: molecular and functional properties of the arachidonic acid monooxygenase. *J. Lipid. Res.* **41**: 163–178.

63. Schwartzman, M.L., Balazy, M., Masferrer, J., Abraham, N.G., McGiff, J.C., Murphy, R.C. (1987) 12(*R*)- hydroxyicosatetraenoic acid: a cytochrome P450-dependent arachidonic acid metabolite that inhibits Na$^+$, K$^+$-ATPase in the cornea. *Proc. Natl. Acad. Sci. USA* **84**: 8125–8131.

64. Kauser, K., Clark, J.E., Masters, B.S.S., Ortiz de Montelano, P.R., Ma, Y.H., Harder, D.R., Roman, R.J. (1991) Inhibition of cytochrome P-450 attenuate the myogenic response to dog renal arcuate arteries. *Circ. Res.* **68**: 1154–1163.

65. Harder, D.R., Gebremedhin, D., Narayanan, J., Jefcoat, C., Falck, J.R., Campbell, W.B., Roman, R.J. (1994) Formation and action of a P-450 4A metabolite of arachidonic acid in cat cerebral vessels. *Am. J. Physiol.* **266**: H2098–H2107.

66. Zou, A.P., Fleming, J.T., Falck, J.R., Jacobs, E.R., Gebremedhin, D., Harder, D.R., Roman, R.J. (1996) 20-HETE is an endogenous inhibitor of the large conductance Ca^{2+}-activated K$^+$ channel in renal arterioles. *Am. J. Physiol.* **270**: R228–R237.

67. Randriamboavonjy, V., Kiss, L., Falck, J.R., Busse, R., Fleming, I. (2005) The synthesis of 20-HETE in small porcine coronary arteries antagonizes EDHF-mediated relaxation. *Cardiovasc. Res.* **65**: 487–494.

68. Gebremedhin, D., Lange, A.R., Lowry, T.F., Taheri, M.R., Birks, E.K., Hudetz, A.G., Narayanan, J., Falck, J.R., Okamoto, H., Roman, R.J., Nithipatikom, K., Campbell, W.B., Harder, D.R. (2000) Production of 20-HETE and its role in the autoregulation of cerebral blood flow. *Circ. Res.* **87**: 60–65.

69. Alonso-Galicia, M., Drummond, H.A., Reddy, K.K., Falck, J.R., Roman, R.J. (1997) Inhibition of 20-HETE production contributes to the vascular responses to nitric oxide. *Hypertension* **29**: 320–325.

70. Jiang, H., Quilley, J., Reddy, L.M., Falck, J.R., Wong, P.Y., McGiff, J.C. (2005) Red blood cells: reservoirs of cis- and trans-epoxyeicosatrienoic acids. *Prostaglandins Other Lipid Mediat.* **75**: 65–78.

71. Gebremedhin, D., Ma, Y.H., Falck, J.R., Roman, R.J., VanRollins, M., Harder, D.R. (1992) Mechanism of action of cerebral epoxyeicosatrienoic acids on cerebral arterial smooth muscle. *Am. J. Physiol.* **263**: H519–H525.

72. Campbell, W.B., Gebremedhin, D., Pratt, P.F., Harder, D.R. (1996) Identification of epoxyeicosatrienoic acids as endothelium-derived hyperpolarizing factor. *Circ. Res.* **78**: 415–423.

73. Eckman, D.M., Hopkins, N.O., McBride, C., Keef, K.D. (1998) Endothelium-dependent relaxation and hyperpolarization in guinea-pig coronary artery: role of epoxyeicosatrienoic acid. *Br. J. Pharmacol.* **124**: 181–189.

74. Ordway, R.W., Walsh, J.V., Singer, J.J. (1989) Arachidonic acid and other fatty acids directly activates potassium channels in vascular smooth muscle cells. *Science* **244**: 1176–1179.

75. Li, P.L., Campbell, W.B. (1997) Epoxyeicosatrienoic acids activate K$^+$ channels in coronary smooth muscle through a guanine nucleotide binding protein. *Cir. Res.* **80**: 877–884.

76. Li, P.L., Chen, C.L., Bortell, R., Campbell, W.B. (1999) Epoxyeicosatrienoic acid stimulates endogenous mono-ADP-ribosylation in bovine coronary artery smooth muscle. *Circ. Res.* **85**: 349–356.

77. Graier, W.F., Simecek, S., Sturek, M. (1995a) Cytochrome P450 mono-oxygenase-regulated signalling of Ca^{2+} entry in human and bovine endothelial cells. *J. Physiol. (Lond.)* **482**: 259–274.

78. Imig, J.D., Navar, L.G., Roman, R.J., Reddy, K.K., Falck, J.R. (1996) Actions of epoxygenase metabolites on the preglomerular vasculature. *J. Am. Soc. Nephrol.* **7**: 2364–2370.

79. Oltman, C.L., Weintraub, N.L., VanRollins, M., Dellsperger, K.C. (1998) Epoxyeicosatrienoic acids and dihydroxyeicosatrienoic acids are potent vasodilators in the canine coronary microcirculation. *Circ. Res.* **83**: 932–939.

80. Fleming, I., Michaelis, U.R., Bredenkotter, D., Fisslthaler, B., Dehghani, F., Brandes, R.P., Busse, R. (2001) Endothelium-derived hyperpolarizing factor synthase (cytochrome P450 2C9) is a functionally significant source of reactive oxygen species in coronary arteries. *Circ. Res.* **88**: 44–51.

81. Rapoport, R.M., Murad, F. (1983) Agonist induced endothelium-dependent relaxation in rat thoracic aorta may be mediated though cyclic GMP. *Circ. Res.* **52**: 352–357.

82. Ignarro, L.J., Burke, K.S., Wood, K.L., Wolin, M.S., Kadowitz, P.J. (1984) Association between cyclic GMP accumulation and acetylcholine-elicited relaxation of bovine intrapulmonary artery. *J. Pharmacol. Exp. Ther.* **228**: 682–690.

83. Murad, F., Mittal, C.R., Arnold, W.P., Katsuki, S., Kimura, H. (1978) Guanylate cyclase: activation by azide, nitro-compounds, nitric acid and hydroxyl radical and inhibition by hemoglobin and myoglobin. *Adv. Cyclic Nucleotide Res.* **9**: 145–158.

84. Griffith, T.M., Edwards, D.H., Lewis, M.J., Newby, A.C., Henderson, A.H. (1984) The nature of endothelium-derived relaxant factor. *Nature* **308**: 645–647.

85. Rubanyi, G.M., Lorenz, R.R., Vanhoutte, P.M. (1985) Bioassay of endothelium-derived relaxing factor(s). Inactivation by catecholamine. *Am. J. Physiol.* **249**: H95–H101.

86. Rubanyi, G.M., Vanhoutte, P.M. (1986) Superoxide anions and hyperoxia inactivate endothelium-derived relaxing factor. *Am. J. Physiol.* **250**: H222–H227.

87. Martin, W., Villani, G.M., Jothianandan, D., Furchgott, R.F. (1985) Selective blockade of endothelium-dependent and glyceryl trinitrate-induced relaxation by hemoglobin and by methylene blue in the rabbit aorta. *J. Pharmacol. Exp. Ther.* **232**: 708–716.

88. Furchgott, R.F. (1988) Studies on relaxation of rabbit aorta by sodium nitrite: the basis for the proposal that the acid-activatable inhibitory factor from bovine retractor penis is inorganic nitrite and the endothelium-derived relaxing factor is nitric oxide. In *Mechanism of Vasodilatation,* P.M. Vanhoutte, ed., Raven Press, New York, pp. 401–414.

89. Ignarro, L.J., Byrns, R.E., Wood, K.S. (1988) Biochemical and pharmacological properties of EDRF and its similarity to nitric oxide radical. In *Mechanism of Vasodilatation,* P.M. Vanhoutte, ed., Raven Press, New York, pp. 427–435.

90. Palmer, R.M.J., Ferridge, A.G., Moncada, S. (1987) Nitric oxide release accounts for the biological activity of endothelium-derived relaxing factor. *Nature* **327**: 524–526.

91. Palmer, R.M.J., Ashton, D.S., Moncada, S. (1988) Vascular endothelial cells synthesize nitric oxide from L-arginine. *Nature* **333**: 664–666.

92. Rees, D.D., Palmer, R.M.J., Hodson, H.F., Moncada, S. (1989) A specific inhibitor of nitric oxide formation from L-arginine attenuates endothelium-dependent relaxation. *Br. J. Pharmacol.* **96**: 418–424.

93. Leiper, J., Vallance, P. (1999) Biological significance of endogenous methylarginines that inhibit nitric oxide synthases. *Cardiovasc. Res.* **43**: 542–548.

94. Stuehr, D.J., Marletta, M.A. (1987) Synthesis of nitrite and nitrate in murine macrophage cell lines. *Cancer Res.* **47**: 5590–5595.

95. Stuehr, D.J., Marletta, M.A. (1987) Induction of nitrite/nitrate synthesis in murine macrophages by BCG infection, lymphokines, or interferon-gamma. *J. Immunol.* **139**: 518–523.

96. Hibbs, J.B., Jr., Vavrin, Z., Taintor, R.R. (1987) L-arginine is required for expression of the activated macrophage effector mechanism causing selective metabolic inhibition. *J. Immunol.* **138**: 550–554.

97. Hibbs, J.B., Jr., Taintor, R.R., Vavrin, Z. (1987) Macrophage cytotoxicity: role for L-arginine deiminase and imino nitrogen oxidation to nitrite. *Science* **235**: 473–476.

98. Marletta, M.A., Yoon, P.S., Iyengar, R., Leaf, C.D., Wishnock, J.S. (1988) Macrophage oxidation of L-arginine to nitrite and nitrate: nitric oxide is an intermediate. *Biochemistry* **27**: 8706–8711.

99. Garthwaite, J., Charles, S.L., Chess-Williams, R. (1988) Endothelium-derived relaxing factor release on activation of NMDA receptors suggests a role as intercellular messenger in the brain. *Nature* **336**: 385–388.

100. Knowles, R.G., Palacios, M., Palmer, R.M., Moncada, S. (1989) Formation of nitric oxide from l-arginine in the central nervous system: a transduction mechanism for stimulation of the soluble guanylate cyclase. *Proc. Natl. Acad. Sci. USA* **86**: 5159–5163.

101. Bredt, D.S., Snyder, S.H. (1990) Isolation of nitric oxide synthetase, a camodulin requiring enzyme. *Proc. Natl. Acad. Sci. USA* **87**: 682–686.

102. Bredt, T.B., Hwang, P.M., Glatt, C.E., Lowenstein, C., Reed, R.R., Snyder, S.H. (1991) Cloned and expressed nitric oxide synthase structurally resembles cytochrome P-450 reductase. *Nature* **351**: 714–717.

103. Yui, Y., Hattori, R., Kosuga, K., Eizawa, H., Hiki, K., Kawai, C. (1991) Purification of nitric oxide synthase from rat macrophages. *J. Biol. Chem.* **266**: 12544–12548.

104. Lyons, C.R., Orloff, G.J., Cunningham, J.M. (1992) Molecular cloning and functional expression of an inducible nitric oxide synthase from a murine macrophage cell line. *J. Biol. Chem.* **267**: 6370–6374.

105. Pollock, J.S., Fostermann, U., Mitchell, J.A., Warner, T.D., Schmidt, H.H., Nakane, M., Murad, F. (1991) Purification and characterization of particulate endothelium-derived relaxing factor synthase from cultured and native bovine endothelial cells. *Proc. Natl. Acad. Sci. USA* **88**: 10480–10484.

106. Sessa, W.C., Harrison, J.K., Barber, C.M., Zeng, D., Durieux, M.E., D'Angelo, D.D., Lynch, K.R., Peach, M.J. (1992) Molecular cloning and expression of a cDNA encoding endothelial cell nitric oxide synthase. *J. Biol. Chem.* **267**: 15274–15278.

107. Gonzales, C., Barosso, C., Martin, C., Gulbenkian, S., Estrada, C. (1997) Neuronal nitric oxide synthase activation by vasoactive intestinal peptide in bovine cerebral arteries. *J. Cereb. Blood Flow Metab.* **17**: 977–984.

108. Papapetropoulos, A., Desai, K.M., Rudic, R.D., Mayer, B., Zhang, R., Ruiz-Torres, M.P., Garcia-Cardena, G., Madri, J.A., Sessa, W.C. (1997) Nitric oxide synthase inhibitors attenuate transforming-growth-factor-beta1-stimulated capillary organization in vitro. *Am. J. Pathol.* **150**: 1835–1844.

109. Boulanger, C.M., Heymes, C., Benessiano, J., Geske, R.S., Levy, B.I., Vanhoutte, P.M. (1998) Neuronal nitric oxide synthase is expressed in rat vascular smooth muscle cells: activation by angiotensin II in hypertension. *Circ. Res.* **83**: 1271–1278.

110. Papapetropoulos, A., Rudic, R.D., Sessa, W.C. (1999) Molecular control of nitric oxide synthases in the cardiovascular system. *Cardiovasc. Res.* **43**: 509–520.

111. Balligand, J.L., Kelly, R.A., Marsden, P.A., Smith, T.W., Michel, T. (1993) Control of cardiac muscle cell function by an endogenous nitric oxide signalling system. *Proc. Natl. Acad. Sci. USA* **90**: 347–351.

112. Sase, K., Michel, T. (1995) Expression of constitutive endothelial nitric oxide synthase in human blood platelets. *Life Sci.* **57**: 2049–2055.

113. Mülsch, A., Bassenge, E., Busse, R. (1989) Nitric oxide synthesis in endothelial cytosol: evidence for a calcium-dependent and a calcium-independent mechanism. *Naunyn Schmiedebergs Arch. Pharmacol.* **340**: 767–770.

114. Ayajiki, K., Kinderman., M., Hecker, M., Fleming, I., Busse, R. (1996) Intracellular pH and tyrosine phosphorylation but not calcium determine shear stress-induced nitric oxide production in native endothelial cells. *Circ. Res.* **78**: 750–758.

115. Fleming, I., Bauersachs, J., Fisslthaler, B., Busse, R. (1998) Calcium-independent activation of the endothelial nitric oxide synthase in response to tyrosine phosphatase inhibitors and fluid shear stress. *Circ. Res.* **81**: 686–695.

116. Dimmeler, S., Fleming, I., Fisslthaler, B., Hermann, C., Busse, R., Zeiher, A. (1999) Activation of nitric oxide synthase in endothelial cells by Akt-dependent phosphorylation. *Nature* **399**: 601–605.

117. Fleming, I., Bauersachs, J., Schafer, A., Scholz, D., Aldershvile, J., Busse, R. (1999) Isometric contraction induces the Ca^{2+}-independent activation of the endothelial nitric oxide synthase. *Proc. Natl. Acad. Sci. USA* **96**: 1123–1128.

118. Fleming, I., Busse, R. (2003) Molecular mechanisms involved in the regulation of the endothelial nitric oxide synthase. *Am. J. Physiol.* **284**: R1–R12.

119. Michell, B.J., Harris, M.B., Chen, Z.P., Ju, H., Venema, V.J., Blackstone, M.A., Huang, W., Venema, R.C., Kemp, B.E. (2002) Identification of regulatory sites of phosphorylation of the bovine endothelial nitric-oxide synthase at serine 617 and serine 635. *J. Biol. Chem.* **44**: 42344–42351.

120. He, H., Venema, V.J., Gu, X., Venema, R.C., Marrero, M.B., Caldwell, R.B. (1999) Vascular endothelial growth factor signals endothelial cell production of nitric oxide and prostacyclin through Flk-1/KDR activation of c-Src. *J. Biol. Chem.* **274**: 25130–25135.

121. Lorenz, M., Wessler, S., Follmann, E., Michaelis, W., Dusterhoft, T., Baumann, G., Stangl, K., Stangl, V. (2004) A constituent of green tea, epigalocatechin-3-gallate, activates endothelial nitric oxide synthase by a phosphatidylinositol-3-OH-kinase, cAMP-dependent protein kinase, and Akt-dependent pathway and leads to endothelium-dependent vasorelaxation. *J. Biol. Chem.* **279**: 6190–6195.

122. Stoclet, J-C., Chataigneau, T., Ndiaye, M., Oak, M-H., El Bedoui, J., Chataigneau, M., Schini-Kerth, V. (2004) Vascular protection by dietary polyphenols. *Eur. J. Pharmacol.* **500**: 299–313.

123. Bredt, D.S., Snyder, S.H. (1994) Nitric oxide: a physiological messenger molecule. *Annu. Rev. Biochem.* **63**: 175–195.

124. Bredt, D.S. (1999) Endogenous nitric oxide synthesis: biological functions and pathophysiology. *Free Radic. Res.* **31**: 577–596.

125. Andrew, P.J., Mayer, B. (1999) Enzymatic function of nitric oxide synthases. *Cardiovasc. Res.* **43**: 521–531.

126. Schmidt, K., Werner, E.R., Mayer, B., Wachter, H., Kukovetz, W.R. (1992) Tetrahydrobiopterin-dependent formation of endothelium-derived relaxing factor (nitric oxide) in aortic endothelial cells. *Biochem. J.* **281**: 297–300.

127. Vasquez-Vivar, J., Kalyanaraman, B., Martasek, P., Hogg, N., Masters, B.S.S., Karoui, H., Tordo, P., Pritchard, K.A., Jr. (1998) Superoxide generation by endothelial nitric oxide synthase: the influence of cofactors. *Proc. Natl. Acad. Sci. USA* **95**: 9220–9225.

128. Xia, Y., Dawson, V.L., Dawson, T.M., Snyder, S.H., Zweier, J.L. (1996) Nitric oxide synthase generates superoxide and nitric oxide in arginine-depleted cells leading to peroxynitrite-mediated cellular injury. *Proc. Natl. Acad. Sci. USA* **93**: 6770–6774.

129. Azuma, H., Ishikawa, M., Sekizakis, S. (1986) Endothelium-dependent inhibition of platelet aggregation. *Br. J. Pharmacol.* **88**: 411–415.

130. Mashimo, H., Goyal, R.K. (1999) Lessons from genetically engineered animal models IV. Nitric oxide synthase knockout mice. *Am. J. Physiol.* **277**: G745–G750.
131. Huang, P.L., Huang, Z., Mashimo, K.D., Bloch, M.A., Moskowitz, J.A., Bevan, J.A., Fishman, M.C. (1995) Hypertension in mice lacking the gene for endothelial nitric oxide synthase. *Nature* **377**: 239–242.
132. Chataigneau, T., Félétou, M., Huang, P.L., Fishman, M.C., Duhault, J., Vanhoutte, P.M. (1999) Acetylcholine-induced relaxation in blood vessels from endothelial nitric oxide synthase knockout mice. *Br. J. Pharmacol.* **126**: 219–226.
133. Brandes, R.P., Schmitz-Winnenthal, F-H., Félétou, M., Gödecke, A., Huang, P-L., Vanhoutte, P.M., Fleming, I., Busse, R. (2000) An endothelium-derived hyperpolarizing factor distinct from NO and prostacyclin is a major endothelium-dependent vasodilator in resistance vessels of wild type and endothelial NO synthase knock-out mice. *Proc. Natl. Acad. Sci. USA* **97**: 9747–9752.
134. Ohashi, Y., Kawashima, S., Hirata, K.I., Yamashita, T., Ishida, T., Inoue, N., Sakoda, H., Kurihara, H., Yazaki, Y., Yokohama, M. (1998) Hypotension and reduced nitric oxide-elicited vasorelaxation in transgenic mice overexpressing endothelial nitric oxide synthase. *J. Clin. Invest.* **102**: 2061–2071.
135. Huang, P.L., Dawson, T.M., Bredt, D.S., Snyder, S.H., Fishman, M.C. (1996) Targeted disruption of the neuronal nitric oxide synthase gene. *Cell* **75**: 1273–1286.
136. Meng, W., Ma, J., Ayata, C., Hara, H., Huang, P.L., Fishman, M.C., Moskowitz, M.A. (1996) Ach dilates pial arterioles in endothelial and neuronal NOS knock-out mice by independent mechanisms. *Am. J. Physiol.* **271**: H1145–H1150.
137. Toda, N., Okamura, T. (2003) The pharmacology of nitric oxide in the peripheral nervous system of blood vessels. *Pharmacol. Rev.* **55**: 271–324.
138. Gross, S.S., Kilbourn, R.G., Griffith, O.W. (1996) NO in septic shock: good, bad or ugly? Learning from the iNOS knockouts. *Trends Microbiol.* **4**: 47–49.
139. Ohara, Y., Peterson, T.E., Harrison, D.G. (1993) Hypercholesterolemia increases endothelial superoxide anion. *J. Clin. Invest.* **91**: 2546–2551.
140. Pritchard, K.A., Jr., Groszek, L., Smalley, D.M., Sessa, W.C., Wu, M., Villalon, P., Wolin, M.S., Stemerman, M.B. (1995) Native low-density lipoprotein increases endothelial cell nitric oxide synthase generation of superoxide anion. *Circ. Res.* **77**: 510–518.
141. Kodja, G., Harrison, D. (1999) Interactions between NO and reactive oxygen species: pathophysiological importance in atherosclerosis, hypertension, diabetes and heart failure. *Cardiovasc. Res.* **43**: 562–571.
142. Cooke, J.P., Andon, N.A., Girerd, X.J., Hirsch, A.T., Creager, M.A. (1991) Arginine restores cholinergic relaxation of hypercholesterolemic rabbit thoracic aorta. *Circulation* **83**: 1057–1062.
143. Arnal, J.F., Munzel, T., Venema, R.C., James, N.L., Bai, C.L., Mitch, W.E., Harrison, D.G. (1995) Interactions between L-arginine and L-glutamine change endothelial NO production. An effect independent of NO-synthase substrate availability. *J. Clin. Invest.* **95**: 2565–2572.
144. Vasquez-Vivar, J., Kalyanaraman, B., Martasek, P. (2003) The role of tetrahydrobiopterin in superoxide generation from eNOS: enzymology and physiological implications. *Free Radic. Res.* **37**: 121–127.
145. Busse, R., Edwards, G., Félétou, M., Fleming, I., Vanhoutte, P.M., Weston, A.H. (2002) Endothelium-dependent hyperpolarization, bringing the concepts together. *Trends Pharmacol. Sci.* **23**: 374–380.
146. Furchgott, R.F., Vanhoutte, P.M. (1989) Endothelium-derived relaxing and contracting factors. *FASEB J.* **3**: 2007–2018.

147. Vanhoutte, P.M., Félétou, M., Taddei, S. (2005) Endothelium-dependent contractions in hypertension. *Br. J. Pharmacol.* **144**: 449–458.

CHAPTER 3

1. De Mey, J.G., Claeys, M., Vanhoutte, P.M. (1982) Endothelium-dependent inhibitory effects of acetylcholine, adenosine triphosphate, thrombin and arachidonic acid in the canine femoral artery. *J. Pharmacol. Exp. Ther.* **222**: 166–173.
2. Moncada, S., Vane, J.R. (1979) Pharmacology and endogenous roles of prostaglandin endoperoxides, thromboxane A2 and prostacyclin. *Pharmacol. Rev.* **30**: 293–331.
3. Rubanyi, G.M., Lorenz, R.R., Vanhoutte, P.M. (1985) Bioassay of endothelium-derived relaxing factor(s). Inactivation by catecholamine. *Am. J. Physiol.* **249**: H95–H101.
4. Rubanyi, G.M., Vanhoutte, P.M. (1987) Nature of endothelium-derived relaxing factor: are there two relaxing mediators? *Circ. Res.* **61**: II61–II67.
5. De Mey, J.G., Claeys, M., Vanhoutte, P.M. (1980) Interaction between Na^+,K^+ exchanges and the direct inhibitory effects of acetylcholine on canine femoral arteries. *Circ. Res.* **46**: 826–836.
6. Rapoport, R.M., Schwartz, K., Murad, F. (1985) Effects of Na^+, K^+ pump inhibitors and membrane depolarising agents on acetylcholine-induced endothelium-dependent relaxation and cyclic-GMP accumulation in rat aorta. *Eur. J. Pharmacol.* **110**: 203–209.
7. Rubanyi, G.M., Vanhoutte, P.M. (1985) Ouabain inhibits endothelium-dependent relaxations to arachidonic acid in canine coronary arteries. *J. Pharmacol. Exp. Ther.* **215**: 81–86.
8. Hoeffner, U., Félétou, M., Flavahan, N.A., Vanhoutte, P.M. (1989) Canine arteries release two different endothelium-derived relaxing factors. *Am. J. Physiol.* **257**: H330–H333.
9. Boulanger, C.M., Hendrickson, H., Lorenz, R.R., Vanhoutte, P.M. (1989) Release of different relaxing factors by cultured porcine endothelial cells. *Circ. Res.* **64**: 1070–1078.
10. Taddei, S., Virdis, A., Mattei, P., Natali, A., Ferrannini, E., Salvetti, A. (1995) Effect of insulin on acetylcholine-induced vasodilatation in normotensive subjects and patients with essential hypertension. *Circulation* **92**: 2911–2918.
11. Taddei, S., Virdis, A., Ghiadoni, L., Versari, D., Salvetti, G., Magagna, A., Salvetti, A. (2003) Calcium antagonist treatment by lercanidipine prevents hyperpolarization in essential hypertension. *Hypertension* **41**: 950–955.
12. Kuriyama, H., Suzuki, H. (1978) The effects of acetylcholine on the membrane and contractile properties of smooth muscle cells of the rabbit superior mesenteric artery. *Br. J. Pharmacol.* **64**: 493–501.
13. Kuriyama, H., Suzuki, H. (1981) Adrenergic transmission in the guinea-pig mesenteric artery and their cholinergic modulation. *J. Physiol.* **317**: 383–396.
14. Kitamura, K., Kuriyama, H. (1979) Effects of acetylcholine on the smooth muscle cells of isolated main coronary artery of the guinea-pig. *J. Physiol.* **293**: 119–133.
15. Bolton, T.B., Lang, R.J., Takewaki, T. (1984) Mechanism of action of noradrenaline and carbachol on smooth muscle of guinea-pig anterior mesenteric artery. *J. Physiol.* **351**: 549–572.
16. Félétou, M., Vanhoutte, P.M. (1985) Endothelium-derived relaxing factor(s) hyperpolarize(s) coronary smooth muscle. *Physiologist* **48**: 325.
17. Félétou, M., Vanhoutte, P.M. (1988) Endothelium-dependent hyperpolarisation of canine coronary smooth muscle. *Br. J. Pharmacol.* **93**: 515–524.

18. Komori, K., Suzuki, H. (1987) Electrical responses of smooth muscle cells during cholinergic vasodilation in the rabbit saphenous artery. *Circ. Res.* **61**: 586–593.
19. Taylor, S.G., Southerton, J.S., Weston, A.H., Baker, J.R.J. (1988) Endothelium-dependent effects of acetylcholine in rat aorta: a comparison with sodium nitroprusside and cromakalim. *Br. J. Pharmacol.* **94**: 853–863.
20. Chen, G., Suzuki, H., Weston, A.H. (1988) Acetylcholine releases endothelium-derived hyperpolarizing factor and EDRF from rat blood vessels. *Br. J. Pharmacol.* **95**: 1165–1174.
21. Nakashima, M., Mombouli, J-V., Taylor, A.A., Vanhoutte, P.M. (1993) Endothelium-dependent hyperpolarisation caused by bradykinin in human coronary arteries. *J. Clin. Invest.* **92**: 2867–2871.
22. Petersson, J., Zygmunt, P.M., Brandt, L., Högestätt, E.D. (1995) Substance P-induced relaxation and hyperpolarisation in human cerebral arteries. *Br. J. Pharmacol.* **115**: 889–894.
23. Rees, D.D., Palmer, R.M.J., Hodson, H.F., Moncada, S. (1989) A specific inhibitor of nitric oxide formation from L-arginine attenuates endothelium-dependent relaxation. *Br. J. Pharmacol.* **96**: 418–424.
24. Rees, D.D., Palmer, R.M.J., Schulz, R., Hodson, H.F., Moncada, S. (1990) Characterization of three inhibitors of endothelial nitric oxide synthase in vitro and in vivo. *Br. J. Pharmacol.* **101**: 746–752.
25. Beny, J-L., Brunet, P.C. (1988) Electrophysiological and mechanical effects of substance P and acetylcholine on rabbit aorta. *J. Physiol. (Lond.)* **398**: 277–289.
26. Richard, V., Tanner, F.C., Tschudi, M.R., Lüscher, T.F. (1990) Different activation of L-arginine pathway by bradykinin, serotonin, and clonidine in coronary arteries. *Am. J. Physiol.* **259**: H1433–H1439.
27. Cowan, C.L., Cohen, R.A. (1991) Two mechanisms mediate relaxation by bradykinin of pig coronary artery: NO-dependent and independent responses. *Am. J. Physiol.* **261**: H830–H835.
28. Mugge, A., Lopez, J.A.G., Piegors, D.J., Breese, K.R., Heistad, D.D. (1991) Acetylcholine-induced vasodilatation in rabbit hindlimb in vivo is not inhibited by analogues of L-arginine. *Am. J. Physiol.* **260**: H242–H247.
29. Hasunuma, K., Yamaguchi, T., Rodman, D., O'Brien, R., McMurtry, I. (1991) Effects of inhibitors of EDRF and EDHF on vasoreactivity of perfused rat lungs. *Am. J. Physiol.* **260**: L97–L104.
30. Illiano, S.C., Nagao, T., Vanhoutte, P.M. (1992) Calmidazolium, a calmodulin inhibitor, inhibits endothelium-dependent relaxations resistant to nitro-L-arginine in the canine coronary artery. *Br. J. Pharmacol.* **107**: 387–392.
31. Nagao, T., Illiano, S.C., Vanhoutte, P.M. (1992) Heterogeneous distribution of endothelium-dependent relaxations resistant to N^G-nitro-L-arginine in rats. *Am. J. Physiol.* **263**: H1090–H1094.
32. Nagao, T., Vanhoutte, P.M. (1992) Characterization of endothelium-dependent relaxations resistant to nitro-L-arginine in the porcine coronary artery. *Br. J. Pharmacol.* **107**: 1102–1107.
33. Pacicca, C., von der Weid, P., Beny, J.L. (1992) Effect of nitro-L-arginine on endothelium-dependent hyperpolarisations and relaxations of pig coronary arteries. *J. Physiol.* **457**: 247–256.
34. Suzuki, H., Chen, G., Yamamoto, Y., Miwa, K. (1992) Nitroarginine-sensitive and insensitive components of the endothelium-dependent relaxation in the guinea-pig carotid artery. *Jpn. J. Physiol.* **42**: 335–347.

35. Mombouli, J.V., Illiano, S., Nagao, T., Vanhoutte, P.M. (1992) The potentiation of bradykinin-induced relaxations by perindoprilat in canine coronary arteries involves both nitric oxide and endothelium-derived hyperpolarizing factor. *Circ. Res.* **71**: 137–144.

36. Zygmunt, P.M., Grundemar, L., Högestätt, E.D. (1994) Endothelium-dependent relaxation resistant to Nω-nitro-L-arginine in the rat hepatic artery and aorta. *Acta Physiol. Scand.* **152**: 107–114.

37. Garcia-Pascual, A., Labadia, A., Jimenez, E., Costa, G. (1995) Endothelium-dependent relaxation to acetylcholine in bovine oviductal arteries: mediation by nitric oxide and changes in apamin-sensitive K+ conductance. *Br. J. Pharmacol.* **115**: 1221–1230.

38. Hayabuchi, Y., Nakaya, Y., Matsukoa, S., Kuroda, Y. (1998) Endothelium-derived hyperpolarizing factor activates Ca^{2+}-activated K^+ channels in porcine coronary artery smooth muscle cells. *J. Cardiovasc. Pharmacol.* **32**: 642–649.

39. Taylor, S.G., Weston, A.H. (1988) Endothelium-derived hyperpolarizing factor: a new endogeneous inhibitor from the vascular endothelium. *Trends Pharmacol. Sci.* **9**: 272–274.

40. Furchgott, R.F., Vanhoutte, P.M. (1989) Endothelium-derived relaxing and contracting factors. *FASEB J.* **3**: 2007–2018.

41. Komori, K., Vanhoutte, P.M. (1990) Endothelium-derived hyperpolarizing factor. *Blood Vessels* **27**: 238–245.

42. Cohen, R.A., Vanhoutte, P.M. (1995) Endothelium-dependent hyperpolarization—beyond nitric oxide and cyclic GMP. *Circulation* **92**: 3337–3349.

43. Garland, C.J., Plane, F., Kemp, B.K., Cocks, T.M. (1995) Endothelium-dependent hyperpolarization: a role in the control of vascular tone. *Trends Pharmacol. Sci.* **16**: 23–30.

44. Félétou, M., Vanhoutte, P.M. (1996) Endothelium-derived hyperpolarizing factor. *Clin. Exp. Pharmacol. Physiol.* **23**: 1082–1090.

45. Moncada, S., Gryglewski, R.J., Bunting, S., Vane, J.R. (1976) An enzyme isolated from arteries transforms prostaglandin endoperoxides to an unstable substance that inhibits platelet aggregation. *Nature* **263**: 663–665.

46. Moncada, S., Herman, A.G., Higgs, E.A., Vane, J.R. (1977) Differential formation of prostacyclin (PGX or PGI_2) by layers of the arterial wall. An explanation for the anti-thrombotic properties of vascular endothelium. *Thromb. Res.* **11**: 323–344.

47. Siegel, G., Stock, G., Schnalke, F., Litza, B. (1987) Electrical and mechanical effects of prostacyclin in canine carotid artery. In *Prostacyclin and Its Stable Analogue Iloprost*, R.J. Gryglewski, G. Stock, eds., Springer-Verlag, Berlin and Heidelberg, pp. 143–149.

48. Jackson, W.F., Konig, A., Dambacher, T., Busse, R. (1993) Prostacyclin-induced vasodilation in rabbit heart is mediated by ATP-sensitive potassium channels. *Am. J. Physiol.* **264**: H238–H243.

49. Parkington, H.C., Tare, M., Tonta, M.A., Coleman, H.A. (1993) Stretch revealed three components in the hyperpolarisation of guinea-pig coronary artery in response to acetylcholine. *J. Physiol.* **465**: 459–476.

50. Parkington, H.C., Tonta, M., Coleman, H., Tare, M. (1995) Role of membrane potential in endothelium-dependent relaxation of guinea-pig coronary arterial smooth muscle. *J. Physiol.* **484**: 469–480.

51. Murphy, M.E., Brayden, J.E. (1995) Apamin-sensitive K^+ channels mediate an endothelium-dependent hyperpolarization in rabbit mesenteric arteries. *J. Physiol.* **489**: 723–734.

52. Corriu, C., Félétou, M., Canet, E., Vanhoutte, P.M. (1996) Endothelium-derived factors and hyperpolarisations of the isolated carotid artery of the guinea-pig. *Br. J. Pharmacol.* **119**: 959–964.

53. Corriu, C., Félétou, M., Edwards, G., Weston, A.H., Vanhoutte, P.M. (2001) Differential effects of prostacyclin and iloprost in the isolated carotid artery of the guinea-pig. *Eur. J. Pharmacol.* **426**: 89–94.

54. Lombard, J.H., Liu, Y., Fredericks, K.T., Bizub, D.M., Roman, R.J., Rusch, N.J. (1999) Electrical and mechanical responses of rat middle cerebral arteries to reduced PO2 and prostacyclin. *Am. J. Physiol.* **276**: H509–H516.

55. Thapaliya, S., Matsuyama, H., Takewaki, T. (2000) Bradykinin causes endothelium-independent hyperpolarisation and neuromodulation by prostanoid synthesis in hamster mesenteric artery. *Eur. J. Pharmacol.* **408**: 313–321.

56. Wise, H., Jones, R.L. (1996) Focus on prostacyclin and its novel mimetics. *Trends Pharmacol. Sci.* **17**: 17–21.

57. Nakashima, M., Vanhoutte, P.M. (1995) Isoproterenol causes hyperpolarization through opening of ATP-sensitive potassium channels in vascular smooth muscle of the canine saphenous vein. *J. Pharmacol. Exp. Ther.* **272**: 379–384.

58. Schubert, R., Serebryakov, N.V., Engel, H., Hopp, H.H. (1996) Iloprost activates KCa channels of vascular smooth muscle cells: role of cyclic-AMP-dependent proteine kinase. *Am. J. Physiol.* **271**: C1203–C1211.

59. Schubert, R., Serebryakov, N.V., Mewes, H., Hopp, H.H. (1997) Iloprost dilates rat small arteries: role of K(ATP) and K(Ca) channel activation by cyclic-AMP-dependent protein kinase. *Am. J. Physiol.* **272**: H1147–H1156.

60. Clapp, L.H., Turcato, S., Hall, S., Baloch, M. (1998) Evidence that Ca^{2+}-activated K^+ channels play a major role in mediating the vascular effects of iloprost and cicaprost. *Eur. J. Pharmacol.* **356**: 215–224.

61. Yamaki, F., Kaga, M., Horinouchi, T., Tanaka, H., Koike, K., Shigenobu, K., Toro, L., Tanaka, Y. (2001) MaxiK channel-mediated relaxation of guinea-pig aorta following stimulation of IP-receptor with beraprost via cyclic AMP-dependent and independent mechanisms. *Naunyn Schmiedebergs Arch. Pharmacol.* **364**: 538–550.

62. Dong, H., Waldron, G.J., Cole, W.C., Triggle, C.R. (1998) Roles of calcium-activated and voltage-gated delayed rectifier potassium channels in endothelium-dependent vasorelaxation in the rabbit middle cerebral artery. *Br. J. Pharmacol.* **123**: 821–832.

63. Zygmunt, P.M., Plane, F., Paulsson, M., Garland, C.J., Högestätt, E.D. (1998) Interactions between endothelium-derived relaxing factors in the rat hepatic artery: focus on regulation of EDHF. *Br. J. Pharmacol.* **124**: 992–1000.

64. Nishiyama, M., Hashitani, H., Fukuta, H., Yamamoto, Y., Suzuki, H. (1998) Potassium channels activated in the endothelium-dependent hyperpolarization in guinea-pig coronary artery. *J. Physiol. (Lond.)* **510**: 455–465.

65. Siegel, G., Emden, J., Wenzel, K., Mironneau, J., Stock, G. (1992) Potassium channel activation and vascular smooth muscle. *Adv. Exp. Med. Biol.* **311**: 53–72.

66. Zhu, S., Han, G., White, R.E. (2002) PGE2 action in human coronary artery smooth muscle: role of potassium channels and signaling cross talk. *J. Vasc. Res.* **39**: 477–488.

67. Barman, S.A., Zhu, S., Han, G., White, R.E. (2003) cAMP activates BK_{Ca} channels in pulmonary arterial smooth muscle via cGMP-dependent protein kinase. *Am. J. Physiol.* **284**: L1004–L1011.

68. Li, P.L., Zou, A.P., Campbell, W.B. (1997) Regulation of potassium channels in coronary arterial smooth muscle by endothelium-derived vasodilators. *Hypertension* **29**: 262–267.

69. Tare, M., Parkington, H.C., Coleman, H.A., Neild, T.O., Dusting, G.J. (1990) Hyperpolarisation and relaxation of arterial smooth muscle caused by nitric oxide derived from the endothelium. *Nature* **346**: 69–71.

70. Siegel, G., Mironneau, J., Schnalke, F., Schroder, G., Schulz, B.G., Grote, J. (1990) Vasodilatation evoked by K+ channel opening. *Prog. Clin. Biol. Res.* **327**: 229–306.

71. Hardy, P., Abran, D., Hou, X., Lahaie, I., Peri, K.G., Asselin, P., Varma, D.R., Chemtob, S. (1998) A major role for prostacyclin in nitric oxide-induced ocular vasorelaxation in the piglet. *Circ. Res.* **83**: 721–729.

72. Shimokawa, H., Flavahan, N.A., Lorenz, R.R., Vanhoutte, P.M. (1988) Prostacyclin releases endothelium-derived relaxing factor and potentiates its action in coronary arteries of the pig. *Br. J. Pharmacol.* **95**: 1197–1203.

73. Yajima, K., Nishiyama, M., Yamamoto, Y., Suzuki, H. (1999) Inhibition of endothelium-dependent hyperpolarization by endothelial prostanoids in guinea-pig coronary artery. *Br. J. Pharmacol.* **126**: 1–10.

74. Félétou, M., Vanhoutte, P.M. (2001) Activation of vascular smooth muscle K+ channels by endothelium-derived factors. In *Potassium Channels in Cardiovascular Biology,* S. Archer, N. Rusch, eds., Kluwer Academic/Plenium Publishers, New York, pp. 691–726.

75. Parkington, H.C., Coleman, H.A., Tare, M. (2004) Prostacyclin and endothelium-dependent hyperpolarization. *Pharmacol. Res.* **49**: 509–514.

76. Borda, E.S., Sterin-Borda, L., Gimeno, M.F., Lazzari, M.A., Gimeno, A.C. (1983) The stimulatory effect of prostacyclin (PGI2) on isolated rabbit and rat aorta is probably associated to the generation of a thromboxane A2 (TXA2) "like material." *Arch. Int. Pharmacodyn. Ther.* **261**: 79–89.

77. Davis, K., Grinsburg, R., Bristow, M., Harrison, D.C. (1980) Biphasic action of prostacyclin in the human coronary artery. *Clin. Res.* **28**: 165A.

78. Pomerantz, K., Sinterose, A., Ramwell, P. (1978) The effect of prostacyclin on the human umbilical artery. *Prostaglandins* **15**: 1035–1044.

79. Levy, J.L. (1980) Prostacyclin-induced contraction of isolated strips from normal and spontaneously hypertensive rats (SHR). *Prostaglandins* **19**: 517–525.

80. Lüscher, T.F., Vanhoutte, P.M. (1986) Endothelium-dependent contractions to acetylcholine in the aorta of spontaneously hypertensive rat. *Hypertension* **8**: 344–348.

81. Williams, S.P., Dorn, G.W., II, Rapoport, R.M. (1994) Prostaglandin I2 mediates contraction and relaxation of vascular smooth muscle. *Am. J. Physiol.* **267**: H796–H803.

82. Zhao, Y.J., Wang, J., Tod, M.L., Rubin, L.J., Yuan, X.J. (1996) Pulmonary vasoconstriction effects of prostacyclin in rats: potential role of thromboxane receptors. *J. Appl. Physiol.* **81**: 2595–2603.

83. Adeagbo, A.S.O., Malik, K.U. (1990) Mechanism of vascular actions of prostacyclin in the rat isolated perfused mesenteric arteries. *J. Pharmacol. Exp. Ther.* **252**: 26–34.

84. Tesfamarian, B., Weisbrod, R.M., Cohen, R.A. (1987) Endothelium inhibits response of rabbit carotid artery to adrenergic nerve stimulation. *Am. J. Physiol.* **253**: H792–H798.

85. Cohen, R.A., Weisbrod, R.M. (1988) Endothelium inhibits norepinephrine release from adrenergic nerves of rabbit carotid artery. *Am. J. Physiol.* **254**: H871–H878.

86. Moncada, S., Palmer, R.J.M., Higgs, E.A. (1991) Nitric oxide: physiology, pathophysiology, and pharmacology. *Pharmacol. Rev.* **43**: 109–142.

87. Haeusler, G., Thorens, M. (1976) The pharmacology of vaso-active antihypertensives. In *Vascular Neuroeffector Mechanisms,* J.A. Bevan et al., eds., Kargel, Basel, Switzerland, pp. 232–241.

88. Ito, Y., Suzuki, H., Kuriyama, K. (1978) Effects of sodium nitroprusside on smooth muscle cells of rabbit pulmonary artery and portal vein. *J. Pharmacol. Exp. Ther.* **207**: 1022–1031.

89. Garland, C.J., Plane, F. (1996) Relative importance of endothelium-derived hyperpolarizing factor for the relaxation of vascular smooth muscle in different arterial beds. In *Endothelium-Derived Hyperpolarizing Factor*, Vol. 1, P.M. Vanhoutte, ed., Harwood Academic Publishers, Amsterdam, pp. 173–179.

90. Cohen, R.A., Plane, F., Najibi, S., Huk, I., Malinski, T., Garland, C.J. (1997) Nitric oxide is the mediator of both endothelium-dependent relaxation and hyperpolarisation of the rabbit carotid artery. *Proc. Natl. Acad. Sci. USA* **94**: 4193–4198.

91. Plane, F., Pearson, T., Garland, C.J. (1995) Multiple pathways underlying endothelium-dependent relaxation in the rabbit in isolated femoral artery. *Br. J. Pharmacol.* **115**: 31–38.

92. Quignard, J-F., Félétou, M., Edwards, G., Duhault, J., Weston, A.H., Vanhoutte, P.M. (2000) Role of endothelial cell hyperpolarization in EDHF-mediated responses in the guinea-pig carotid artery. *Br. J. Pharmacol.* **129**: 1103–1112.

93. Murphy, M.E., Brayden, J.E. (1995) Nitric oxide hyperpolarisation of rabbit mesenteric arteries via ATP-sensitive potassium channels. *J. Physiol.* **486**: 47–58.

94. Ito, Y., Kitamura, K., Kuriyama, K. (1980) Actions of nitroglycerin on the membrane and mechanical properties of smooth muscle cells of the coronary artery of the pig. *Br. J. Pharmacol.* **70**: 197–204.

95. Ito, Y., Kitamura, K., Kuriyama, K. (1980) Nitroglycerin and catecholamine actions on smooth muscle cells of the canine coronary artery. *J. Physiol. (Lond.)* **309**: 171–183.

96. Komori, K., Lorenz, R.R., Vanhoutte, P.M. (1988) Nitric oxide, ACh and electrical and mechanical properties of canine arterial smooth muscle. *Am. J. Physiol.* **255**: H207–H212.

97. Félétou, M., Hoeffner, U., Vanhoutte, P.M. (1989) Endothelium-dependent relaxing factors do not affect the smooth muscle of portal mesenteric veins. *Blood Vessels* **26**: 21–32.

98. Félétou, M., Vanhoutte, P.M. (1996) Bioassay of endothelium-derived hyperpolarizing factor in canine arteries. In *Endothelium-Derived Hyperpolarizing Factor*, Vol. 1, P.M. Vanhoutte, ed., Harwood Academic Publishers, Amsterdam, pp. 25–32.

99. Brayden, J.E., Murphy, M.E. (1996) Potassium channels activated by endothelium-derived factors in mesenteric and cerebral resistance arteries. In *Endothelium-Derived Hyperpolarizing Factor*, Vol. 1, P.M. Vanhoutte, ed., Harwood Academic Publishers, Amsterdam, pp. 137–142.

100. Shimamura, K., Zou, L.B., Matsuda, K., Sekeguchi, F., Yamamoto, K., Sunano, S. (2000) Role of nitric oxide in the contraction of the circular muscle of the portal vein. *Pflugers Arch.* **440**: 435–439.

101. Robertson, B.E., Schubert, R., Hescheler, J., Nelson, M.T. (1993) Cyclic-GMP-dependent protein kinase activates Ca-activated K channels in cerebral artery smooth muscle cells. *Am. J. Physiol.* **265**: C299–C303.

102. Bolotina, V.M., Najibi, S., Palacino, J.J., Pagano, P.J., Cohen, R.A. (1994) Nitric oxide directly activates calcium-dependent potassium channels in vascular smooth muscle cells. *Nature* **368**: 850–853.

103. Quignard, J-F., Chataigneau, T., Corriu, C., Duhault, J., Félétou, M., Vanhoutte, P.M. (1999) Effects of SIN-1 on potassium channels of vascular smooth muscle cells of the rabbit aorta and guinea-pig carotid artery. In *Endothelium-Dependent Hyperpolarizations*, P.M. Vanhoutte, ed., Harwood Academic Publishers, Amsterdam, pp. 193–200.

104. Archer, S.L., Huang, J.M.C., Hampl, V., Nelson, D.P., Shultz, P.J., Weir, E.K. (1994) Nitric oxide and cyclic-GMP cause vasorelaxation by activation of a charybdotoxin-sensitive K channel by cyclic-GMP-dependent protein kinase. *Proc. Natl. Acad. Sci. USA* **91**: 7583–7587.

105. Wellman, G.C., Bonev, A.D., Nelson, M.T., Brayden, J.E. (1996) Gender differences in coronary artery diameter involve estrogen, nitric oxide and Ca^{2+}-dependent K^{+} channels. *Circ. Res.* **79**: 1024–1030.

106. Carrier, G.O., Fuchs, L.C., Winecoff, A.P., Giulumian, A.D., White, R.E. (1997) Nitrovasodilators relax mesenteric microvessels by cyclic-GMP-induced stimulation of Ca-activated K channels. *Am. J. Physiol.* **42**: H76–H84.

107. Mistry, D.K., Garland, C.J. (1998) Nitric oxide (NO)-induced activation of large conductance Ca^{2+}-dependent K^{+} channels (BKCa) in smooth muscle cells isolated from the rat mesenteric artery. *Br. J. Pharmacol.* **124**: 1131–1140.

108. Hoang, L.M., Mathers, D.A. (1998) Internally applied endotoxins and the activation of BK channels in cerebral artery smooth muscle via a nitric oxide-like pathway. *Br. J. Pharmacol.* **123**: 5–12.

109. Peng, W., Hoidal, J.R., Farrukh, I.S. (1996) Regulation of Ca^{2+}activated K^{+} channels in pulmonary vascular smooth muscle cells—role of nitric oxide. *J. Appl. Physiol.* **81**: 1264–1272.

110. Bychkov, R., Gollasch, M., Steinke, T., Ried, C., Luft, F.C., Haller, H. (1998) Calcium-activated potassium channels and nitrate-induced vasodilation in human coronary arteries. *J. Pharmacol. Exp. Ther.* **285**: 293–298.

111. Taniguchi, J., Furukawa, K.I., Shigekawa, M. (1993) Maxi K^{+} channels are stimulated by cyclic guanosine monophosphate-dependent protein kinase in canine coronary artery smooth muscle cells. *Pflügers Arch. Eur. J. Physiol.* **423**: 167–172.

112. Li, P.L., Jin, M.W., Campbell, W.B. (1998) Effect of soluble inhibition of soluble guanylate cyclase on the K(Ca) channel activity in coronary smooth muscle. *Hypertension* **31**: 303–308.

113. Nara, M., Dhulipala, P.D., Ji, G.J., Kamasani, U.R., Wang, Y.X., Matalon, S., Kotlikoff, M.I. (2000) Guanylyl cyclase stimulatory coupling to K(Ca) channels. *Am. J. Physiol.* **279**: C1938–C1945.

114. Shin, J.H., Chung, S., Park, E.J., Uhm, D.Y., Suh, C.K. (1997) Nitric oxide directly activates calcium-activated potassium channels from rat brain reconstituted into planar lipid bilayer. *FEBS Lett.* **415**: 299–302.

115. Abdelrrahmane, A., Salvail, D., Dumoulin, M., Garon, J., Cadieux, A., Rousseau, E. (1998) Direct activation of KCa channel in airway smooth muscle by nitric oxide: involvement of a nitrothiosylation mechanism? *Am. J. Respir. Cell. Mol. Biol.* **19**: 485–497.

116. Lang, R.J., Harvey, J.R., McPhee, G.J., Klemm, M.F. (2000) Nitric oxide and thiol reagent modulation of Ca^{2+}-activated K^{+} (BKCa) channels in myocytes of the guinea-pig taenia caeca. *J. Physiol.* **525**: 363–376.

117. Lang, R.J., Harvey, J.R., Mulholland, E.L. (2003) Sodium (2-sulfonatoethyl) methanethiosulfonate prevents S-nitroso-L-cysteine activation of Ca^{2+}-activated K^{+} (BK_{Ca}) channels in myocytes of the guinea-pig taenia caeca. *Br. J. Pharmacol.* **139**: 1153–1163.

118. Nossaman, B.D., Kaye, A.D., Feng, C.J., Kadowitz, P.J. (1997) Effects of charybdotoxin on responses to nitrovasodilators and hypoxia in the rat lung. *Am. J. Physiol.* **16**: L787–L791.

119. Zhao, Y.J., Wang, J., Rubin, L.J., Yuan, X.J. (1997) Inhibition of K-V and K-Ca channels antagonizes NO-induced relaxation in pulmonary artery. *Am. J. Physiol.* **41**: H904–H912.

120. Plane, F., Hurrell, A., Jeremy, J.Y., Garland, C.J. (1996) Evidence that potassium channels make a major contribution to SIN-1-evoked relaxation of rat isolated mesenteric artery. *Br. J. Pharmacol.* **119**: 1557–1562.

121. Price, J.M., Hellerman, A. (1997) Inhibition of cyclic-GMP mediated relaxation in small rat coronary arteries by block of Ca^{++}-activated K^+ channels. *Life Sci.* **61**: 1185–1192.

122. Kitazono, T., Ibayashi, S., Nagao, T., Fujii, K., Fujishima, M. (1997) Role of Ca^{2+}-activated K^+ channels in acetylcholine-induced dilatation of the basilar artery in vivo. *Br. J. Pharmacol.* **120**: 102–106.

123. Taguchi, H., Heistad, D.D., Chu, Y., Rios, C.D., Ooboshi, H., Faraci, F.M. (1996) Vascular expression of inducible nitric oxide synthase isoform associated with activation of Ca^{++}-dependent K^+ channels. *J. Pharmacol. Exp. Ther.* **279**: 1514–1519.

124. Khan, S.A., Mathews, S.R., Meisheri, K.D. (1993) Role of calcium-activated K^+ channels in vasodilatation induced by nitroglycerin, acetylcholine and nitric oxide. *J. Pharmacol. Exp. Ther.* **267**: 1327–1335.

125. Najibi, S., Cowan, C.L., Palacino, J.J., Cohen, R.A. (1994) Enhanced role of K^+ channels in relaxations of hypercholesterolemic rabbit carotid artery to NO. *Am. J. Physiol.* **266**: H2061–H2067.

126. Plane, F., Wiley, K.E., Jeremy, J.Y., Cohen, R.A., Garland, C.J. (1998) Evidence that different mechanisms underlie smooth muscle relaxation to nitric oxide and nitric oxide donors in the rabbit isolated carotid artery. *Br. J. Pharmacol.* **123**: 1351–1358.

127. Bialecki, R.A., Stinson-Fisher, C. (1995) Kca channel antagonists reduce NO donor-mediated relaxation of vascular and tracheal smooth muscle. *Am. J. Physiol.* **12**: L152–L159.

128. Paταricza, J., Toth, G.K., Penke, B., Hohn, J., Papp, J.G. (1995) Effect of selective inhibition of potassium channels on vasorelaxing response to cromakalim, nitroglycerin and nitric oxide of canine coronary arteries. *J. Pharm. Pharmacol.* **47**: 921–925.

129. Node, K., Kitazake, M., Kosaka, H., Minamino, T., Sato, H., Kuzuya, T., Hori, M. (1997) Roles of NO and Ca^{2+}-activated K^+ channels in coronary vasodilatation induced by 17-beta-estradiol in ischemic heart failure. *FASEB J.* **11**: 793–799.

130. Onoue, H., Katuzic, Z.S. (1997) Role of potassium channels in relaxations of canine middle cerebral arteries induced by nitric oxide donors. *Stroke* **28**: 1264–1270.

131. Darkow, D.J., Lu, L., White, R.E. (1997) Estrogen relaxation of coronary artery smooth muscle is mediated by nitric oxide and cyclic-GMP. *Am. J. Physiol.* **41**: H2765–H2773.

132. Zanzinger, J., Czachurski, J., Seller, H. (1996) Role of calcium-dependent K^+ channels in the regulation of arterial and venous tone by nitric oxide in pigs. *Pflügers Arch. Eur. J. Physiol.* **432**: 671–677.

133. Plane, F., Sampson, L.J., Smith, J.J., Garland, C.J. (2001) Relaxation to authentic nitric oxide and SIN-1 in rat isolated mesenteric arteries: variable role for smooth muscle hyperpolarization. *Br. J. Pharmacol.* **133**: 665–672.

134. Sampson, L.J., Plane, F., Garland, C.J. (2001) Involvement of cyclic GMP and potassium channels in relaxation evoked by the nitric oxide donor diethylamine NONOate, in the rat small isolated mesenteric artery. *Naunyn Schmiedebergs Arch. Pharmacol.* **364**: 220–225.

135. Edwards, G., Weston, A.H. (1995) Potassium channels in the regulation of vascular smooth muscle tone. In *Pharmacologicol Control of Calcium and Potassium Homeostasis: Biological, Therapeutical and Clinical Aspects,* T. Godfraind et al., eds., Kluwer Academic Press, Dordrecht, pp. 85–93.

136. Yu, M., Sun, C.W., Maier, K.G., Harder, D.R., Roman, R.J. (2002) Mechanism of cGMP contribution to the vasodilator response to NO in rat middle cerebral arteries. *Am. J. Physiol.* **282**: H1724–H1731.

137. Sun, C.W., Alonso-Galicia, M., Taheri, M.R., Falck, J.R., Harder, D.R., Roman, R.J. (1998) Nitric oxide-20-hydroxyeicosatetrienoic acid interaction in the regulation of K+ channel activity and vascular tone in renal arterioles. *Circ. Res.* **83**: 1069–1079.

138. Zou, A.P., Fleming, J.T., Falck, J.R., Jacobs, E.R., Gebremedhin, D., Harder, D.R., Roman, R.J. (1996) 20-HETE is an endogenous inhibitor of the large conductance Ca^{2+}-activated K+ channel in renal arterioles. *Am. J. Physiol.* **270**: R228–R237.

139. Alonso-Galicia, M., Drummond, H.A., Reddy, K.K., Falck, J.R., Roman, R.J. (1997) Inhibition of 20-HETE production contributes to the vascular responses to nitric oxide. *Hypertension* **29**: 320–325.

140. Liu, Y., Terata, K., Chai, Q., Li, H., Kleinman, L.H., Gutterman, D.D. (2002) Peroxynitrite inhibits Ca^{2+}-activated K+ channel activity in smooth muscle of human coronary arterioles. *Circ. Res.* **91**: 1070–1076.

141. Miyoshi, H., Nakaya, Y., Moritoki, H. (1994) Non-endothelial-derived nitric oxide activates the ATP-sensitive K channel of vascular smooth muscle cells. *FEBS Lett.* **345**: 47–49.

142. Miyoshi, H., Nakaya, Y. (1994) Endotoxin-induced non-endothelial nitric oxide activates the Ca^{2+}-activated K+ channel in cultured vascular smooth muscle cells. *J. Mol. Cell. Cardiol.* **26**: 1487–1495.

143. Yuan, X.J., Tod, M.L., Rubin, L.J., Blaustein, M.P. (1996) NO hyperpolarizes pulmonary artery smooth muscle cells and decreases the intracellular Ca^{2+} concentration by activating voltage-gated K+ channels. *Proc. Natl. Acad. Sci. USA* **93**: 10489–10494.

144. Schubert, R., Krien, U., Wulfsen, I., Schiemann, D., Lehmann, G., Ulfig, N., Veh, R.W., Schwarz, J.R., Gago, H. (2004) Nitric oxide donor sodium nitroprusside dilates rat small arteries by activation of inward rectifier potassium channels. *Hypertension* **43**: 891–896.

145. Quignard, J.-F., Félétou, M., Duhault, J., Vanhoutte, P.M. (1999) Potassium ions as endothelium-derived hyperpolarizing factors in the isolated carotid artery of the guinea-pig. *Br. J. Pharmacol.* **127**: 27–34.

146. Quignard, J.-F., Félétou, M., Corriu, C., Chataigneau, T., Edwards, G., Weston, A.H., Vanhoutte, P.M. (2000) 3-Morpholinosydnonimine (SIN-1) and K+ channels in smooth muscle cells of the rabbit and guinea-pig carotid arteries. *Eur. J. Pharmacol.* **399**: 9–16.

147. Koh, S.D., Monaghan, K., Sergeant, G.P., Ro, S., Walker, R.L., Sanders, K.M., Horowitz, B. (2001) TREK-1 regulation by nitric oxide and cGMP-dependent protein kinase. An essential role in smooth muscle inhibitory neurotransmission. *J. Biol. Chem.* **276**: 44338–44346.

148. Weidelt, T., Boldt, W., Markwardt, F. (1997) Acetylcholine-induced K+ currents in smooth muscle of intact rat small arteries. *J. Physiol. (Lond.)* **500**: 617–630.

149. von der Weid, P-Y. (1998) ATP-sensitive K+ channels in smooth muscle cells of guinea-pig lymphatics: role in nitric oxide and β-adrenoceptor agonist-induced hyperpolarizations. *Br. J. Pharmacol.* **125**: 17–22.

150. Si, J.Q., Zhao, H., Yang, Y., Jiang, Z.G., Nuttall, A.L. (2002) Nitric oxide induces hyperpolarization by opening ATP-sensitive K(+) channels in guinea pig spiral modiolar artery. *Hear. Res.* **171**: 167–176.

151. Berg, T., Koteng, O. (1997) Signalling pathway in bradykinin- and nitric oxide-induced hypotension in the normotensive rat; role of K+ channels. *Br. J. Pharmacol.* **121**: 1113–1120.

152. Bracamonte, M.P., Burnett, J.C., Jr., Miller, V.M. (1999) Activation of soluble gua-nylate cyclase and potassium channels contribute to relaxations to nitric oxide in smooth muscle derived from canine femoral veins. *J. Cardiovasc. Pharmacol.* **34**: 407–413.

153. Simonsen, U., Garcia-Sacristan, A., Prieto, D. (1997) Apamin-sensitive K^+ channels involved in the inhibition of acetylcholine-induced contractions in lamb coronary small arteries. *Eur. J. Pharmacol.* **329**: 153–163.

154. Plane, F., Garland, C.J. (1994) Smooth muscle hyperpolarization and relaxation to acetylcholine in the rabbit basilar artery. *J. Autonom. Nerv. Syst.* **49**: S15–S18.

155. Archer, S.L., Huang, J.M.C., Reeve, H.L., Hampl, V., Tolarova, S., Michelakis, E., Weir, E.K. (1996) Differential distribution of electrophysiologically distinct myocytes in conduit and resistance arteries determines their response to nitric oxide and hypoxia. *Circ. Res.* **78**: 431–442.

156. Ellershaw, D.C., Greenwood, I.A., Large, W.A. (2000) Dual modulation of swelling-activated chloride current by NO and NO donors in rabbit portal vein myocytes. *J. Physiol.* **528**: 15–24.

157. Yamazaki, J., Kitamura, K. (2001) Cell-to-cell communication via nitric oxide mod-ulation of oscillatory Cl(-) currents in rat intact cerebral arterioles. *J. Physiol.* **536**: 67–78.

158. Hirakawa, Y., Gericke, M., Cohen, R.A., Bolotina, V.M. (1999) Ca(2+)-dependent Cl(–) channels in mouse and rabbit aortic smooth muscle cells: regulation by intra-cellular Ca(2+) and NO. *Am. J. Physiol.* **277**: H1732–H1744.

159. Sakagami, K., Kawamura, H., Wu, D.M., Puro, D.G. (2001) Nitric oxide/cGMP-induced inhibition of calcium and chloride currents in retinal pericytes. *Microvasc. Res.* **62**: 196–203.

160. Lamb, F.S., Barna, T.J. (1998) The endothelium modulates the contribution of chloride current to norepinephrine-induced vascular contraction. *Am. J. Physiol.* **275**: H161–H168.

161. Graves, J.E., Greenwood, I.A., Large, W.A. (2000) Tonic regulation of vascular tone by nitric oxide and chloride ions in rat isolated small coronary arteries. *Am. J. Physiol.* **279**: H2604–H2611.

162. Minowa, T., Miwa, S., Kobayashi, S., Enoki, T., Zhang, X.F., Komuro, T., Iwamuro, Y., Masaki, T. (1997) Inhibitory effects of nitrovasodilators and cyclic GMP on ET-1 activated Ca(2+)-permeable nonselective cation channel in rat aortic smooth muscle cells. *Br. J. Pharmacol.* **120**: 1536–1544.

163. Wayman, C.P., McFadzean, I., Gibson, A., Tucker, J.F. (1996) Inhibition of a calcium store depletion activated non-selective cation current in smooth muscle cells of the mouse anococcygeus. *Br. J. Pharmacol.* **118**: 2001–2008.

164. Brakemier, S., Eichler, I., Knorr, A., Fassheber, T., Kohler, R., Hoyer, J. (2003) Modulation of Ca^{2+}-activated K^+ channel in renal artery endothelium in situ by nitric oxide and reactive oxygen species. *Kidney Int.* **64**: 199–207.

165. Busse, R., Edwards, G., Félétou, M., Fleming, I., Vanhoutte, P.M., Weston, A.H. (2002) Endothelium-dependent hyperpolarization, bringing the concepts together. *Trends Pharmacol. Sci.* **23**: 374–380.

166. Koh, S.D., Campbell, J.D., Carl, A., Sanders, K.M. (1995) Nitric oxide activates multiple potassium channels in canine colonic smooth muscle. *J. Physiol.* **489**: 735–743.

167. Suh, S.H., Watanabe, H., Droogmans, G., Nilius, B. (2002) ATP and nitric oxide modulate a Ca^{2+}-activated non-selective cation current in macrovascular endothelial cells. *Pflügers Arch. Eur. J. Physiol.* **444**: 438–445.

168. Haburcak, M., Wei, L., Viana, F., Prenen, J., Droogmans, G., Nilius, B. (1997) Calcium-activated potassium channels in cultured human endothelial cells are not directly modulated by nitric oxide. *Cell Calcium* **21**: 291–300.

169. Kwan, H.Y., Huang, Y., Yao, X. (2000) Store-operated calcium entry in vascular endothelial cells is inhibited by cGMP via a protein kinase G-dependent mechanism. *J. Biol. Chem.* **275**: 6758–6763.

170. Malinski, T., Taha, Z. (1992) Nitric oxide release from a single cell measure in situ by a porphyrinic-based microsensor. *Nature* **358**: 676–678.

171. Simonsen, U., Wadsworth, R.M., Buus, N.H., Mulvany, M.J. (1999) In vitro simultaneous measurements of relaxation and nitric oxide concentration in rat superior mesenteric artery. *J. Physiol.* **516**: 271–282.

172. Stoen, R., Lossius, K., Karlsson, J.O. (2003) Acetylcholine-induced vasodilatation depends entirely upon NO in the femoral artery of young piglets. *Br. J. Pharmacol.* **138**: 39–46.

173. Ge, Z.D., Zhang, X.H., Fung, P.C., He, G.W. (2000) Endothelium-dependent hyperpolarization and relaxation: resistance to N(G)-nitro-L-arginine and indomethacin in coronary circulation. *Cardiovasc. Res.* **46**: 547–556.

174. Buus, N.H., Simonsen, U., Pilegaard, H.K., Mulvany, M.J. (2000) Nitric oxide, prostanoid and non NO, non prostanoid involvement in acetylcholine relaxation of isolated human small arteries. *Br. J. Pharmacol.* **129**: 184–192.

175. Kemp, B.K., Cocks, T.M. (1997) Evidence that mechanisms dependent and independent of nitric oxide mediate endothelium-dependent relaxation to bradykinin in human small resistance-like coronary arteries. *Br. J. Pharmacol.* **120**: 757–762.

176. Quignard, J-F., Chataigneau, T., Corriu, C., Edwards, G., Weston, A.H., Félétou, M., Vanhoutte, P.M. (2002) Endothelium-dependent hyperpolarization and lipoxygenase derived metabolites of arachidonic acid in the carotid artery of the guinea pig. *J. Cardiovasc. Pharmacol.* **40**: 467–477.

177. He, G.W., Liu, Z.G. (2001) Comparison of nitric oxide release and endothelium-derived hyperpolarizing factor-mediated hyperpolarization between human radial and internal mammary arteries. *Circulation* **104**: I344–I349.

178. Waldron, G.J., Ding, H., Lovren, F., Kubes, P., Triggle, C.R. (1999) Acetylcholine-induced relaxation of peripheral arteries isolated from mice lacking endothelial nitric oxide synthase. *Br. J. Pharmacol.* **128**: 653–658.

179. Brandes, R.P., Schmitz-Winnenthal, F-H., Félétou, M., Gödecke, A., Huang, P-L., Vanhoutte, P.M., Fleming, I., Busse, R. (2000) An endothelium-derived hyperpolarizing factor distinct from NO and prostacyclin is a major endothelium-dependent vasodilator in resistance vessels of wild type and endothelial NO synthase knock-out mice. *Proc. Natl. Acad. Sci. USA* **97**: 9747–9752.

180. Ding, H., Kubes, P., Triggle, C. (2000) Potassium and acetylcholine-induced vasorelaxation in mice lacking endothelial nitric oxide synthase. *Br. J. Pharmacol.* **129**: 1194–1200.

181. Huang, A., Sun, D., Smith, C.J., Connetta, J.A., Shesely, E.G., Koller, A., Kaley, G. (2000) In eNOS knockout mice skeletal muscle, arteriolar dilation is mediated by EDHF. *Am. J. Physiol.* **278**: H762–H768.

182. Huang, A., Sun, D., Carroll, M.A., Jiang, H., Smith, C.J., Connetta, J.A., Falck, J.R., Shesely, E.G., Koller, A., Kaley, G. (2001) EDHF mediates flow-induced dilation in skeletal muscle arterioles of female eNOS-KO mice. *Am. J. Physiol.* **280**: H2462–H2469.

183. Stamler, J.S., Jaraki, O., Osborne, J., Simon, D.I., Keaney, J., Vita, J., Singel, D., Valeri, C.R., Loscalzo, J. (1992) Nitric oxide circulates in mammalian plasma primarily as an nitroso-adduct of serum albumin. *Proc. Natl. Acad. Sci. USA* **89**: 7674–7677.

184. Keaney, J.F., Jr., Simon, D.L., Stamler, J.S., Jaraki, O., Scharfstein, J., Vita, J.A., Loscalzo, J. (1993) NO forms an adduct with serum albumin that has endothelium-derived relaxing properties. *J. Clin. Invest.* **91**: 1582–1589.

185. Jia, L., Bonaventura, C., Bonaventura, J., Stamler, J.S. (1996) S-nitrosohaemoglobin: a dynamic activity of blood involved in the vascular control. *Nature* **380**: 221–226.

186. Hobbs, A.J., Gladwin, M.T., Patel, R.P., Williams, D.L., Butler, A.R. (2002) Haemoglobin: NO transporter, NO inactivator or NOne of the above. *Trends Pharmacol. Sci.* **23**: 406–411.

187. Furchgott, R.F., Jothianandan, D. (1991) Endothelium-dependent and -independent vasodilatation involves cyclic GMP: relaxation induced by nitric oxide, carbon monoxide and light. *Blood Vessels* **28**: 52–61.

188. Flitney, F.W., Megson, I.L. (2003) Nitric oxide and the mechanism of rat vascular smooth muscle photorelaxation. *J. Physiol.* **550**: 819–828.

189. Andrews, K.L., McGuire, J.J., Triggle, C.R. (2003) A photosensitive vascular smooth muscle store of nitric oxide in mouse aorta: no dependence on expression of endothelial nitric oxide synthase. *Br. J. Pharmacol.* **138**: 932–940.

190. Lovren, F., Triggle, C.R. (1998) Involvement of nitrosothiols, nitric oxide and voltage-gated K^+ channels in photorelaxation of vascular smooth muscle cells. *Eur. J. Pharmacol.* **347**: 215–221.

191. Kakuyama, M., Vallence, P., Ahluwalia, A. (1998) Endothelium-dependent sensory NANC vasodilatation: involvement of ATP, CGRP and a possible NO store. *Br. J. Pharmacol.* **123**: 310–316.

192. Chauhan, S., Rahman, A., Nilsson, H., Clapp, L., MacAllister, R., Ahluwalia, A. (2003) NO contributes to EDHF-like responses in rat small arteries: role of NO stores. *Cardiovasc. Res.* **57**: 207–216.

193. Muller, B., Kleschyov, A.L., Stoclet, J.C. (1996) Evidence for N-acetylcysteine-sensitive nitric oxide storage as dinitrosyl iron complexes in lipopolysaccharide-treated rat aorta. *Br. J. Pharmacol.* **119**: 1281–1285.

194. Muller, B., Kleschyov, A.L., Alencar, J.L., Vanin, A., Stoclet, J.C. (2002) Nitric oxide transport and storage in the cardiovascular system. *Ann. NY Acad. Sci.* **962**: 131–139.

195. Batenburg, W.W., Popp, R., Fleming, I., Vries, Rd.R., Garrelds, I.M., Saxena, P.R., Danser, A.H. (2004) Bradykinin-induced relaxation of coronary microarteries: S-nitrosothiols as EDHF? *Br. J. Pharmacol.* **142**: 125–135.

196. Batenburg, W.W., De Vries, R., Saxena, P.R., Jan Danser, A.H. (2004) L-S nitrosothiols: endothelium-derived hyperpolarizing factors in porcine coronary arteries. *J. Hypertens.* **22**: 1927–1936.

197. Jewell, R.P., Saundry, C.M., Bonev, A.D., Tranmer, B.I., Wellman, G.C. (2004) Inhibition of Ca^{++} sparks by oxyhemoglobin in rabbit cerebral arteries. *J. Neurosurg.* **100**: 295–302.

198. Vanheel, B., Van de Voorde, J. (2000) EDHF and residual NO: different factors. *Cardiovasc. Res.* **46**: 370–375.

199. Chen, G., Suzuki, H. (1989) Some electrical properties of the endothelium-dependent hyperpolarisation recorded from rat arterial smooth muscle cells. *J. Physiol.* **410**: 91–106.

200. Chen, G., Suzuki, H. (1989) Direct and indirect action of acetylcholine and histamine on intrapulmonary artery and vein smooth muscles of the rat. *Jpn. J. Physiol.* **39**: 51–65.

201. Nagao, T., Vanhoutte, P.M. (1992) Hyperpolarisation as a mechanism for endothelium-dependent relaxations in the porcine coronary artery. *J. Physiol.* **445**: 355–367.

202. Corriu, C., Félétou, M., Canet, E., Vanhoutte, P.M. (1996) Inhibitors of the cyto-chrome P450-monooxygenase and endothelium-dependent hyperpolarisations in the guinea-pig isolated carotid artery. *Br. J. Pharmacol.* **117**: 607–610.

203. Quignard, J-F., Chataigneau, T., Corriu, C., Duhault, J., Félétou, M., Vanhoutte, P.M. (1999) Potassium channels involved in EDHF-induced hyperpolarization of the smooth muscle cells of the isolated guinea-pig carotid artery. In *Endothelium-Dependent Hyperpolarizations*, P.M. Vanhoutte, ed, Harwood Academic Publishers, Amsterdam, pp. 201–208.

204. Peach, M.J., Singer, H.A., Izzo, N.J., Loeb, A.L. (1987) Role of calcium in endothelium-dependent relaxation of arterial smooth muscle. *Am. J. Cardiol.* **59**: 35–43.

205. Chen, G., Suzuki, H. (1990) Calcium dependency of the endothelium-dependent hyperpolarisation in smooth muscle cells of the rabbit carotid artery. *J. Physiol.* **421**: 521–534.

206. Gluais, P., Edwards, G., Weston, A.H., Falck, J.R., Vanhoutte, P.M., Félétou, M. (2005) SK_{Ca} and IK_{Ca} in the endothelium-dependent hyperpolarization of the guinea-pig isolated carotid artery. *Br. J. Pharmacol.* **144**: 477–485.

207. Luckhoff, A., Pohl, U., Mulsch, A., Busse, R. (1988) Differential role of extra- and intracellular calcium in the release of EDRF and prostacyclin from cultured endo-thelial cells. *Br. J. Pharmacol.* **95**: 189–196.

208. Johns, A., Freay, A.D., Adams, D.J., Lategan, T.W., Ryan, U.S., Van Breemen, C. (1988) Role of calcium in the activation of endothelial cells. *J. Cardiovasc. Pharma-col.* **12**: S119–S123.

209. Fleming, I., Fisslthaler, B., Busse, R. (1996) Interdependence of calcium signaling and protein tyrosine phosphorylation in human endothelial cells. *J. Biol. Chem.* **271**: 11009–11015.

210. Nilius, B., Droogmans, G., Wondergen, R. (2003) Transient receptor potential chan-nels in endothelium: solving the calcium entry puzzle? *Endothelium* **10**: 5–15.

211. Weintraub, N.L., Stephenson, A.H., Sprague, R.S., McMurdo, L., Lonigro, A.J. (1995) Relationship of arachidonic acid release to porcine coronary artery relaxation. *Hyper-tension* **26**: 684–690.

212. Fulton, D., McGiff, J.C., Quilley, J. (1996) Role of phospholipase C and phospholi-pase A2 in the nitric oxide-independent coronary vasodilator effect of bradykinin in the rat perfused heart. *J. Pharmacol. Exp. Ther.* **286**: 1146–1151.

213. Fukao, M., Hattori, Y., Kanno, M., Sakuma, I., Kitabatake, A. (1997) Sources of Ca^{2+} in relation to generation of acetylcholine-induced endothelium-dependent hyperpo-larization in rat mesenteric artery. *Br. J. Pharmacol.* **120**: 439–446.

214. Hutcheson, L.R., Chaytor, A.T., Evans, W.H., Griffith, T.M. (1999) NO-independent relaxations to acetylcholine and A23187 involve different routes of heterocellular communications: role of gap junction and phospholipase A2. *Circ. Res.* **84**: 53–63.

215. You, J., Marrelli, S.P., Bryan, R.M., Jr., (2002) Role of cytoplasmic phospholipase A2 in endothelium-derived hyperpolarizing factor dilations of rat middle cerebral arteries. *J. Cereb. Blood Flow Metab.* **22**: 1239–1247.

216. Cheung, D.W., Chen, G. (1992) Calcium activation of hyperpolarization response to acetylcholine in coronary endothelial cells. *J. Cardiovasc. Pharmacol.* **12**: S120–S123.

217. von der Weid, P.Y., Beny, J-L. (1992) Effect of Ca^{2+} ionophores on membrane potential of pig coronary endothelial cells. *Am. J. Physiol.* **262**: H1823–H1831.

218. Pasyk, E., Inazu, M., Daniel, E.E. (1995) CPA enhances Ca^{2+} entry in cultured bovine pulmonary arterial endothelial cells in an IP3-independent maner. *Am. J. Physiol.* **268**: H138–H146.

219. Vaca, L., Licea, A., Possani, L.D. (1996) Modulation of cell membrane potential in cultured vascular endothelium. *Am. J. Physiol.* **270**: C819–C824.

220. Wang, X., Chu, W., van Breemen, C. (1996) Potentiation of acetylcholine-induced responses in freshly isolated rabbit aortic endothelial cells. *J. Vasc. Res.* **33**: 414–424.

221. Ohashi, M., Satoh, K., Itoh, T. (1999) Acetylcholine-induced membrane potential changes in endothelial cells of rabbit aortic valve. *Br. J. Pharmacol.* **126**: 19–26.

222. Fukao, M., Hattori, Y., Kanno, M., Sakuma, I., Kitabatake, A. (1995) Thapsigargin- and cyclopiazonic acid-induced endothelium-dependent hyperpolarization in rat mesenteric artery. *Br. J. Pharmacol.* **115**: 987–992.

223. Brayden, J.E. (1990) Membrane hyperpolarization is a mechanism of endothelium-dependent cerebral vasodilation. *Am. J. Physiol.* **259**: H668–H673.

224. Lacza, Z., Puskar, M., Kis, B., Perciaccante, J.V., Miller, A.W., Busija, D.W. (2002) Hydrogen peroxide acts as an EDHF in the piglet pial vasculature in response to bradykinin. *Am. J. Physiol.* **283**: H406–H411.

225. Chen, G., Suzuki, H. (1991) Endothelium-dependent hyperpolarization elicited by adenine compounds in rabbit carotid artery. *Am. J. Physiol.* **260**: H1037–H1042.

226. Van de Voorde, J., Vanheel, B., Leusen, I. (1992) Endothelium-dependent relaxation and hyperpolarisation in aorta from control and renal hypertensive rats. *Circ. Res.* **70**: 1–8.

227. Zygmunt, P.M., Högestätt, E.D. (1996) Endothelium-dependent hyperpolarization and relaxation in the hepatic artery of the rat. In *Endothelium-Derived Hyperpolarizing Factor*, Vol. 1, P.M. Vanhoutte, ed., Harwood Academic Publishers, Amsterdam, pp. 191–202.

228. Zygmunt, P.M., Edwards, G., Weston, A.H., Larsson, B., Högestätt, E.D. (1997) Involvement of voltage-dependent potassium channels in the EDHF-mediated relaxation of rat hepatic artery. *Br. J. Pharmacol.* **121**: 141–149.

229. Petersson, J., Zygmunt, P.M., Högestätt, E.D. (1997) Characterization of the potassium channels involved in EDHF-mediated relaxation in cerebral arteries. *Br. J. Pharmacol.* **120**: 1344–1350.

230. Chataigneau, T., Félétou, M., Duhault, J., Vanhoutte, P.M. (1998) Epoxyeicosatrienoic acids, potassium channel blockers and endothelium-dependent hyperpolarisation in the guinea-pig carotid artery. *Br. J. Pharmacol.* **123**: 574–580.

231. Yamanaka, A., Ishikawa, K., Goto, K. (1998) Characterization of endothelium-dependent relaxation independent of NO and prostaglandins in guinea pig coronary artery. *J. Pharmacol. Exp. Ther.* **285**: 480–489.

232. Andersson, D.A., Zygmunt, P.M., Movahed, P., Andersson, T.L., Hogestatt, E.D. (2000) Effects of inhibitors of calcium-activated potassium channels, inwardly-rectifying potassium channels and Na(+)/K(+)-ATPase on EDHF relaxation in the rat hepatic artery. *Br. J. Pharmacol.* **129**: 1490–1496.

233. Ding, H., Jiang, Y., Triggle, C.R. (2003) The contribution of D-tubocurarine and apamin-sensitive potassium channels to endothelium-derived hyperpolarizing factor-mediated relaxation of small arteries from e-NOS-/- mice. In *EDHF 2002*, P.M. Vanhoutte, ed., Taylor and Francis, London, pp. 283–296.

234. Zygmunt, P.M., Högestätt, E.D., Waldeck, K., Edwards, G., Kirkup, A.J., Weston, A.H. (1997) Studies on the effects of anandamide in rat hepatic artery. *Br. J. Pharmacol.* **122**: 1679–1686.

235. Jensen, B.S., Strobaek, D., Christophersen, P., Jorgensen, T.D., Hansen, C., Silahtaroglu, A., Olesen, S.P., Ahring, P.K. (1998) Characterization of the cloned human intermediate-conductance Ca^{2+}-activated K^+ channel. *Am. J. Physiol.* **275**: C848–C856.

236. Edwards, G., Gardener, M.J., Félétou, M., Brady, G., Vanhoutte, P.M., Weston, A.H. (1999) Further investigation of endothelium-derived hyperpolarizing factor (EDHF) in rat hepatic artery: studies using 1-EBIO and ouabain. *Br. J. Pharmacol.* **128**: 1064–1070.

237. Pedersen, K.A., Schroder, R.L., Skaaning-Jensen, B., Strobaek, D., Olesen, S.P., Christophersen, P. (1999) Activation of the human intermediate-conductance Ca(2+)-activated K(+) channel by 1-ethyl-2-benzimidazolinone is strongly Ca(2+)-dependent. *Biochim. Biophys. Acta.* **1420**: 231–240.

238. Syme, C.A., Gerlach, A.C., Singh, A.K., Devor, D.C. (2000) Pharmacological activation of cloned intermediate and small conductance Ca(2+)-activated K(+) channels. *Am. J. Physiol.* **278**: C570–C581.

239. Cao, Y., Dreixler, J.C., Roizen, J.D., Roberts, M.T., Houamed, K.M. (2001) Modulation of recombinant small-conductance Ca(2+)-activated K(+) channels by the muscle relaxant chlorzoxazone and structurally related compounds. *J. Pharmacol. Exp. Ther.* **296**: 683–689.

240. Coleman, H.A., Tare, M., Parkington, H.C. (2001) EDHF is not K⁺ but may be due to spread of current from the endothelium in guinea pig arterioles. *Am. J. Physiol.* **280**: H2478–H2483.

241. Pedarzani, P., Mosbacher, J., Rivard, A., Cingolani, L.A., Oliver, D., Stocker, M., Adelman, J.P., Fakler, B. (2001) Control of electrical activity in central neurons by modulating the gating of small conductance Ca^{2+}-activated K⁺ channels. *J. Biol. Chem.* **276**: 9762–9769.

242. Walker, S.D., Dora, K.A., Ings, N.T., Crane, G.J., Garland, C.J. (2001) Activation of endothelial IK(CA) with 1-ethyl-2-benzimidazolinone evokes smooth muscle hyperpolarization in rat isolated mesenteric artery. *Br. J. Pharmacol.* **134**: 1548–1554.

243. Bychkov, R., Burnham, M.P., Richards, G.R., Edwards, G., Weston, A.H., Félétou, M., Vanhoutte, P.M. (2002) Characterization of a charybdotoxin-sensitive intermediate conductance Ca^{2+}-activated K⁺ channel in porcine coronary endothelium: relevance to EDHF. *Br. J. Pharmacol.* **138**: 1346–1354.

244. Edwards, G., Félétou, M., Gardener, M.J., Thollon, C., Vanhoutte, P.M., Weston, A.H. (1999) Role of gap junctions in the responses to EDHF in rat and guinea-pig small arteries. *Br. J. Pharmacol.* **128**: 1788–1794.

245. Crane, G.J., Gallagher, N., Dora, K.A., Garland, C.J. (2003) Small- and intermediate-conductance calcium-activated K⁺ channels provide different facets of endothelium-dependent hyperpolarization in rat mesenteric artery. *J. Physiol.* **553**: 183–189.

246. Brakemier, S., Kersten, A., Eichler, I., Grgic, I., Zakrzewicz, A., Hopp, H., Kohler, R., Hoyer, J. (2003) Shear stress-induced up-regulation of the intermediate-conductance Ca(2+) activated K(+) channel in human endothelium. *Cardiovasc. Res.* **60**: 488–496.

247. Marrelli, S.P., Eckmann, M.S., Hunte, M.S. (2003) Role of endothelial intermediate conductance K_{Ca} channels in cerebral EDHF-mediated dilations. *Am. J. Physiol.* **285**: H1590–H1599.

248. Edwards, G., Thollon, C., Gardener, M.J., Félétou, M., Vilaine, J-P., Vanhoutte, P.M, Weston, A.H. (2000) Role of gap junctions and EETs in endothelium-dependent hyperpolarization of porcine coronary artery. *Br. J. Pharmacol.* **129**: 1145–1162.

249. Wulff, H., Miller, M.J., Haensel, W., Grissner, S., Cahalan, M.D., Chandy, K.G. (2000) Design of potent and selective inhibitor of the intermediate-conductance Ca^{2+}-activated K⁺ channel, IKCa1: a potential immunosuppressant. *Proc. Natl. Acad. Sci. USA* **97**: 8151–8156.

250. Eichler, I., Wibawa, J., Grgic, I., Knorr, A., Brakemier, S., Pries, A.R., Hoyer, J., Kohler, R. (2003) Selective blockade of endothelial Ca^{2+}-activated small- and intermediate-conductance K^+ channels suppresses EDHF-mediated vasodilation. *Br. J. Pharmacol.* **138**: 594–601.

251. Hinton, J.M., Langton, P.D. (2003) Inhibition of EDHF by two new combinations of K^+-channel inhibitors in rat isolated mesenteric arteries. *Br. J. Pharmacol.* **138**: 1031–1035.

252. Ding, H., Triggle, C.R. (2000) Novel endothelium-derived relaxing factors. Identification of factors and cellular targets. *J. Pharmacol. Toxicol. Methods* **44**: 441–452.

253. Ishii, T.M., Silvia, C., Hirschberg, B., Bond, C.T., Adelman, J.P., Maylie, J. (1997) A human intermediate conductance calcium-activated potassium channel. *Proc. Natl. Acad. Sci. USA* **94**: 11651–11656.

254. Monaghan, A.S., Benton, D.C., Bahia, P.K., Hosseini, R., Shah, Y.A., Haylett, D.G., Moss, G.W. (2004) The SK3 subunit of small conductance Ca^{2+}-activated K^+ channels interacts with both SK1 and SK2 subunits in a heterologous expression system. *J. Biol. Chem.* **279**: 1003–1009.

255. Castle, N.A. (1999) Recent advances in the biology of small conductance calcium-activated potassium channels. *Perspect. Drug Discov. Design* **15/16**: 131–154.

256. Gebremedhin, D., Kaldunski, M., Jacobs, E.R., Harder, D.R., Roman, R.J. (1996) Coexistence of two types of calcium activated potassium channels in rat renal arterioles. *Am. J. Physiol.* **270**: F69–F81.

257. Latorre, R., Oberhauser, A., Labarca, P., Alvarez, O. (1989) Varieties of calcium-activated potassium channels. *Annu. Rev. Physiol.* **51**: 385–399.

258. Neylon, C.B., Lang, R.J., Fu, Y., Bobik, A., Reinhart, P.H. (1999) Molecular cloning and characterization of the intermediate-conductance $Ca(2+)$-activated $K(+)$ channel in vascular smooth muscle: relationship between K(Ca) channel diversity and smooth muscle function. *Circ. Res.* **85**: 33–43.

259. Kohler, R., Wulff, H., Eichler, I., Kneifel, M., Neumann, D., Knorr, A., Grgi, I., Kampfe, D., Si, H., Wibawa, J., Real, R., Borner, K., Brakemeier, S., Orzechowski, H.D., Reusch, H.P., Paul, M., Chandy, K.G., Hoyer, J. (2003) Blockade of intermediate-conductance calcium-activated potassium channel as a new therapeutic strategy for restenosis. *Circulation* **108**: 1119–1125.

260. Marchenko, S.M., Sage, S.O. (1996) Calcium-activated potassium channels in the endothelium of intact aorta. *J. Physiol.* **492**: 53–60.

261. Frieden, M., Sollini, M., Beny, J-L. (1999) Substance P and bradykinin activate different types of Kca currents to hyperpolarize cultured porcine coronary artery endothelial cells. *J. Physiol.* **519**: 361–371.

262. Köhler, R., Degenhardt, C., Kühn, M., Runkel, N., Paul, M., Hoyer, J. (2000) Expression and function of endothelial Ca^{2+}-activated K^+ channels in human mesenteric artery—a single cell reverse transcriptase-polymerase chain reaction and electrophysiological study in situ. *Circ. Res.* **87**: 496–503.

263. Sollini, M., Frieden, M., Bény, J-L. (2002) Charybdotoxin-sensitive small conductance KCa channel activated by bradykinin and substance P in endothelial cells. *Br. J. Pharmacol.* **136**: 1201–1209.

264. Burnham, M.P., Bychkov, R., Félétou, M., Richards, G.R., Vanhoutte, P.M., Weston, A.H., Edwards, G. (2002) Characterization of an apamin-sensitive small conductance Ca^{2+}-activated K^+ channel in porcine coronary artery endothelium: relevance to EDHF. *Br. J. Pharmacol.* **135**: 1133–1143.

265. Bychkov, R., Burnham, M.P., Richards, G.R., Thollon, C., Edwards, G., Weston, A.H., Vanhoutte, P.M., Félétou, M. (2003) Small and intermediate conductance Ca^{2+}-activated K^+channels (SK_{Ca} and IK_{Ca}) in porcine coronary endothelium: relevance to EDHF. In *EDHF 2002,* P.M. Vanhoutte, ed., Taylor and Francis., London, pp. 261–273.

266. Doughty, J.M., Plane, F., Langton, P.D. (1999) Charybdotoxin and apamin block EDHF in rat mesenteric artery if selectively applied to the endothelium. *Am. J. Physiol.* **276**: H1107–H1112.

267. Edwards, G., Dora, K.A., Gardener, M.J., Garland, C.J., Weston, A.H. (1998) K^+ is an endothelium-derived hyperpolarizing factor in rat arteries. *Nature* **396**: 269–272.

268. Busse, R., Fichtner, H., Luckhoff, A., Kohlhardt, M. (1988) Hyperpolarisation and increased free calcium in acetylcholine-stimulated endothelial cells. *Am. J. Physiol.* **255**: H965–H969.

269. Luckhoff, A., Busse, R. (1990) Calcium influx into endothelial cells and formation of endothelium-derived relaxing factor is controlled by the membrane potential. *Pflügers Arch.* **416**: 305–311.

270. Kamouchi, M., Droogmans, G., Nilius, B. (1999) Membrane potential as a modulator of the free intracellular Ca^{2+} concentration in agonist-activated endothelial cells. *Gen. Physiol. Biophys.* **18**: 199–208.

271. Marchenko, S.M., Sage, S.O. (1993) Electrical properties of resting and acetylcholine-stimulated endothelium in intact rat aorta. *J. Physiol.* **462**: 735–751.

272. Sedova, M., Klishin, A., Huser, J., Blater, L.A. (2000) Capacitive Ca^{2+} entry is graded with degree of intracellular Ca^{2+} store depletion in bovine vascular cells. *J. Physiol.* **523**: 549–559.

273. Nilius, B., Droogmans, G. (2001) Ion channels and their functional role in vascular endothelium. *Physiol. Rev.* **81**: 1416–1459.

274. Ghisdal, P., Morel, N. (2001) Cellular targets of voltage and calcium-dependent K^+ channel blockers involved in EDHF-mediated responses in rat superior mesenteric artery. *Br. J. Pharmacol.* **134**: 1021–1028.

275. Marrelli, S.P. (2001) The primary role of endothelial Ca^{2+} in the EDHF response of cerebral arteries is to stimulate endothelial K_{Ca} channels. *J. Vasc. Res.* **38**: 17.

276. Xia, X.M., Fakler, B., Rivard, A., Wayman, G., Johnson-Pais, T., Keen, J.E., Ishii, T., Hirschberg, B., Bond, C.T., Lutsenko, S., Maylie, J., Adelman, J.P. (1998) Mechanism of calcium-gating in small-conductance calcium-activated potassium channels. *Nature* **395**: 503–507.

277. Fanger, C.M., Ghanshani, S., Logsdon, N.J., Rauer, H., Kalman, K., Zhou, J., Beckingham, K., Chandy, K.G., Cahalan, M.D., Aiyar, J. (1999) Calmodulin mediates calcium-dependent activation of the intermediate conductance KCa channel, IKCa1. *J. Biol. Chem.* **274**: 5746–5754.

278. Schumacher, M.A., Rivard, A.F., Bachinger, H.P., Adelman, J.P. (2001) Structure of the gating domain of a Ca^{2+}-activated K^+ channel complexed with Ca^{2+}/calmodulin. *Nature* **410**: 1120–1124.

279. Takahata, T., Hayashi, M., Ishikawa, T. (2003) SK4/IK1-like channels mediate TEA-insensitive, Ca^{2+}-activated K^+ currents in bovine parotid acinar cells. *Am. J. Physiol.* **284**: C127–C144.

280. Kong, I.D., Koh, S.D., Bayguinov, O., Sanders, K.M. (2000) Small conductance Ca^{2+}-activated K^+ channels are regulated by Ca^{2+}-calmodulin-dependent protein kinase II in murine colonic myocytes. *J. Physiol.* **524**: 331–337.

281. Nagao, T., Illiano, S., Vanhoutte, P.M. (1992) Calmodulin antagonists inhibit endothelium-dependent hyperpolarization in the canine coronary artery. *Br. J. Pharmacol.* **107**: 382–386.

282. Wood, P.G., Gillespie, J.I. (1998) In permeabilised endothelial cells IP3-induced Ca^{2+} release is dependent on the cytoplasmic concentration of monovalent cations. *Cardiovasc. Res.* **37**: 263–270.

283. Félétou, M., Vanhoutte, P.M. (2000) Endothelium-dependent hyperpolarization of vascular smooth muscle cells. *Acta Pharmacol. Sin.* **21**: 1–18.

284. McGuire, J.J., Ding, H., Triggle, C.R. (2001) Endothelium-derived relaxing factors: a focus on endothelium-derived hyperpolarizing factor(s). *Can. J. Physiol. Pharmacol.* **79**: 443–470.

285. Christ, G.J., Spray, D.C., el Sabban, M., Moore, L.K., Brink, P.R. (1996) Gap junctions in vascular tissues. Evaluating the role of intercellular communication in the modulation of vasomotor tone. *Circ. Res.* **79**: 631–646.

286. Brink, P.R., Beyer, E.C., Christ, G.J. (2001) What do gap junctions do anyway? In *EDHF 2000*, P.M. Vanhoutte, ed., Taylor and Francis, New York, pp. 1–13.

287. Griffith, T.M. (2004) Endothelium-dependent smooth muscle hyperpolarization: do gap junctions provide a unifying hypothesis? *Br. J. Pharmacol.* **141**: 881–903.

288. Sandow, S.L., Hill, C.E. (2000) Incidence of myo-endothelial gap junctions in the proximal and distal mesenteric arteries of the rat is suggestive of a role in endothelium-derived hyperpolarizing factor-mediated responses. *Circ. Res.* **86**: 341–346.

289. Beny, J-L., Brunet, P.C., Huggel, H. (1986) Effect of mechanical stimulation, substance P and vasoactive intestinal polypeptide on the electrical and mechanical activity of the circular smooth muscles from pig coronary arteries contracted with acetylcholine: role of the endothelium. *Pharmacology* **33**: 61–68.

290. Beny, J-L., Brunet, P.C., Huggel, H. (1987) Interaction of bradykinin and Des-Arg9-Bradykinin with isolated pig coronary arteries: mechanical and electrophysiological events. *Regul. Pept.* **17**: 181–190.

291. Chen, G., Yamamoto, Y., Miwa, K., Suzuki, H. (1991) Hyperpolarisation of arterial smooth muscle induced by endothelial humoral substances. *Am. J. Physiol.* **260**: H1888–H1892.

292. Popp, R., Bauersachs, J., Sauer, E., Hecker, M., Fleming, I., Busse, R. (1996) A transferable, β-naphtoflavone-inducible, hyperpolarizing factor is synthesized by native and cultured porcine coronary endothelial cells. *J. Physiol (Lond.)* **497**: 699–709.

293. Gebremedhin, D., Harder, D.R., Pratt, P.F., Campbell, W.B. (1998) Bioassay of an endothelium-derived hyperpolarizing factor from bovine coronary arteries: role of a cytochrome P450 metabolites. *J. Vasc. Res.* **35**: 274–284.

294. Mombouli, J-V., Bissiriou, I., Vanhoutte, P.M. (1996) Bioassay of endothelium-derived hyperpolarizing factor: is endothelium-derived depolarizing factor a confounding element? In *Endothelium-Derived Hyperpolarizing Factor*, Vol. 1, P.M. Vanhoutte, ed., Harwood Academic Publishers, Amsterdam, pp. 51–57.

295. Liu, Z.G., Ge, Z.D., He, G.W. (2000) Difference in endothelium-derived hyperpolarizing factor-mediated hyperpolarization and nitric oxide release between human internal mammary artery and saphenous vein. *Circulation* **102**: III296–III301.

296. Kagota, S., Yamaguchi, Y., Nakamura, K., Kunitomo, M. (1999) Characterization of nitric oxide and prostaglandin-independent relaxation in response to acetylcholine in rabbit renal artery. *Clin. Exp. Pharmacol. Physiol.* **26**: 790–796.

297. Chaytor, A.Y., Evans, W.H., Griffith, T.M. (1998) Central role of heterocellular gap junction communication in endothelium-dependent relaxations of rabbit arteries. *J. Physiol. (Lond.)*, **508**: 561–73.

298. Chaytor, A.Y., Taylor, H.J., Griffith, T.M. (2002) Gap-junction-dependent and -independent EDHF-type relaxations may involve smooth muscle cAMP accumulation. *Am. J. Physiol.* **282**: H1548–H1555.

299. Daut, J., Mehrke, G., Nees, S., Newma, W.H. (1988) Passive electrical properties and electrogenic sodium pump transport in cultured guinea-pig coronary endothelial cells. *J. Physiol.* **402**: 237–254.

300. Davies, P.F., Oleson, S.P., Clapham, D.E., Morel, E.M., Schoen, F.J. (1988) Endothelial communication: state of the art lecture. *Hypertension* **11**: 563–572.

301. Daut, J., Standen, N.B., Nelson, M.T. (1994) The role of the membrane potential of endothelial and smooth muscle cells in the regulation of coronary blood flow. *J. Cardiovasc. Electrophysiol.* **5**: 154–181.

302. Beny, J-L., Pacicca, C. (1994) Bidirectional electrical communication between smooth muscle and endothelial cells in the pig coronary artery. *Am. J. Physiol.* **266**: H1465–H1472.

303. Kühlberger, E., Groschner, K., Kukovetz, W.R., Brunner, F. (1994) The role of myoendothelial cell contact in non-nitric oxide-, non-prostanoid-mediated endothelium-dependent relaxation of porcine coronary artery. *Br. J. Pharmacol.* **113**: 1289–1294.

304. Beny, J-L. (1990) Endothelial and smooth muscle cells hyperpolarized by bradykinin are not dye coupled. *Am. J. Physiol.* **258**: H836–H841.

305. Kristek, F., Gerova, M. (1992) Myoendothelial relations in the conduit coronary artery of the dog and rabbit. *J. Vasc. Res.* **29**: 29–32.

306. Hill, C.E., Rummery, N., Hickey, H., Sandow, S.L. (2002) Heterogeneity in the distribution of vascular gap junctions and connexins: implications for function. *Clin. Exp. Pharmacol. Physiol.* **29**: 620–625.

307. Hwa, J.J., Ghibaudi, L., Williams, P., Chaterjee, M. (1994) Comparison of acetylcholine-dependent relaxation in large and small arteries of rat mesenteric vascular bed. *Am. J. Physiol.* **266**: H952–H958.

308. Shimokawa, H., Yasutake, H., Fujii, K., Owada, M.K., Nakaike, R., Fukumoto, Y., Takayanagani, T., Nagao, T., Egashira, K., Fujishima, M., Takeshita, A. (1996) The importance of the hyperpolarizing mechanism increases as the vessel size decreases in endothelium-dependent relaxations in rat mesenteric circulation. *J. Cardiovasc. Pharmacol.* **28**: 703–711.

309. Berman, R.S., Griffith, T.M. (1997) Differential actions of charybdotoxin on central and daughter arteries of the rabbit isolated ear. *Br. J. Pharmacol.* **120**: 639–646.

310. Berman, R.S., Griffith, T.M. (1998) Spatial heterogeneity in the mechanisms contributing to acetylcholine-induced dilatation in the rabbit isolated ear. *Br. J. Pharmacol.* **124**: 1245–1253.

311. Tomioka, H., Hattori, Y., Fukao, M., Sato, A., Lui, M., Sakuma, I., Kitabatake, A., Kanno, M. (1999) Relaxation in different-sized blood vessels mediated by endothelium-derived hyperpolarizing factor: importance of processes mediating precontractions. *J. Vasc. Res.* **36**: 311–320.

312. Miura, H., Liu, Y., Gutterman, D.D. (1999) Human coronary artery arteriolar dilation to bradykinin depends on membrane hyperpolarization: role of nitric oxide and Ca^{2+}-activated K^+ channels. *Circulation* **99**: 3132–3138.

313. Sandow, S.L., Tare, M., Coleman, H.A., Hill, C.E., Parkington, H.C. (2002) Involvement of myoendothelial gap junctions in the action of endothelium-derived hyperpolarizing factor. *Circ. Res.* **90**: 1108–1113.

314. Sandow, S.L., Goto, K., Rummery, N.M., Hill, C.E. (2004) Developmental changes in myoendothelial gap junction mediated vasodilator activity in the rat saphenous artery. *J. Physiol.* **556**: 875–886.

315. Dora, K.A., Sandow, S.L., Gallagher, N.T., Takano, H., Rummery, N.M., Hill, C.E., Garland, C.J. (2003) Myoendothelial gap junctions may provide the pathway for EDHF in mouse mesenteric artery. *J. Vasc. Res.* **40**: 480–490.

316. Sandow, S.L., Branich, N.J., Bandy, H.P., Rummery, N.M., Hill, C.E. (2003) Structure, function, and endothelium-derived hyperpolarizing factor in the caudal artery of the SHR and WKY rat. *Arterioscler. Thromb. Vasc. Biol.* **23**: 822–828.

317. Dora, K.A., Doyle, M.P., Duling, B.R. (1997) Elevation of intracellular calcium in smooth muscle causes endothelial cell generation of NO in arterioles. *Proc. Natl. Acad. Sci. USA* **94**: 6529–6534.

318. Yashiro, Y., Duling, B.R. (2000) Integrated Ca(2+) signaling between smooth muscle and endothelium of resistance vessels. *Circ. Res.* **87**: 1048–1054.

319. Marchenko, S.M., Sage, S.O. (1994) Smooth muscle cells affect endothelial membrane potential in rat aorta. *Am. J. Physiol.* **267**: H804–H811.

320. Beach, J.M., McGahren, E.D., Duling, B.R. (1998) Capillaries and arterioles are electrically coupled in hamster cheek pouch. *Am. J. Physiol.* **275**: H1489–H1496.

321. Yamamoto, Y., Fukuta, H., Nakahira, Y., Suzuki, H. (1998) Blockade by 18β-glycyrrhetinic acid of intercellular electrical coupling in guinea-pig arterioles. *J. Physiol.* **511**: 501–518.

322. Yamamoto, Y., Imaeda, K., Suzuki, H. (1999) Endothelium-dependent hyperpolarization and intercellular electrical coupling in guinea-pig mesenteric arterioles. *J. Physiol.* **514**: 505–513.

323. Emerson, G.G., Segal, S.S. (2000) Electrical coupling between endothelial cells and smooth muscle cells in hamster feed arteries: role in vasomotor control. *Circ. Res.* **87**: 474–479.

324. Tare, M., Coleman, H.A., Parkington, H.C. (2002) Glycyrrhetinic acid derivatives inhibit hyperpolarization in endothelial cells of guinea pig and rat arteries. *Am. J. Physiol.* **282**: H335–H341.

325. Baker, M.F., Fanestil, D.D. (1991) Licorice, computer-based analyses of dehydrogenase sequences, and the regulation of steroid and prostaglandin action. *Mol. Cell. Endocrinol.* **78**: C99–C102.

326. Davidson, J.S., Baumgarten, I.M., Harley, E.H. (1986) Reversible inhibition of intercellular junctional communication by glycyrrhetinic acid. *Biochem. Biophys. Res. Comm.* **134**: 29–36.

327. Davidson, J.S., Baumgarten, I.M. (1988) Glycyrrhetinic acid derivatives: a novel class of inhibitors of gap junctional intercellular communication. Structure activity relationship. *J. Pharmacol. Exp. Ther.* **246**: 1104–1107.

328. Taylor, H.J., Chaytor, A.T., Evans, W.H., Griffith, T.M. (1998) Inhibition of the gap junctional component of endothelium-dependent relaxations in rabbit iliac artery by 18β-glycyrrhetinic acid. *Br. J. Pharmacol.* **125**: 1–3.

329. Chaytor, A.Y., Marsh, W.L., Hutcheson, I.R., Griffith, T.M. (2000) Comparison of glycyrrhetinic acid isoforms and carbenoxolone as inhibitors of EDHF-type relaxations mediated by gap junctions. *Endothelium* **7**: 265–278.

330. Kenny, L.C., Baker, P.N., Kendall, D.A., Randall, M.D., Dunn, W.R. (2002) The role of gap junctions in mediating endothelium-dependent responses to bradykinin in myometrial small arteries isolated from pregnant women. *Br. J. Pharmacol.* **136**: 1085–1088.

331. Goto, K., Fujii, K., Kansui, Y., Abe, I., Iida, M., (2002) Critical role of gap-junctions in endothelium-dependent hyperpolarization in rat mesenteric arteries. *Clin. Exp. Pharmacol. Physiol.* **29**: 592–602.

332. Sandow, S.L., Looft-Wilson, R., Doran, B., Grayson, T.H., Segal, S.S., Hill, C.E. (2003) Expression of homocellular and heterocellular gap junctions in hamster arterioles and feed arteries. *Cardiovasc. Res.* **60**: 643–653.

333. Luksha, L., Nisell, H., Kublickiene, K. (2004) The mechanism of EDHF-mediated responses in subcutaneous small arteries of healthy pregnant women. *Am. J. Physiol.* **286**: R1102–R1109.

334. Kilarski, W.M., Fu, X., Roomans, G.M., Backstrom, T., Ulmsten, U. (1995) In vitro effects of Hepes buffer on maintaining the number of gap junction plaques in human myometrium at term. *Folia Histochem. Cytobiol.* **33**: 151–155.

335. Bevans, C.G., Harris, A.L. (1999) Regulation of connexin channels by pH. Direct action of the protonated form of taurine and other aminosulfonates. *J. Biol. Chem.* **274**: 3711–3719.

336. Edwards, G., Félétou, M., Gardener, M.J., Glen, C.D., Richards, G.R., Vanhoutte, P.M., Weston, A.H. (2001) Further investigations into the endothelium-dependent hyperpolarizing effects of bradykinin and substance P in porcine coronary artery. *Br. J. Pharmacol.* **133**: 1145–1153.

337. Berman, R.S., Martin, P.E.M., Evans, W.H., Griffith, T.M. (2002) Relative contributions of NO and gap junctional communication to endothelium-dependent relaxations of rabbit resistance arteries vary with vessel size. *Microvasc. Res.* **63**: 115–128.

338. Chaytor, A.Y., Martin, P.E., Edwards, D.H., Griffith, T.M. (2001) Gap junctional communication underpins EDHF-type relaxations evoked by ACh in the rat hepatic artery. *Am. J. Physiol.* **280**: H2441–H2450.

339. Dora, K.A., Martin, P.E., Chaytor, A.T., Evans, W.H., Garland, C.J., Griffith, T.M. (1999) Role of heterocellular gap junctional communication in endothelium-dependent smooth muscle hyperpolarization: inhibition by a connexin-mimetic peptide. *Biochem. Biophys. Res. Commun.* **254**: 27–31.

340. Doughty, J.M., Boyle, J.P., Langton, P.D. (2000) Potassium does not mimic EDHF in rat mesenteric arteries. *Br. J. Pharmacol.* **130**: 1174–1182.

341. Ujiie, H., Chaytor, A.T., Bakker, L.M., Griffith, T.M. (2003) Essential role for gap junctions in NO- and prostanoid-independent relaxations evoked by acetylcholine in rabbit intracerebral arteries. *Stroke* **34**: 544–550.

342. De Vriese, A.S., Van de Voorde, J., Lameire, N.H. (2002) Effects of connexin-mimetic peptides on nitric oxide synthase- and cyclooxygenase-independent renal vasodilatation. *Kidney Int.* **61**: 177–185.

343. Coleman, H.A., Tare, M., Parkington, H.C. (2001) K^+ currents underlying the action of endothelium-derived hyperpolarizing factor in guinea-pig, rat and human blood vessels. *J. Physiol.* **531**: 359–373.

344. Emerson, G.G., Segal, S.S. (2001) Electrical activation of endothelium evokes vasodilatation and hyperpolarization along hamster feed arteries. *Am. J. Physiol.* **280**: H160–H167.

345. Emerson, G.G., Segal, S.S. (2000) Endothelial cell pathway for conduction of hyperpolarization and vasodilation along hamster feed artery. *Circ. Res.* **86**: 94–100.

346. Gustafsson, F., Mikkelsen, H.B., Arensbak, B., Thuneberg, L., Neve, S., Jensen, L.J., Holstein-Rathlou, N.H. (2003) Expression of connexin 37, 40 and 43 in rat mesenteric arterioles and resistance arteries. *Histochem. Cell Biol.* **119**: 139–148.

347. Martin, P.E., Wall, C., Griffith, T.M. (2005) Effects of connexin-mimetic peptides on gap junction functionality and connexin expression in cultured vascular cells. *Br. J. Pharmacol.* **144**: 617–627.

348. Li, X., Simard, J.M. (1999) Multiple connexins form gap junction channel in rat basilar artery smooth muscle cells. *Circ. Res.* **84**: 1277–1284.

349. Welsh, D.G., Segal, S.S. (2000) Role of EDHF in conduction of vasodilation along hamster cheek pouch arterioles in vivo. *Am. J. Physiol.* **278**: 1832–1839.

350. Bartlett, I.S., Segal, S.S. (2000) Resolution of smooth muscle and endothelial pathway for conduction along hamster cheek pouch arterioles. *Am. J. Physiol.* **278**: H604–H612.

351. Budel, S., Bartlett, I.S., Segal, S.S. (2003) Homocellular conduction along endothelium and smooth muscle of arterioles in hamster cheek pouch. Unmasking an NO wave. *Circ. Res.* **93**: 61–68.

352. Grasby, D.J., Morris, J.L., Segal, S.S. (1999) Heterogeneity of vascular innervation in the hamster cheek pouch and retractor muscle. *J. Vasc. Res.* **36**: 465–476.

353. Looft-Wilson, R.C., Haug, S.J., Neufer, P.D., Segal, S.S. (2004) Independence of connexin expression and vasomotor conduction from sympathetic innervation in hamster feed arteries. *Microcirculation* **11**: 397–408.

354. Selemidis, S., Cocks, T.M. (2001) Myoendothelial and circumferential spread of endothelium-dependent hyperpolarization in coronary arteries. In *EDHF 2000*, P.M. Vanhoutte, ed., Taylor and Francis, London, pp. 75–86.

355. Selemidis, S., Cocks, T.M. (2002) Endothelium-dependent hyperpolarization as a remote anti-atherogenic mechanism. *Trends Pharmacol. Sci.* **23**: 213–220.

356. Chaytor, A.T., Bakker, L.M., Edwards, D.H., Griffith, T.M. (2005) Connexin-mimetic peptides dissociate electrotonic EDHF-type signaling via myo-endothelial and smooth muscle gap junctions in the rabbit iliac artery. *Br. J. Pharmacol.* **144**: 108–114.

357. Griffith, T.M., Chaytor, A.T., Taylor, H.J., Giddings, B.D., Edwards, D.H. (2002) cAMP facilitates EDHF-type relaxations in conduit arteries by enhancing electrotonic conduction via gap junctions. *Proc. Natl. Acad. Sci. USA* **99**: 6392–6397.

358. Taylor, H.J., Chaytor, A.T., Edwards, D.H., Griffith, T.M. (2001) Gap-junction-dependent increases in smooth muscle cAMP underpin the EDHF phenomenon in rabbit arteries. *Biochem. Biophys. Res. Comm.* **283**: 583–589.

359. Matsumoto, T., Wakabayashi, K., Kobayashi, T., Kamata, K. (2004) Diabetes-related changes in cAMP-dependent protein kinase activity and decrease in relaxation response in rat mesenteric artery. *Am. J. Physiol.* **287**: H64–H71.

360. Griffith, T.M., Chaytor, A.T., Edwards, D.H. (2004) The obligatory link: role of gap junctional communication in endothelium-dependent smooth muscle hyperpolarization. *Pharmacol. Res.* **49**: 551–564.

361. Simon, A.M., McWhorter, A.R. (2003) Decreased intercellular dye-transfer and downregulation of non-ablated connexins in aortic endothelium deficient in connexin37 or connexin40. *J. Cell Sci.* **116**: 2223–2236.

362. Simon, A.M., McWhorter, A.R. (2002) Vascular abnormalities in mice lacking the endothelial gap junction protein connexin37 and connexin40. *Dev. Biol.* **251**: 206–220.

363. De Wit, C., Roos, F., Bolz, S.S., Kirchhoff, S., Kruger, O., Willecke, K., Pohl, U. (2000) Impaired conduction of vasodilatation along arterioles in connexin 40-deficient mice. *Circ. Res.* **86**: 649–655.

364. De Wit, C., Roos, F., Bolz, S.S., Pohl, U. (2003) Lack of vascular connexin 40 is associated with hypertension and irregular arteriolar vasomotion. *Physiol. Genomics.* **13**: 169–177.

365. Figueroa, X.F., Paul, D.L., Simon, A.M., Goodenough, D.A., Day, K.H., Damon, D.N., Duling, B.R. (2003) Central role of connexin 40 in the propagation of electrically activated vasodilation in mouse cremasteric arterioles in vivo. *Circ. Res.* **92**: 793–800.

366. Theis, M., de Wit, C., Shlaeger, T.M., Eckardt, D., Kruger, O., Doring, B., Risau, W., Deutsch, U., Pohl, U., Willecke, K. (2001) Endothelium-specific replacement of the connexin 43 coding region by a lacZ reporter gene. *Genesis* **29**: 1–13.

367. Liao, Y., Day, K.H., Damon, D.N., Duling, B.R. (2001) Endothelial cell-specific knockout of connexin 43 causes hypotension and bradycardia in mice. *Proc. Natl. Acad. Sci. USA* **98**: 9989–9994.

368. Kruger, O., Beny, J-L., Chabaud, F., Traub, O., Theis, M., Brix, K., Kirchhoff, S., Willecke, K. (2002) Altered dye diffusion and upregulation of connexin 47 in mouse aortic endothelium-deficient in connexin 40. *J. Vasc. Res.* **39**: 160–172.

369. Cohen, S.E. (1936) The influence of the Ca and K ions on toners and adrenaline response of the coronary arteries. *Arch. Int. Pharmacodyn.* **54**: 1–6.

370. Dawes, G.S. (1941) The vasodilator action of K^+. *J. Physiol.* **99**: 224–238.

371. Kjellmer, I. (1965) The potassium ion as a vasodilator during muscular exercise. *Acta. Physiol. Scand.* **63**: 460–468.

372. Skinner, N.S., Powell, W.J. (1967) Action of oxygen and potassium on vascular resistance of dog skeletal muscle. *Am. J. Physiol.* **212**: 533–540.

373. Haddy, F.J. (1975) Potassium and blood vessels. *Life Sci.* **16**: 1489–1497.

374. Shepherd, J.T., Vanhoutte, P.M. (1979) *The Human Cardiovascular System. Facts and Concepts.* Raven Press, New York, pp. 1–351.

375. Hendrickx, H., Casteels, R. (1974) Electrogenic sodium pump in arterial smooth muscle cells. *Pflugers Arch.* **346**: 299–306.

376. Toda, N. (1974) Responsiveness to potassium and calcium ions of isolated cerebral arteries. *Am. J. Physiol.* **227**: 1206–1211.

377. Nelson, M.T., Quayle, J.M. (1995) Physiological roles and properties of potassium channels in arterial smooth muscle. *Am. J. Physiol.* **268**: C799–C822.

378. Prior, H.M., Webster, N., Quinn, K., Beech, D.J., Yates, M.S. (1998) K(+)-induced dilation of a small renal artery: no role for inward rectifier K^+ channels. *Cardiovasc. Res.* **37**: 780–790.

379. Makino, A., Ohuchi, K., Kamata, K. (2000) Mechanisms underlying the attenuation of endothelium-dependent vasodilatation in the mesenteric arterial bed of the streptozotocin-induced rat. *Br. J. Pharmacol.* **130**: 549–556.

380. Dora, K.A., Garland, C.J. (2001) Properties of smooth muscle hyperpolarization and relaxation to K^+ in the rat isolated mesenteric artery. *Am. J. Physiol.* **280**: H2424–H2429.

381. De Clerck, I., Boussery, K., Pannier, J.L., Van De Voorde, J. (2003) Potassium potently relaxes small rat skeletal arteries. *Med. Sci. Sports Exerc.* **35**: 2005–2012.

382. Savage, D., Perkins, J., Hong Lim, C., Bund, S.J. (2003) Functional evidence that K^+ is the non-nitric oxide, non-prostanoid endothelium-derived relaxing factor in rat femoral arteries. *Vascul. Pharmacol.* **40**: 23–28.

383. McGuire, J.J., Hollenberg, M.D., Andrade-Gordon, P., Triggle, C.R. (2002) Multiple mechanisms of vascular smooth muscle relaxation by the activation of proteinase-activated receptor 2 in mouse mesenteric arterioles. *Br. J. Pharmacol.* **135**: 155–169.

384. Beny, J-L., Schaad, O. (2000) An evaluation of potassium ions as endothelium-derived hyperpolarizing factor in porcine coronary arteries. *Br. J. Pharmacol.* **131**: 965–973.

385. Nelli, S., Wilson, W.S., Laidlaw, H., Llano, A., Middleton, S., Price, A.G., Martin, W. (2003) Evaluation of potassium ions as the endothelium-derived hyperpolarizing factor (EDHF) in the bovine coronary artery. *Br. J. Pharmacol.* **139**: 982–988.

386. Randriamboavonjy, V., Kiss, L., Falck, J.R., Busse, R., Fleming, I. (2005) The synthesis of 20-HETE in small porcine coronary arteries antagonizes EDHF-mediated relaxation. *Cardiovasc. Res.* **65**: 487–494.

387. Büssemaker, E., Wallner, C., Fisslthaler, B., Fleming, I. (2002) The Na$^+$K$^+$-ATPase is a target for an EDHF displaying characteristics similar to potassium ions in the porcine renal interlobar arteries. *Br. J. Pharmacol.* **137**: 647–654.

388. Büssemaker, E., Popp, R., Binder, J., Busse, R., Fleming, I. (2003) Characterization of the endothelium-derived hyperpolarizing factor (EDHF) response in the human interlobar artery. *Kidney Int.* **63**: 1749–1755.

389. Torondel, B., Vila, J.M., Segarra, G., Lluch, P., Medina, P., Martinez-Leon, J., Ortega, J., Lluch, S. (2004) Endothelium-dependent responses in human isolated thyroid arteries from donors. *J. Endocrinol.* **181**: 379–384.

390. Bradley, K.K., Jaggar, J.H., Bonev, A.D., Heppner, T.J., Flynn, E.R., Nelson, M.T., Horowitz, B. (1999) Kir2.1 encodes the inward rectifier potassium channel in rat arterial smooth muscle cells. *J. Physiol.* **515**: 639–651.

391. Edwards, G., Richards, G.R., Gardener, M.J., Félétou, M., Vanhoutte, P.M., Weston, A.H. (2003) Role of the inward-rectifier K$^+$ channel and Na$^+$/K$^+$-ATPase in the hyperpolarization to K$^+$ in rat mesenteric arteries. In *EDHF 2002*, P.M. Vanhoutte, ed., Taylor and Francis, London, pp. 309–317.

392. Zaritsky, J.J., Eckman, D.M., Wellman, G.C., Nelson, M.T., Schwarz, T.L. (2000) Targeted disruption of Kir2.1 and Kir2.2 genes reveal the essential role of the inwardly rectifying K$^+$ current in K$^+$-mediated vasodilation. *Circ. Res.* **87**: 160–166.

393. Juhaszova, M., Blaustein, M.P. (1997) Distinct distribution of different Na$^+$ pump α subunit isoforms in plasmalemma. Physiological implications. In *Na/K-ATPase and Related Transport ATPases,* Vol 834., L.A. Beaugé et al., eds., Annals of the New York Academy of Sciences, New York, pp. 524–536.

394. Weston, A.H., Richards, G.R., Burnham, M.P., Félétou, M., Vanhoutte, P.M., Edwards, G. (2002) K$^+$-induced hyperpolarization in rat mesenteric artery: identification, localization and role for Na$^+$/K$^+$-ATPases. *Br. J. Pharmacol.* **136**: 918–926.

395. Lansbery, K., Mendenhall, M.L., Vehige, L.C., Taylor, J.A., Sanchez, G., Blanco, G., Mercer, R.W. (2003) Isoforms of the Na,K-ATPase. In *EDHF 2002*, P.M. Vanhoutte, ed., Taylor and Francis, London, pp. 27–34.

396. Quayle, J.M., McCarron, J.G., Brayden, J.E., Nelson, M.T. (1993) Inward rectifier K$^+$ currents in smooth muscle from rat resistance-sized cerebral arteries. *Am. J. Physiol.* **265**: C1363–C1370.

397. Robertson, B.E., Bonev, A.D., Nelson, M.T. (1996) Inward rectifier K+ currents in smooth muscle cells from rat coronary arteries: block by Mg^{2+}, Ca^{2+}, and Ba^{2+}. *Am. J. Physiol.* **271**: H696–H705.

398. Zygmunt, P.M., Sorgard, M., Petersson, J., Johansson, R., Hoggestatt, E.D. (2000) Differential actions of anandamide, potassium ions and endothelium-derived hyperpolarizing factor in guinea-pig basilar artery. *Naunyn Schmiedebergs Arch. Pharmacol.* **361**: 535–542.

399. Dong, H., Jiang, Y., Cole, W.C., Triggle, C.R. (2000) Comparison of the pharmacological properties of EDHF-mediated vasorelaxation in the guinea-pig cerebral and mesenteric resistance vessels. *Br. J. Pharmacol.* **130**: 1983–1991.

400. Van de Voorde, J., Vanheel, B. (2000) EDHF-mediated relaxation in rat gastric small arteries: influence of ouabain/Ba^{2+} and relation to potassium ions. *J. Cardiovasc. Pharmacol.* **35**: 543–548.

401. Jiang, F., Dusting, G.J. (2001) Endothelium-dependent vasorelaxation independent of nitric oxide and K(+) release in isolated renal arteries of rat. *Br. J. Pharmacol.* **132**: 1558–1564.

402. Pratt, P.F., Li, P., Hillard, C.J., Kurian, J., Campbell, W.B. (2001) Endothelium-independent, ouabain-sensitive relaxation of bovine coronary arteries by EETs. *Am. J. Physiol.* **280**: H1113–H1121.

403. Matoba, T., Shimokawa, H., Morikawa, K., Kubota, H., Kunihiro, I., Urakami-Harasawa, L., Mukai, Y., Hirakawa, Y., Akaike, T., Takeshita, A. (2003) Electron-spin resonance detection of hydrogen peroxide as an endothelium-derived hyperpolarizing factor in porcine coronary arteries microvessels. *Arterioscler. Thromb. Vasc. Biol.* **23**: 1224–1230.

404. Drummond, G.R., Selemidis, S., Cocks, T.M. (2000) Apamin-sensitive non-nitric oxide (NO) endothelium-dependent relaxations to bradykinin in the bovine isolated coronary artery: no role for cytochrome P450 and K^+. *Br. J. Pharmacol.* **129**: 811–819.

405. Marchetti, J., Praddaude, F., Rajerison, R., Ader, J.L., Alhenc-Gelas, F. (2001) Brady-kinin attenuates the [Ca(2+)](i) response to angiotensin II of renal juxtamedullary efferent arterioles via an EDHF. *Br. J. Pharmacol.* **132**: 749–759.

406. Coats, P., Johnston, F., MacDonald, J., McMurray, J.J., Hillier, C. (2001) Endothelium-derived hyperpolarizing factor: identification and mechanisms of action in human subcutaneous resistance arteries. *Circulation* **103**: 1702–1708.

407. McIntyre, C.A., Buckley, C.H., Jones, G.C., Sandeep, T.C., Andrews, R.C., Elliott, A.I., Gray, G.A., Williams, B.C., McKnight, J.A., Walker, B.R., Hadoke, P.W. (2001) Endothelium-derived hyperpolarizing factor and potassium use different mechanisms to induce relaxation of human subcutaneous resistance arteries. *Br. J. Pharmacol.* **133**: 902–908.

408. Seol, G.H., Ahn, S.C., Kim, J.A., Nilius, B., Suh, S.H. (2004) Inhibition of endothelium-dependent vasorelaxation by extracellular K^+: a novel controlling signal for vascular contractility. *Am. J. Physiol.* **286**: H329–H339.

409. Vanheel, B., Van de Voorde, J. (1999) Barium decreases endothelium-dependent smooth muscle responses to transient but not to more prolonged acetylcholine applications. *Pflugers Arch.* **439**: 123–129.

410. Lacy, P.S., Pilkington, G., Hanvesakul, R., Fish, H.J., Boyle, J.P., Thurston, H. (2000) Evidence against potassium as an endothelium-derived hyperpolarizing factor in rat mesenteric small arteries. *Br. J. Pharmacol.* **129**: 605–611.

411. Doughty, J.M., Boyle, J.P., Langton, P.D. (2001) Blockade of chloride channels reveals relaxations of rat small mesenteric arteries to raised potasssium. *Br. J. Pharmacol.* **132**: 293–301.

412. Harris, D., Martin, P.E., Evans, W.H., Kendall, D.A., Griffith, T.M., Randall, M.D. (2000) Role of gap junctions in endothelium-derived hyperpolarizing factor responses and mechanisms of K(+)-relaxation. *Eur. J. Pharmacol.* **402**: 119–128.

413. Richards, G.R., Weston, A.H., Burnham, M.P., Félétou, M., Vanhoutte, P.M., Edwards, G. (2001) Suppression of K^+-induced hyperpolarization by phenylephrine in rat mesenteric artery: relevance to studies of endothelium-derived hyperpolarizing factor. *Br. J. Pharmacol.* **134**: 1–5.

414. Dawes, M., Sieniawska, C., Delves, T., Dwivedi, R., Chowienczyk, P.J., Ritter, J.M. (2002) Barium reduces blood flow and inhibits potassium-induced vasodilation in the human forearm. *Circulation* **105**: 1323–1328.

415. Fujita, T., Ito, Y. (1993) Salt loads attenuate potassium-induced vasodilation of forearm vasculature in humans. *Hypertension* **21**: 772–778.

416. Taddei, S., Ghiadoni, L., Virdis, A., Buralli, S., Salvetti, A. (1999) Vasodilatation to bradykinin is mediated by an ouabain-sensitive pathway as a compensatory mechanism for impaired nitric oxide availability in essential hypertensive patients. *Circulation* **100**: 1400–1405.

417. Jiang, Z.G., Nuttall, A.L., Zhao, H., Dai, C.F., Guan, B.C., Si, J.Q., Yang, Y.Q. (2005) Electrical coupling and release of K^+ from endothelial cells co-mediate ACh-induced smooth muscle hyperpolarization in inner ear artery. *J. Physiol. (Lond.)* **564**: 475–487.

418. Pinto, A., Abraham, N.G., Mullane, K.M. (1987) Arachidonic acid-induced endothelial-dependent relaxations of canine coronary arteries: contribution of a cytochrome P450 dependent pathway. *J. Pharmacol. Exp. Ther.* **240**: 856–862.

419. Hecker, M., Bara, A.T., Bauersachs, J., Busse, R. (1994) Characterization of endothelium-derived hyperpolarizing factor as a cytochrome P_{450}-derived arachidonic acid metabolite in mammals. *J. Physiol.* **481**: 407–414.

420. Adeagbo, A.S. (1997) Endothelium-derived hyperpolarizing factor: characterization as a cytochrome P450 1A-linked metabolite of arachidonic acid in perfused rat mesenteric prearteriolar bed. *Am. J. Hypertens.* **10**: 763–771.

421. Graziani, A., Bricko, V., Carmignani, M., Graier, W.F., Groschner, K. (2004) Cholesterol- and caveolin-rich membrane domains are essential for phospholipase A_2-dependent EDHF formation. *Cardiovasc. Res.* **64**: 234–242.

422. Quilley, J., McGiff, J.C. (2000) Is EDHF an epoxyeicosatrienoic acid? *Trends Pharmacol. Sci.* **21**: 121–124.

423. Campbell, W.B., Gebremedhin, D., Pratt, P.F., Harder, D.R. (1996) Identification of epoxyeicosatrienoic acids as endothelium-derived hyperpolarizing factor. *Circ. Res.* **78**: 415–423.

424. Widmann, M.D., Weintraub, N.L., Fudge, J.L., Brooks, L.A., Dellsperger, K.C. (1998) Cytochrome P-450 pathway in acetylcholine-induced canine coronary microvascular vasodilation in vivo. *Am. J. Physiol.* **43**: H283–H289.

425. Gebremedhin, D., Ma, Y.H., Falck, J.R., Roman, R.J., VanRollins, M., Harder, D.R. (1992) Mechanism of action of cerebral epoxyeicosatrienoic acids on cerebral arterial smooth muscle. *Am. J. Physiol.* **263**: H519–H525.

426. Graier, W.F., Holzmann, S., Hoebel, B.G., Kukovetz, W.R., Kostner, G.M. (1996) Mechanisms of L-N^G nitroarginine/indomethacin-resistant relaxation in bovine and porcine coronary arteries. *Br. J. Pharmacol.* **119**: 1177–1186.

427. Eckman, D.M., Hopkins, N.O., McBride, C., Keef, K.D. (1998) Endothelium-dependent relaxation and hyperpolarization in guinea-pig coronary artery: role of epoxyeicosatrienoic acid. *Br. J. Pharmacol.* **124**: 181–189.

428. Fulton, D., McGiff, J.C., Quilley, J. (1998) Pharmacological evaluation of an epoxide as the putative hyperpolarizing factor mediating the nitric oxide-independent vasodilator effect of bradykinin in the rat heart. *J. Pharmacol. Exp. Ther.* **287**: 497–503.

429. Oltman, C.L., Weintraub, N.L., VanRollins, M., Dellsperger, K.C. (1998) Epoxyeicosatrienoic acids and dihydroxyeicosatrienoic acids are potent vasodilators in the canine coronary microcirculation. *Circ. Res.* **83**: 932–939.

430. Fisslthaler, B., Popp, R., Kiss, L., Potente, M., Harder, D.R., Fleming, I., Busse, R. (1999) Cytochrome P4540 2C is an EDHF synthase in coronary arteries. *Nature* **401**: 493–497.

431. Hu, S., Kim, H.S. (1993) Activation of K^+ channel in vascular smooth muscles by cytochrome P450 metabolites of arachidonic acid. *Eur. J. Pharmacol.* **230**: 215–221.

432. Dumoulin, M., Salvail, D., Gaudreault, S.B., Cadieux, A., Rousseau, E. (1998) Epoxyeicosatrienoic acids relax airway smooth muscles and directly activate reconstituted KCa channels. *Am. J. Physiol.* **275**: L423–L431.

433. Li, P.L., Campbell, W.B. (1997) Epoxyeicosatrienoic acids activate K^+ channels in coronary smooth muscle through a guanine nucleotide binding protein. *Cir. Res.* **80**: 877–884.

434. Fukao, M., Mason, H.S., Kenyon, J.L., Horowitz, B., Keef, K.D. (2001) Regulation of BK(Ca) channels expressed in human embryonic kidney 293 cells by epoxyeicosatrienoic acid. *Mol. Pharmacol.* **59**: 16–23.

435. Li, P.L., Chen, C.L., Bortell, R., Campbell, W.B. (1999) Epoxyeicosatrienoic acid stimulates endogenous mono-ADP-ribosylation in bovine coronary artery smooth muscle. *Circ. Res.* **85**: 349–356.

436. Li, P.L., Zhang, D.X., Ge, Z.D., Campbell, W.B. (2002) Role of ADP-ribose in 11,12-EET-induced activation of K(Ca) channels in coronary arterial smooth muscle cells. *Am. J. Physiol.* **282**: H1229–H1236.

437. Lu, T., Hoshi, T., Weintraub, N.L., Spector, A.A., Lee, H.C. (2001) Activation of ATP-sensitive K(+) channels by epoxyeicosatrienoic acids in rat cardiac ventricular myocytes. *J. Physiol.* **537**: 811–827.

438. Ye, D., Zhou, W., Lee, H.C. (2005) Activation of rat mesenteric arteric arterial K_{ATP} channels by 11,12-epoxyeicosatrienoic acid. *Am. J. Physiol.* **288**: H358–H364.

439. Wong, P-Y.K., Lai, P-S., Kalck, J.R. (2000) Mechanism and signal transduction of 14(R), 15(S)-epoxyeicosatrienoic acid (14,15-EET) binding in guinea pig monocytes. *Prostaglandin Other Lipid Mediat.* **62**: 321–333.

440. Gauthier, K.M., Deeter, C., Krishna, U.M., Reddy, Y.K., Bondlela, M., Falck, J.R., Campbell, W.B. (2002) 14,15-Epoxyeicosa-5(Z)-enoic acid: a selective epoxyeicosatrienoic acid antagonist that inhibits endothelium-dependent hyperpolarization and relaxation in coronary arteries. *Circ. Res.* **90**: 1028–1036.

441. Gauthier, K.M., Jagadeesh, S.G., Falck, J.R., Campbell, W.B. (2003) 14,15-Epoxyeicosa-5(Z)-enoic-mSI: a 14,15- and 5,6-EET antagonist in bovine coronary arteries. *Hypertension* **42**: 555–561.

442. Gauthier, K.M., Falck, J.R., Reddy, J.R., Campbell, W.B. (2004) 14,15-EET analogs: characterization of structural requirements for agonist and antagonist activity in bovine coronary arteries. *Pharmacol. Res.* **49**: 515–524.

443. Gauthier, K.M., Spitzbarth, N., Edwards, E.M., Campbell, W.B. (2004) Apamin-sensitive K^+ currents mediate arachidonic acid-induced relaxation of rabbit aorta. *Hypertension* **43**: 413–419.

444. Rosolowski, M., Campbell, W.B. (1993) Role of PGI2 and EETs in the relaxation of bovine coronary arteries to arachidonic acid. *Am. J. Physiol.* **264**: H327–H335.

445. Bauersachs, J., Hecker, M., Busse, R. (1994) Display of the characteristics of endothelium-derived hyperpolarizing factor by a cytochrome P450-derived arachidonic acid metabolite in the coronary microcirculation. *Br. J. Pharmacol.* **113**: 1548–1553.

446. Fulton, D., McGiff, J.C., Quilley, J. (1992) Contribution of NO and cytochrome P_{450} to the vasodilator effect of bradykinin in the rat kidney. *Br. J. Pharmacol.* **107**: 722–725.

447. Fulton, D., Mahboudi, K., McGiff, J.C., Quilley, J. (1995) Cytochrome P450-dependent effects of bradykinin in the rat heart. *Br. J. Pharmacol.* **114**: 99–102.

448. Kessler, P., Lischke, V., Hecker, M. (1996) Etomidate and thiopental inhibit the release of endothelium-derived-hyperpolarizing factor in the human renal artery. *Anesthesiology* **84**: 1485–1488.

449. Miura, H., Gutterman, D.D. (1998) Human coronary arteriolar dilation to arachidonic acid depends on cytochrome P450 monooxygenase and Ca^{2+}-activated K^+ channels. *Circ. Res.* **83**: 501–507.

450. Halcox, J.P., Narayanan, S., Cramer-Joyce, L., Mincemoyer, R., Quyyumi, A.A. (2001) Characterization of endothelium-derived hyperpolarizing factor in the human forearm microcirculation. *Am. J. Physiol.* **280**: H2470–H2477.

451. Oltman, C.L., Kane, C.L., Fudge, J.L., Weintraub, N.L., Dellsperger, K.C. (2001) Endothelium-derived hyperpolarizing factor in coronary microcirculation: responses to arachidonic acid. *Am. J. Physiol.* **281**: H1553–H1560.

452. Archer, S.L., Gragasin, F.S., Wu, X., Wang, S., McMurthry, S., Kim, D.H., Platonov, M., Koshal, A., Hashimoto, K., Campbell, W.B., Falck, J.R., Michelakis, E.D. (2003) Endothelium-derived hyperpolarizing factor in human internal mammary artery is 11,12-epoxieicosatrienoic acid and causes relaxation by activating BK(Ca) channels. *Circulation* **107**: 769–776.

453. Tanaka, M., Kanatsuka, H., Ong, B.H., Tanikawa, T., Urano, A., Komaru, T., Koshida, R., Shirato, K. (2003) Cytochrome P-450 metabolites but not NO, PGI_2 and H_2O_2 contribute to Ach-induced hyperpolarization of pressurized canine microvessels. *Am. J. Physiol.* **285**: H1939–H1948.

454. Zhang, D.X., Gauthier, K.M., Campbell, W.B. (2004) Acetylcholine-induced relaxation and hyperpolarization in small bovine adrenal cortical arteries: role of cytochrome P450 metabolites. *Endocrinology* **145**: 4532–4539.

455. Alvarez, J., Montero, M., Garcia-Sancho, J. (1992) High affinity inhibition of Ca^{2+}-dependent K^+ channels by cytochrome P-450 inhibitors. *J. Biol. Chem.* **267**: 11789–11793.

456. Edwards, G., Zygmunt, P.M., Högestätt, E.D., Weston, A.H. (1996) Effects of cytochrome P450 inhibitors on potassium currents in mechanical activity in rat portal vein. *Br. J. Pharmacol.* **119**: 691–701.

457. Vanheel, B., Calders, P., Van den Bossche, I., Van de Voorde, J. (1999) Influence of some phospholipase A2 and cytochrome P 450 inhibitors on rat arterial smooth muscle K^+ currents. *Can. J. Physiol. Pharmacol.* **77**: 481–489.

458. Chataigneau, T., Félétou, M., Thollon, C., Villeneuve, N., Vilaine, J-P., Duhault, J., Vanhoutte, P.M. (1998) Cannabinoid CB_1 receptor and endothelium-dependent hyperpolarisation in guinea-pig carotid, rat mesenteric and porcine coronary arteries. *Br. J. Pharmacol.* **123**: 968–974.

459. Graier, W.F., Holzmann, S., Hoebel, B.G., Kukovetz, W.R. (1995) L$^\Omega$N-nitro-arginine resistant vessel relaxation is mediated via a pertussis toxin sensitive pathway but not via cytochrome P450 mono-oxygenase in bovine coronary arteries. *Circulation* **92**: 751.

460. Eckman, D.M., Hopkins, N.O., Keef, K.D. (1995) Effects of inhibitors of cytochrome P450 pathway on relaxation and hyperpolarisation induced with acetylcholine and lemakalim. *Circulation* **92**: i–751.

461. Van de Voorde, J., Vanheel, B. (1997) Evidence against the involvement of cytochrome P450 metabolites in endothelium-dependent hyperpolarization of the rat mesenteric artery. *J. Physiol.* **501**: 331–341.

462. Van de Voorde, J., Vanheel, B. (1997) Influence of cytochrome P450 inhibitors on endothelium-dependent nitro-L-arginine-resistant relaxation and cromakalim-induced relaxation in rat mesenteric arteries. *J. Cardiovasc. Pharmacol.* **29**: 827–832.

463. Capdevilla, J., Gil, L., Orellana, M., Marnett, L.J., Yadagiri, P., Falck, J.R. (1988) Inhibitors of cytochrome P-450-dependent arachidonic acid metabolism. *Arch. Biochem. Biophys.* **261**: 257–263.

464. Jensen, B.S., Strobaek, D., Christophersen, P. (2001) The intermediate-conductance Ca^{2+}-activated K^+ channel: a molecular target for novel treatments? *Curr. Drug Targets* **2**: 401–422.

465. Imig, J.D., Falck, J.R., Wei, S., Capdevila, J.H. (2001) Epoxygenase metabolites contribute to nitric oxide-independent afferent arteriolar vasodilation in response to bradykinin. *J. Vasc. Res.* **38**: 247–255.

466. Harrington, L.S., Falck, J.R., Mitchell, J.A. (2004) Not so EEZE: the "EDHF" antagonist 14,15 epoxyeicosa-5(Z)-enoic acid has vasodilator properties in mesenteric arteries. *Eur. J. Pharmacol.* **506**: 165–168.

467. Wulff, H., Gutman, G.A., Cahalan, M.D., Chandy, K.G. (2001) Delineation of the clotrimazole/TRAM-34 binding site on the intermediate conductance calcium-activated potassium channel, $IK_{Ca}1$. *J. Biol. Chem.* **276**: 32040–32045.

468. Bolz, S.S., Fisslthaler, B., Pieperhoff, S., De Wit, C., Fleming, I., Busse, R., Pohl, U. (2000) Antisense oligonucleotides against cytochrome P450 attenuate EDHF-mediated Ca(2+) changes and dilation in isolated resistance arteries. *FASEB J.* **14**: 255–260.

469. Rosolowski, M., Campbell, W.B. (1996) Synthesis of hydroxyeicosatetraenoic (HETEs) and epoxyeicosatrienoic acids (EETs) by cultured bovine coronary endothelial cells. *Biochem. Biophys. Acta* **1299**: 267–277.

470. Popp, R., Fleming, I., Busse, R. (1998) Pulsatile stretch in coronary arteries elicits release of endothelium-derived hyperpolarizing factor: a modulator of arterial compliance. *Circ. Res.* **82**: 696–703.

471. Gauthier, K.M., Edwards, E.M., Falck, J.R., Reddy, D.S., Campbell, W.B. (2005) 14,15-Epoxyeicosatrienoic acid represents a transferable endothelium-dependent relaxing factor in bovine coronary arteries. *Hypertension* **45**: 666–671.

472. Weston, A.H., Félétou, M., Vanhoutte, P.M., Falck, J.R., Campbell, W.B., Edwards, G. (2005) Bradykinin-induced, endothelium-dependent responses in porcine coronary arteries: involvement of potassium channel activation and epoxyeicosatrienoic acids. *Br. J. Pharmacol.* **145**: 775–784.

473. Yang, W., Gauthier, K.M., Reddy, L.M., Sangras, B., Sharma, K.K., Nithipatikom, K., Falck, J.R., Campbell, W.B. (2005) Stable 5,6-epoxyeicosatrienoic acid analog relaxes coronary arteries through potassium channel activation. *Hypertension* **45**: 681–686.

474. Zygmunt, P.M., Edwards, G., Weston, A.H., Davis, S.C., Högestätt, E.D. (1996) Effects of cytochrome P450 inhibitors on EDHF-mediated relaxation in the rat hepatic artery. *Br. J. Pharmacol.* **118**: 1147–1152.

475. Fukao, M., Hattori, Y., Kanno, M., Sakuma, I., Kitabatake, A. (1997) Evidence against a role of cytochrome P450-derived arachidonic acid metabolites in endothelium-dependent hyperpolarisation by acetylcholine in rat isolated mesenteric artery. *Br. J. Pharmacol.* **120**: 439–446.

476. Petersson, J., Zygmunt, P.M., Jonsson, P., Högestätt, E.D. (1998) Characterization of endothelium-dependent relaxation in guinea pig basilar artery—effect of hypoxia and role of cytochrome P450 monooxygenase. *J. Vasc. Res.* **35**: 285–294.

477. Ungvari, Z., Koller, A. (2001) Mediation of EDHF-induced reduction of smooth muscle [Ca(2+)](i) and arteriolar dilation by K(+) channels, 5,6-EET, and gap junctions. *Microcirculation* **8**: 265–274.

478. Wallerstedt, S.M., Bodelsson, M. (1997) Endothelium-dependent relaxations by substance P in human isolated omental arteries and veins: relative contribution of prostanoids, nitric oxide and hyperpolarisation. *Br. J. Pharmacol.* **120**: 25–30.

479. Ohlmann, P., Martinez, M.C., Schneider, F., Stoclet, J.C., Andriantsitohaina, R. (1997) Characterization of endothelium-derived relaxing factors released by bradykinin in human resistance arteries. *Br. J. Pharmacol.* **121**: 657–664.

480. Urakami-Harasawa, L., Shimokawa, H., Nakashima, M., Egashira, K., Takeshita, A. (1997) Importance of endothelium-derived hyperpolarizing factor in human arteries. *J. Clin. Invest.* **100**: 2793–2799.
481. Matoba, Y., Shimokawa, H., Kubota, H., Morikawa, K., Fujiki, T., Kunihiro, I., Mukai, Y., Hirakawa, Y., Takeshita, A. (2002) Hydrogen peroxide is an endothelium-derived hyperpolarizing factor in human mesenteric artery. *Biochem. Biophys. Res. Commun.* **290**: 909–913.
482. Hatoum, O.A., Binion, D.G., Miura, H., Telford, G., Otterson, M.F., Gutterman, D.D. (2004) The role of hydrogen peroxide in ACh-induced dilation of human sumucosal intestinal microvessels. *Am. J. Physiol.* **288**: H48–H54.
483. Fujimoto, S., Ikegami, Y., Isaka, M., Kato, T., Nishimura, K., Itoh, T. (1999) K(+) channel blockers and cytochrome P450 inhibitors on acetylcholine-induced, endothelium-dependent relaxations in rabbit mesenteric artery. *Eur. J. Pharmacol.* **384**: 7–15.
484. Jiang, F., Li, C.G., Rand, M.J. (2000) Mechanisms of nitric oxide-independent relaxations induced by carbachol and acetylcholine in rat renal arteries. *Br. J. Pharmacol.* **130**: 1191–1200.
485. Hoebel, B.G., Kostner, G.M., Graier, W.F. (1997) Activation of microsomal P450 mono-oxygenase by Ca^{2+} store depletion and its contribution to Ca^{2+} entry in porcine aortic endothelial cells. *Br. J. Pharmacol.* **121**: 1579–1588.
486. Mombouli, J-V., Holzmann, S., Kostner, G.M., Graier, W.F. (1999) Potentiation of Ca^{2+} signaling in endothelial cells by 11,12-epoxyeicosatrienoic acid. *J. Cardiovasc. Pharmacol.* **33**: 779–784.
487. Kuroiwa-Matsumoto, M., Hirano, K., Ahmed, A., Kawasaki, J., Nishimura, J., Kanaide, H. (2000) Mechanisms of the thapsigargin-induced Ca(2+) entry in *in situ* endothelial cells of the porcine aortic valve and the endothelium-dependent relaxation in the porcine coronary artery. *Br. J. Pharmacol.* **131**: 115–123.
488. Takeuchi, K., Watanabe, H., Tran, Q.K., Ozeki, M., Uehara, A., Katoh, H., Satoh, H., Terasa, H., Ohashi, K., Hayashi, H. (2003) Effects of cytochrome P450 inhibitors on agonist-induced Ca^{2+} responses and production of NO and PGI2 in vascular endothelial cells. *Mol. Cell Biochem.* **248**: 129–134.
489. Xie, Q., Zhang, Y., Zhai, C., Bonanno, J.A. (2002) Calcium influx factor from cytochrome P450 metabolism and secretion-like coupling mechanisms for capacitive calcium entry in corneal endothelial cells. *J. Biol. Chem.* **277**: 16559–16566.
490. Alonso-Torre, S.R., Alvarez, J., Montero, M., Sanchez, A., Garcia-Sancho, J. (1993) Control of Ca^{2+} entry into HL60 and U937 human leukaemia cells by the filling state of the intracellular Ca^{2+} stores. *Biochem. J.* **289**: 761–766.
491. Rzigalinski, B.A., Blackmore, P.F., Rosenthal, M.D. (1996) Arachidonate mobilization is coupled to depletion of intracellular calcium stores and influx of extracellular calcium in differentiated U937 cells. *Biochem. Biophys. Acta* **1299**: 342–352.
492. Gailly, P. (1998) Ca^{2+} entry in CHO cells, after Ca^{2+} stores depletion, is mediated by arachidonic acid. *Cell Calcium* **24**: 293–304.
493. Graier, W.F., Simecek, S., Sturek, M. (1995) Cytochrome P450 mono-oxygenase-regulated signalling of Ca^{2+} entry in human and bovine endothelial cells. *J. Physiol. (Lond.)* **482**: 259–274.
494. Rzigalinski, B.A., Willoughby, K.A., Hoffman, S.W., Falck, J.R., Ellis, E.F. (1999) Calcium influx factor, further evidence it is 5, 6-epoxyeicosatrienoic acid. *J. Biol. Chem.* **274**: 175–182.
495. Nilius, B. (2003) From TRPs to SOCs, CCEs and CRACs: consensus and controversies. *Cell Calcium* **33**: 293–298.

496. Watanabe, H., Vriens, J., Prenen, J., Droogmans, G., Voets, T., Nilius, B. (2003) Anandamide and arachidonic acid use epoxyeicosatrienoic acids to activate TRPV4 channels. *Nature* **424**: 434–438.

497. Baron, A., Frieden, M., Beny, J.–L. (1997) Epoxyeicosatrienoic acids activate a high-conductance, Ca^{2+}-dependent K^+ channel on pig coronary artery endothelial cells. *J. Physiol.* **504**: 537–543.

498. Popp, R., Brandes, R.P., Ott, G., Busse, R., Fleming, I. (2002) Dynamic modulation of inter-endothelial gap junctional communication by 11,12-epoxyeicosatrienoic acid. *Circ. Res.* **90**: 800–806.

499. Krotz, F., Riexinger, T., Buerkle, M.A., Nithipatikom, K., Gloe, T., Sohn, H.Y., Campbell, W.B., Pohl, U. (2004) Membrane potential-dependent inhibition of platelet adhesion to endothelial cells by epoxyeicosatrienoic acids. *Arterioscler. Thromb. Vasc. Biol.* **24**: 595–600.

500. Node, K., Ruan, X., Dai., J., Yang, B., Graham, L., Zeldin, D.C., Liao, J.K. (2001) Activation of Galpha s mediates induction of tissue-type plasminogen activator gene transcription by epoxyeicosatrienoic acids. *J. Biol. Chem.* **276**: 15983–15989.

501. Node, K., Huo, Y., Ruan, X., Yang, B., Spiecker, M., Ley, K., Zeldin, D.C., Liao, J.K. (1999) Anti-inflammatory properties of cytochrome P450 epoxygenase-derived eicosanoids. *Science* **285**: 1276–1279.

502. Hoebel, B.G., Steyrer, E., Graier, W.F. (1998) Origin and function of epoxyeicosa-trienoic acids in vascular endothelial cells: more than just endothelium-derived hyper-polarizing factor? *Clin. Exp. Pharmacol. Physiol.* **25**: 826–830.

503. Hoebel, B.G., Graier, W.F. (1998) 11,12-epoxyeicosatrienoic acid stimulates tyrosine kinase activity in porcine aortic endothelial cells. *Eur. J. Pharmacol.* **346**: 115–117.

504. Fleming, I., Fisslthaler, B., Michaelis, U.R., Kiss, L., Popp, R., Busse, R. (2001) The coronary endothelium-derived hyperpolarizing factor (EDHF) stimulates multiple signaling pathways and proliferation in vascular cells. *Pflugers Arch.* **442**: 511–518.

505. Chen, J.K., Wang, D.W., Falck, J.R., Capdevila, J., Harris, R.C. (1999) Transfection of an active cytochrome P450 arachidonic acid epoxygenase indicates that 14,15-epoxyeicosatrienoic acid functions as an intracellular second messenger in response to epidermal growth factor. *J. Biol. Chem.* **274**: 4764–4769.

506. Potente, M., Michaelis, U.R., Fisslthaler, B., Busse, R., Fleming, I. (2002) Cyto-chrome P450 2C9-induced endothelial cell proliferation involves induction of mitogen-activated protein (MAP) kinase phosphatase-1, inhibition of the c-Jun N-terminal kinase and regulation of cyclin D1. *J. Biol. Chem.* **277**: 15671–15676.

507. Potente, M., Fisslthaler, B., Busse, R., Fleming, I. (2003) 11,12-epoxyeicosatrienoic acid-induced inhibition of FOXO factors promotes endothelial proliferation by down-regulation of p27Kip1. *J. Biol. Chem.* **278**: 29619–29625.

508. Michaelis, U.R., Fisslthaler, B., Medhora, M., Harder, D., Fleming, I., Busse, R. (2003) Cytochrome P450 2C9-derived epoxyeicosatrienoic acids induce angiogenesis via cross-talk with the epidermal growth factor receptor (EGFR). *FASEB J.* **17**: 770–772.

509. Medhora, M., Daniels, J., Mundey, K., Fisslthaler, B., Busse, R., Jacobs, E.R., Harder, D.R. (2003) Epoxygenase-driven angiogenesis in human microvascular endothelial cells. *Am. J. Physiol.* **284**: H215–H224.

510. Zhang, C., Harder, D.R. (2002) Cerebral capillary endothelial cell mitogenesis and morphogenesis induced by astrocytic epoxyeicosatrienoic acid. *Stroke* **33**: 2957–2964.

511. Sun, J., Sui, X., Bradbury, J.A., Zeldin, D.C., Conte, M.S., Lia, J.K. (2002) Inhibition of vascular smooth muscle migration by cytochrome P450 epoxygenase-derived eicosanoids. *Circ. Res.* **90**: 1020–1027.

512. Fleming, I., Michaelis, U.R., Bredenkotter, D., Fisslthaler, B., Dehghani, F., Brandes, R.P., Busse, R. (2001) Endothelium-derived hyperpolarizing factor synthase (cytochrome P450 2C9) is a functionally significant source of reactive oxygen species in coronary arteries. *Circ. Res.* **88**: 44–51.

513. Yang, B., Graham, L., Dikakov, S., Mason, R.P., Falck, J.R., Liao, J.K., Zeldin, D.C. (2001) Overexpression of cytochrome P450 CYP 2J2 protects against hypoxia-reoxygenation injury in cultured bovine aortic endothelial cells. *Mol. Pharmacol.* **60**: 310–320.

514. Fleming, I. (2004) Cytochrome P450 epoxygenases as EDHF synthase(s). *Pharmacol. Res.* **49**: 525–533.

515. Vanhoutte, P.M., Félétou, M., Taddei, S. (2005) Endothelium-dependent contractions in hypertension. *Br. J. Pharmacol.* **144**: 449–458.

516. Zhang, Y.Y., Walker, J.L., Huang, A., Keaney, J.F., Clish, C.B., Serhan, C.N., Loscalzo, J. (2002) Expression of 5-lipoxygenase in pulmonary endothelial cells. *Biochem. J.* **361**: 267–276.

517. Gu, J.L., Pei, H., Thomas, L., Nadler, J.L., Rossi, J.J., Lanting, L., Natarajan, R. (2001) Ribozyme-mediated inhibition of rat leukocyte-type 12-lipoxygenase prevents intimal hyperplasia in balloon-injured rat carotid arteries. *Circulation* **103**: 1446–1452.

518. Kim, J.A., Gu, J.L., Natarajan, R., Berliner, J.A., Nadler, J.L. (1995) A leukocyte-type of 12-lipoxygenase is expressed in human vascular and mononuclear cells. Evidence for upregulation by angiotensin II. *Arterioscler. Thromb. Vasc. Biol.* **15**: 942–948.

519. Lee, Y.W., Kuhn, H., Kaiser, S., Hennig, B., Daugherty, A., Toborek, M. (2001) Interleukin 4 induces transcription of the 15-lipoxygenase I type gene in human endothelial cells. *J. Lipid Res.* **42**: 783–791.

520. McLean, P.G., Aston, D., Sarkar, D., Ahluwalia, A. (2002) Protease-activated receptor-2 activation causes EDHF-like coronary vasodilation: selective preservation in ischemia/reperfusion injury: involvement of lipoxygenase products, VR1 receptors and C-fibers. *Circ. Res.* **90**: 465–472.

521. Faraci, F.M., Sobey, C.G., Chrissobolis, S., Lund, D.D., Heistad, D.D., Weintraub, N.L. (2001) Arachidonate dilates basilar artery by lipoxygenase-dependent mechanism and activation of K^+ channels. *Am. J. Physiol.* **281**: R246–R253.

522. Zinc, M.H., Oltman, C.L., Lu, T., Katakam, P.V., Kaduce, T.L., Lee, H., Dellsperger, K.C., Spector, A.A., Myers, P.R., Weintraub, N.L. (2001) 12-lipoxygenase in porcine coronary circulation: implications for coronary vasoregulation. *Am. J. Physiol.* **280**: H693–H704.

523. Barlow, R.S., El-Mowafy, A.M., White, R.E. (2000) H_2O_2 opens BK_{Ca} channels via the PLA_2-arachidonic acid signaling cascade in coronary artery smooth muscle. *Am. J. Physiol.* **279**: H475–H483.

524. Kurachi, Y., Ito, H., Sugimoto, T., Shimizu, T., Miki, I., Ui, M. (1989) Arachidonic acid metabolites as intracellular modulators of the G-protein-gated cardiac K^+ channel. *Nature* **337**: 1176–1179.

525. Pfister, S.L., Spitzbarth, N., Edgemont, W., Campbell, W.B. (1996) Vasorelaxation by an endothelium-derived metabolite of arachidonic acid. *Am. J. Physiol.* **270**: H1021–H1030.

526. Tang, X., Spitzbarth, N., Kuhn, H., Chaitidis, P., Campbell, W.B. (2003) Interleukin-13 upregulates vasodilatory 15-lipoxygenase eicosanoids in rabbit aorta. *Arterioscler. Thromb. Vasc. Biol.* **23**: 1768–1774.

527. Zhang, D.X., Gauthier, K.M., Chawengsub, Y., Homes, B.B., Campbell, W.B. (2005) Cyclooxygenase- and lipoxygenase-dependent relaxation to arachidonic acid in rabbit small mesenteric arteries. *Am. J. Physiol.* **288**: H302–H309.

528. Pfister, S.L., Spitzbarth, N., Nithipatikom, K., Edgemont, W., Campbell, W.B. (1999) Endothelium-derived eicosanoids from lipoxygenase relax the rabbit aorta by opening the potassium channel. In *Endothelium-Dependent Hyperpolarizations,* Vol 2., P.M. Vanhoutte, ed., Harwood Academic Publishers, The Netherlands, pp. 17–28.

529. Pfister, S.L., Spitzbarth, N., Nithipatikom, K., Edgemont, W., Falk, J.R., Campbell, W.B. (1998) Identification of 11,14,15- and 11,12,15-trihydroxyeicosatrienoic acids as endothelium-derived relaxing factors of rabbit aorta. *J. Biol. Chem.* **273**: 30879–30887.

530. Campbell, W.B., Spitzbarth, N., Gauthier, K.M., Pfister, S.L. (2003) 11,12,15-trihydroxyeicosatrienoic acid mediates Ach-induced relaxations in rabbit aorta. *Am. J. Physiol.* **285**: H2648–H2656.

531. Fulep, E.E., Vedernikov, Y.P., Saade, G.R., Garfield, R.E. (2001) The role of endothelium-derived hyperpolarizing factor in the regulation of the uterine circulation in pregnant rats. *Am. J. Obstet. Gynecol.* **185**: 638–642.

532. Weintraub, N.L., Joshi, N.L., Branch, C.A., Stephenson, A.H., Sprague, R.S., Lonigro, A.J. (1994) Relaxation of porcine coronary artery to bradykinin. Role of arachidonic acid. *Hypertension* **23**: 976–981.

533. Pomposiello, S., Rhaleb, N.E., Alva, M., Caretero, O.A. (1999) Reactive oxygen species: role in the relaxation induced by bradykinin or arachidonic acid via EDHF in isolated porcine coronary arteries. *J. Cardiovasc. Pharmacol.* **34**: 567–574.

534. Clark, D.L., Linden, J. (1986) Modulation of guanylate cyclase by lipoxygenase inhibitors. *Hypertension* **8**: 947–950.

535. Yamamura, I.I., Nagano, N., Iirano, M., Muraki, K., Watanabe, M., Imaizumi, Y. (1999) Activation of Ca^{2+}-dependent K^+ current by nordihydroguaiaretic acid in porcine coronary arterial smooth muscle cells. *J. Pharmacol. Exp. Ther.* **291**: 140–146.

536. Devane, W.A., Hanus, L., Breuer, A., Pertwee, R.G., Stevenson, L.A., Griffin, G., Gibson, D., Mandelbaum, A., Etinger, A., Mechoulam, R. (1992) Isolation and structure of a brain constituent that binds to the cannabinoid receptor. *Science* **258**: 1946–1949.

537. Di Marzo, V., Fontana, A., Cadas, H., Schinelli, S., Cimino, G., Schwartz, J.C., Piomelli, D. (1994) Formation and inactivation of endogenous cannabinoid anandamide in central neurons. *Nature* **372**: 686–691.

538. De Petrocellis, L., Cascio, M.G., Di Marzo, V. (2004) The endocannabinoid system: a general view and latest additions. *Br. J. Pharmacol.* **141**: 765–774.

539. Randall, M.D., Kendall, D.A., O'Sullivan, S. (2004) The complexities of the cardiovascular actions of cannabinoids. *Br. J. Pharmacol.* **142**: 20–26.

540. Rinaldi-Carmona, M., Barth, F., Héaulme, M., Shire, D., Calandra, B., Congy, C., Martinez, S., Maruani, J., Néliat, G., Caput, D., Ferrara, P., Soubrié, P., Breliere, J.-C., Le Fur, G. (1994) SR141716A, a potent and selective antagonist of the brain cannabinoid receptors. *FEBS Lett.* **350**: 240–244.

541. Randall, M.D., Alexander, S.P.H., Bennett, T., Boyd, E.A., Fry, J.R., Gardiner, S.M., Kemp, P.A., McCulloch, A.I., Kendall, D.A. (1996) An endogenous cannabinoid as an endothelium-derived vasorelaxant. *Biochem. Biophys. Res. Commun.* **229**: 114–120.

542. Randall, M.D., McCulloch, A.I., Kendall, D.A. (1997) Comparative pharmacology of endothelium-derived hyperpolarizing factor and anandamide in rat isolated mesentery. *Eur. J. Pharmacol.* **333**: 191–197.

543. Randall, M.D., Kendall, D.A. (1997) Involvement of a cannabinoid in endothelium-derived hyperpolarizing factor-mediated coronary vasorelaxation. *Eur J. Pharmacol.* **335**: 205–209.

544. Randall, M.D., Kendall, D.A. (1998) Anandamide and endothelium-derived hyper-polarizing factor act via a common vasorelaxant mechanism in rat mesentery. *Eur. J. Pharmacol.* **346**: 51–53.

545. Plane, F., Holland, M., Waldron, G.J., Garland, C.J., Boyle, J.P. (1997) Evidence that anandamide and EDHF act via different mechanisms in the rat isolated mesenteric arteries. *Br. J. Pharmacol.* **121**: 1509–1511.

546. White, R., Hiley, C.R. (1997) A comparison of EDHF-mediated responses and anandamide-induced relaxations in the rat isolated mesenteric artery. *Br. J. Pharmacol.* **122**: 1573–1584.

547. Pratt, P.F., Edgemont, W.S., Hillard, C.J., Campbell, W.B. (1998) N-Arachidonyleth-anolamide relaxation of bovine coronary arteries is not mediated by CB1 cannabinoid receptor. *Am. J. Physiol.* **274**: H375–H381.

548. Ellis, E.F., Moore, S.F., Willoughby, K.A. (1995) Anandamide and delta 9-THC dilation is blocked by indomethacin. *Am. J. Physiol.* **269**: H1859–H1864.

549. Chaytor, A.Y., Martin, P.E., Evans, W.H., Randall, M.D., Griffith, T.M. (1999) The endothelial component of cannabinoid-induced relaxation in rabbit mesenteric artery depends on gap junctional communications. *J. Physiol.* **520**: 539–550.

550. Fleming, I., Schermer, B., Popp, R., Busse, R. (1999) Inhibition of the production of endothelium-derived hyperpolarizing factor by cannabinoid receptor agonists. *Br. J. Pharmacol.* **126**: 949–960.

551. Hillard, C.J. (2000) Endocannabinoids and vascular function. *J. Pharmacol. Exp. Ther.* **294**: 27–32.

552. Kunos, G., Jarai, Z., Varga, K., Liu, J., Wang, L., Wagner, J.A. (2000) Cardiovascular effects of endocannabinoids—the plot thickens. *Prostaglandins Other Lipid Mediat.* **61**: 71–84.

553. Ralevic, V., Kendall, D.A., Randall, M.D., Smart, D. (2002) Cannabinoid modulation of sensory neurotransmission via cannabinoid and vanilloid receptors: role in regulation of cardiovascular function. *Life Sci.* **71**: 2577–2594.

554. Jarai, Z., Wagner, J.A., Vzarga, K., Lake, K.D., Compton, D.R., Martin, B.R., Zimmer, A.M., Bonner, T.I., Buckley, N.E., Mezey, E., Razdan, R.K., Zimmer, A., Kontos, G. (1999) Cannabinoid-induced mesenteric vasodilation through an endothelial site distinct from CB1 and CB2 receptors. *Proc. Natl. Acad. Sci. USA* **96**: 14136–14141.

555. Di Marzo, V., De Petrocellis, L., Fezza, F., Ligresti, A., Bisogno, T. (2002) Anandamide receptors. *Prostaglandins Leukot. Essent. Fatty Acids* **66**: 377–391.

556. Zygmunt, P.M., Petersson, J., Andersson, D.A., Chuang, H.H., Di Marzo, V., Julius, D., Högestätt, E.D. (1999) Vanilloid receptors on sensory nerves mediate the vasodilator action of anandamide. *Nature* **400**: 452–457.

557. Chemin, J., Monteil, A., Perez-Reyes, E., Nargeot, J., Lory, P. (2001) Direct inhibition of the T-type calcium channels by the endogenous cannabinoid anandamide. *EMBO J.* **20**: 7033–7040.

558. Maingret, F., Patel, A.J., Lazdunski, M., Honore, E. (2001) The endocannabinoid anandamide is a direct and selective blocker of the background K^+ channel TASK-1. *EMBO J.* **20**: 47–54.

559. Arroyo, C.M., Carmichael, A.J., Bouscarel, B., Liang, J.H., Weglicki, W.B. (1990) Endothelial cells as a source of oxygen-free radicals. An ESR study. *Free Radic. Res. Commun.* **9**: 287–296.

560. Sundquist, T. (1991) Bovine aortic endothelial cells release hydrogen peroxide. *J. Cell. Physiol.* **148**: 152–156.

561. Heinzel, B., John, M., Klatt, P., Böhme, E., Mayer, B. (1992) Ca^{2+}/calmodulin-dependent formation of hydrogen peroxide by brain nitric oxide synthase. *Biochem. J.* **281**: 627–630.

562. Brandes, R.P., Barton, M., Philippens, K.M., Schweitzer, G., Mugge, A. (1997) Endothelium-derived superoxide anions in pig coronary arteries: evidence from lucigenin chemiluminoscence and histochemical techniques. *J. Physiol.* **500**: 331–342.

563. Chaytor, A.T., Edwards, D.H., Bakker, L.M., Griffith, T.M. (2003) Distinct hyperpolarizing and relaxant roles for gap junctions and endothelium-derived H_2O_2 in NO-independent relaxations of rabbit arteries. *Proc. Natl. Acad. Sci. USA* **100**: 15212–15217.

564. Nishio, E., Watanabe, Y. (1997) The involvement of reactive oxygen species and arachidonic acid in alpha 1-adrenoceptor-induced smooth muscle cell proliferation and migration. *Br. J. Pharmacol.* **121**: 665–670.

565. Zafari, A.M., Ushio-Fukai, M., Akers, M., Yin, Q., Shah, A., Harrison, D.G., Taylor, W.R., Griendling, K.K. (1998) Role of NADH/NADPH oxidase-derived H_2O_2 in angiotensin II-induced vascular hypertrophy. *Hypertension* **32**: 488–495.

566. Kinoshita, H., Kakutani, T., Iranami, H., Hatano, Y. (2001) The role of oxygen-derived free radicals in augmented relaxations to levcromakalim in the aorta from hypertensive rats. *Jpn. J. Pharmacol.* **85**: 29–33.

567. Itoh, T., Kajikuri, J., Hattori, T., Kusanuma, N., Yamamoto, T. (2003) Involvement of H_2O_2 in superoxide-dismutase-induced enhancement of endothelium-dependent relaxation in rabbit mesenteric resistance arteries. *Br. J. Pharmacol.* **139**: 444–456.

568. Yang, H., Shi, M., Vanremmen, H., Chen, X., Vijg, J., Richardson, A., Guo, Z. (2003) Reduction of pressor response to vasoconstrictor agents by overexpression of catalase in mice. *Am. J. Hypertens.* **16**: 1–5.

569. Needleman, P., Jakshik, B., Johnson, E.M. (1973) Sulfhydryl requirement for relaxation of vascular smooth muscle. *J. Pharmacol. Exp. Ther.* **187**: 324–331.

570. Rubanyi, G.M., Vanhoutte, P.M. (1986) Superoxide anions and hyperoxia inactivate endothelium-derived relaxing factor. *Am. J. Physiol.* **250**: H222–H227.

571. Rubanyi, G.M., Vanhoutte, P.M. (1986) Oxygen-derived free radicals, endothelium, and responsiveness of vascular smooth muscle. *Am. J. Physiol.* **250**: H815–H821.

572. Beny, J-L., von der Weid, P.Y. (1991) Hydrogen peroxide: an endogenous smooth muscle cell hyperpolarizing factor. *Biochem. Biophys. Res. Commun.* **176**: 378–384.

573. Consentino, F., Katusic, Z.S. (1995) Tetrahydrobiopterin and dysfunction of endothelial nitric oxide synthase in coronary arteries. *Circulation* **91**: 139–144.

574. Barlow, R.S., White, R.E. (1998) Hydrogen peroxide relaxes porcine coronary arteries by stimulating BK_{Ca} channel activity. *Am. J. Physiol.* **275**: H1283–H1289.

575. Hayabuchi, Y., Nakaya, Y., Matsukoa, S., Kuroda, Y. (1998) Hydrogen peroxide-induced vascular relaxation in porcine coronary arteries is mediated by Ca^{2+}-activated K^+ channels. *Heart Vessels* **13**: 9–17.

576. Miura, H., Bosnjak, J.J., Ning, G., Salto, T., Miura, M., Gutterman, D.D. (2003) Role for hydrogen peroxide in flow-induced dilation of human coronary arterioles. *Circ. Res.* **92**: e31–e40.

577. Sato, A., Sakuma, I., Gutterman, D.D. (2003) Mechanism of dilation to reactive oxygen species in human coronary arterioles. *Am. J. Physiol.* **285**: H2345–H2354.

578. Thengchaisri, N., Kuo, L. (2003) Hydrogen peroxide induces endothelium-dependent and -independent coronary arteriolar dilation: role of cyclooxygenase and potassium channels. *Am. J. Physiol.* **285**: H2255–H2263.

579. Matoba, Y., Shimokawa, H., Nakashima, M., Hirakawa, Y., Mukai, Y., Hirano, K., Kanaide, H., Takeshita, A. (2000) Hydrogen peroxide is an endothelium-derived hyperpolarizing factor in mice. *J. Clin. Invest.* **106**: 1521–1530.

580. Fujimoto, S., Asano, T., Sakai, M., Sakurai, K., Takagi, D., Yoshimoto, N., Itoh, T. (2001) Mechanism of hydrogen peroxide-induced relaxation in rabbit mesenteric artery. *Eur. J. Pharmacol.* **412**: 291–300.

581. Ellis, A., Pannirselvam, M., Anderson, T.J., Triggle, C.R. (2003) Catalase has negligible effects on endothelium-dependent relaxations in mouse isolated aorta and small mesenteric artery. *Br. J. Pharmacol.* **140**: 1193–1200.

582. Gao, Y.J., Hirota, S., Zhang, D.W., Janssen, L.J., Lee, R.M. (2003) Mechanisms of hydrogen-peroxide-induced biphasic response in rat mesenteric artery. *Br. J. Pharmacol.* **138**: 1085–1092.

583. Hattori, T., Kajikuri, J., Katsuya, H., Itoh, T. (2003) Effects of H_2O_2 on membrane potential of smooth muscle cells in rabbit mesenteric resistance artery. *Eur. J. Pharmacol.* **464**: 101–109.

584. Wei, E.P., Kontos, H.A. (1990) H_2O_2 and endothelium-dependent cerebral arteriolar dilation. Implications for the identity of endothelium-derived relaxing factor generated by acetylcholine. *Hypertension* **16**: 162–169.

585. Yang, Z.W., Zhang, A., Altura, B.T., Altura, B.M. (1998) Endothelium-dependent relaxation to hydrogen peroxide in canine basilar artery: a potential new cerebral dilator mechanism. *Brain Res. Bull.* **47**: 257–263.

586. Iida, Y., Katusic, Z.S. (2000) Mechanisms of cerebral arterial relaxations to hydrogen peroxide. *Stroke* **31**: 2224–2230.

587. Bharadwadj, L., Prasad, K. (1995) Mediation of H_2O_2-induced vascular relaxation by endothelium-derived relaxing factor. *Mol. Cell Biochem.* **149–150**: 267–270.

588. Iesaki, T., Gupte, S.A., Kaminski, P.M., Wolin, M.S. (1999) Inhibition of guanylate cyclase stimulation by NO and bovine arterial relaxation to peroxynitrite and H_2O_2. *Am. J. Physiol.* **277**: H978–H985.

589. Yang, Z., Zhang, A., Altura, B.T., Altura, B.M. (1999) Hydrogen peroxide-induced endothelium-dependent relaxation of rat aorta involvement of Ca^{2+} and other cellular metabolites. *Gen. Pharmacol.* **33**: 325–336.

590. Fujimoto, S., Mori, M., Tsushima, H. (2003) Mechanisms underlying the hydrogen peroxide-induced, endothelium-dependent relaxation of the norepinephrine-contraction in guinea pig aorta. *Eur. J. Pharmacol.* **459**: 65–73.

591. Rabelo, L.A., Cortes, S.F., Alvarez-Leite, J.I., Lemos, V.S. (2003) Endothelium dysfunction in LDL receptor knock-out mice: a role for H_2O_2. *Br. J. Pharmacol.* **138**: 1215–1220.

592. Oeckler, R.A., Kaminski, P.M., Wolin, M.S. (2003) Stretch enhances contraction of bovine coronary arteries via an NAD(P)H oxidase-mediated activation of the extracellular signal-regulated kinase mitogen-activated protein kinase cascade. *Circ. Res.* **92**: 23–31.

593. Katusic, Z.S., Schugel, J., Cosentino, F., Vanhoutte, P.M. (1993) Endothelium-dependent contractions to oxygen-derived free radicals in the canine basilar artery. *Am. J. Physiol.* **264**: H859–H864.

594. Rodriguez-Martinez, M.A., Garcia-Cohen, E.C., Baena, A.B., Gonzales, R., Salaices, M., Marin, J. (1998) Contractile responses elicited by hydrogen peroxide in aorta from normotensive and spontaneously hypertensive rats. Endothelial modulation and mechanism involved. *Br. J. Pharmacol.* **125**: 1329–1335.

595. Sotlikova, R. (1998) Investigations of the mechanisms underlying H_2O_2-evoked contraction in the isolated rat aorta. *Gen. Pharmacol.* **31**: 115–119.

596. Yang, Z.W., Zheng, T., Zhang, A., Altura, B.T., Altura, B.M. (1998) Mechanisms of hydrogen peroxide-induced contractions of rat aorta. *Eur. J. Pharmacol.* **344**: 169–181.

597. Shen, J.Z., Zheng, X.F., Kwan, C.Y. (2000) Evidence for P(2)-purinoceptors contribution in H(2)O(2)-induced contraction of rat aorta in the absence of endothelium. *Cardiovasc. Res.* **47**: 574–585.

598. Gao, Y.J., Lee, R.M. (2001) Hydrogen peroxide induces a greater contraction in mesenteric arteries of spontaneously hypertensive rats through thromboxane A(2) production. *Br. J. Pharmacol.* **134**: 1639–1646.

599. Torrecillas, G., Boyano-Adanez, M.C., Medina, J., Parra, T., Griera, M., Lopez-Ongil, S., Arilla, E., Rodriguez-Puyol, M., Rodriguez-Puyol, D. (2001) The role of hydrogen peroxide in the contractile response to angiotensin II. *Mol. Pharmacol.* **59**: 104–112.

600. Lucchesi, P.A., Belmadani, S., Matrougui, K. (2005) Hydrogen peroxide acts as both vasodilator and vasoconstrictor in the control of perfused mesenteric resistance arteries. *J. Hypertens.* **23**: 571–579.

601. Tang, X.D., Garcia, M.L., Heinemann, S.H., Hoshi, T. (2004) Reactive oxygen species impair Slo1 BK channel function by altering cystein-mediated calcium sensing. *Nat. Struct. Mol. Biol.* **11**: 171–178.

602. Michelakis, E.D., Rebeyka, I., Wu, X., Nsair, A., Thebaud, B., Hashimoto, K., Dyck, J.R., Haromy, A., Harry, G., Barr, A., Archer, S.L. (2002) O_2 sensing in the human ductus arteriosus: regulation of voltage-gated K^+ channel in smooth muscle by mitochondrial redox sensor. *Circ. Res.* **91**: 478–486.

603. Greenwood, I.A., Leblanc, N., Gordienko, D.V., Large, W.A. (2002) Modulation of ICl_{Ca} in vascular smooth muscle cells by oxidising and cystein-reactive reagents. *Pflugers Arch.* **443**: 473–482.

604. Bychkov, R., Pieper, K., Ried, C., Milosheva, M., Bychkov, E., Luft, F.C., Haller, H. (1999) Hydrogen peroxide, potassium currents, and membrane potential in human endothelial cells. *Circulation* **99**: 1719–1725.

605. Ji, G., O'Brien, C.D., Feldman, M., Manevich, Y., Lim, P., Sun, J., Albelda, S.M., Kotlikoff, M.I. (2002) PECAM-1 (CD31) regulates a hydrogen peroxide-activated nonselective cation channel in endothelial cells. *J. Cell Biol.* **157**: 173–184.

606. Hu, Q., Corda, S., Zweier, J.L., Capogrossi, M.C., Ziegelstein, R.C. (1998) Hydrogen peroxide induces intracellular calcium oscillations in human aortic endothelial cells. *Circulation* **97**: 268–275.

607. Cai, R., Sauve, R. (1997) Effects of thiol-modifying agents on a KCa^{2+} channel of intermediate conductance in bovine aortic endothelial cells. *J. Membr. Biol.* **158**: 147–158.

608. Krippeit-Drews, P., Haberland, C., Fingerle, J., Drews, G., Lang, F. (1995) Effects of H_2O_2 on membrane potential and $[Ca^{2+}]_i$ on cultured rat arterial smooth muscle cells. *Biochem. Biophys. Res. Commun.* **209**: 139–145.

609. Sobey, C.G., Heistad, D.D., Faraci, F.M. (1997) Mechanisms of bradykinin-induced cerebral vasodilatation in rats. Evidence that reactive oxygen species activate K^+ channels. *Stroke* **28**: 2290–2294.

610. Wei, E.P., Kontos, H.A., Beckman, J.S. (1996) Mechanisms of cerebral vasodilation by superoxide, hydrogen peroxide, and peroxynitrite. *Am. J. Physiol.* **271**: H1262–H1266.

611. Gluais, P., Edwards, G., Weston, A.H., Vanhoutte, P.M., Félétou, M. (2005) Hydrogen peroxide and the endothelium-dependent hyperpolarization of the guinea-pig isolated carotid artery. *Eur. J. Pharmacol.* 513: 219–224..

612. Shimokawa, H., Matoba, T. (2004) Hydrogen peroxide as an endothelium-derived hyperpolarizing factor. *Pharmacol. Res.* **49**: 543–549.

613. Kimura, K., Tsuda, K., Moriwaki, C., Kawabe, T., Hamada, M., Obana, M., Baba, A., Hano, T., Nishio, I. (2002) Leukemia inhibitory factor relaxes arteries through endothelium-dependent mechanism. *Biochem. Biophys. Res. Commun.* **294**: 359–362.

614. Morikawa, K., Shimokawa, H., Matoba, T., Kubota, H., Akaike, T., Talukder, M.A., Hatanaka, M., Fujiki, T., Maeda, H., Takahashi, S., Takeshita, A. (2003) Pivotal role of Cu,Zn-superoxide dismutase in endothelium-dependent hyperpolarization. *J. Clin. Invest.* **112**: 1871–1879.

615. Yada, T., Shimokawa, H., Hiramatsu, O., Kajita, T., Shigeto, F., Goto, M., Ogasawara, Y., Kajiya, F. (2003) Hydrogen peroxide, an endogenous endothelium-derived hyperpolarizing factor, plays an important role in coronary autoregulation in vivo. *Circulation* **107**: 1040–1045.

616. Fulton, D., McGiff, J.C., Wolin, M.S., Kaminski, P., Quilley, J. (1997) Evidence against a cytochrome P450-derived reactive oxygen species as the mediator of the nitric oxide-independent vasodilator effect of bradykinin in the perfused heart of the rat. *J. Pharmacol. Exp. Ther.* **280**: 702–709.

617. Hamilton, C.A., McPhaden, A.R., Berg, G., Pathi, V., Dominiczak, A.F. (2001) Is hydrogen peroxide an EDHF in human radial arteries? *Am. J. Physiol.* **280**: H2451–H2455.

618. Ellis, A., Triggle, C.R. (2003) Endothelium-derived reactive oxygen species: their relationship to endothelium-dependent hyperpolarization and vascular tone. *Can. J. Physiol. Pharmacol.* **81**: 1013–1028.

619. Cai, W.Q., Bodin, P., Loesch, A., Sexton, A., Burnstock, G. (1993) Endothelium of human umbilical blood vessels — ultrastructural immunolocalization of neuropeptides. *J. Vasc. Res.* **30**: 348–355.

620. Stingo, A.J., Clavell, A.L., Heublein, D.M., Wei, C.M., Pittlekow, M.R., Burnett, J.C., Jr. (1992) Presence of C-type natriuretic peptide in cultured human endothelial cells and plasma. *Am. J. Physiol.* **263**: H1318–H1321.

621. Sugo, S., Minamino, N., Kangawa, K., Miyamoto, K., Kitamura, K., Sakata, J., Eto, T., Matsuo, H. (1994) Endothelial cells actively synthesize and secrete adrenomedullin. *Biochem. Biophys. Res. Commun.* **201**: 719–726.

622. Koller, K.J., Goeddel, D.V. (1992) Molecular biology of the natriuretic peptides and their receptors. *Circulation* **86**: 1081–1088.

623. Nakao, K., Itoh, H., Saito, Y., Mukoyama, M., Ogawa, Y. (1996) The natriuretic peptide family. *Curr. Opin. Nephrol. Hypertens.* **5**: 4–11.

624. Suga, S., Itoh, H., Komatsu, Y., Ogawa, Y., Hama, N., Yoshimasa, T., Nakao, K. (1993) Cytokine-induced c-type natriuretic peptide (CNP) secretion from vascular endothelial cells—evidence for CNP as a novel autocrine/paracrine regulator from endothelial cells. *Endocrinology* **133**: 3038–3041.

625. Nazario, B., Hu, R.M., Pedram, A., Prins, B., Levin, E.R. (1995) Atrial and brain natriuretic peptides stimulate the production and secretion of C-type natriuretic peptide from bovine aortic endothelial cells. *J. Clin. Invest.* **95**: 1151–1157.

626. Wei, C.M., Hu, S., Miller, V.M., Burnett, J.C., Jr. (1994) Vascular actions of C-type natriuretic peptide in isolated porcine coronary arteries and coronary vascular smooth muscle cells. *Biochem. Biophys. Res. Comm.* **205**: 765–771.

627. Nakamura, M., Arakawa, N., Yoshida, H., Makita, S., Hiramori, K. (1994) Vasodilatory effects of C-type natriuretic peptide on forearm resistance vessels are distinct from those of natriuretic peptide in chronic heart failure. *Circulation* **90**: 1210–1214.

628. Banks, M., Wei, C.M., Kim, C.H., Burnett, J.C., Jr., Miller, V.M. (1996) Mechanism of relaxations to C-type natriuretic peptide in veins. *Am. J. Physiol.* **271**: H1907–H1911.

629. Wennberg, P.W., Miller, V.M., Rabelink, T., Burnett, J.C., Jr. (1999) Further attenuation of endothelium-dependent relaxation imparted by natriuretic peptide receptor antagonism. *Am. J. Physiol.* **277**: H1618–H1821.

630. Honing, M.L., Smits, P., Morrison, P.J., Burnett, J.C., Jr., Rabelink, T.J. (2001) C-type natriuretic peptide-induced vasodilation is dependent on hyperpolarization in human forearm resistance vessels. *Hypertension* **37**: 1179–1183.

631. Barton, M., Beny, J-L., Uscio, L.V., Wyss, T., Noll, G., Luscher, T.F. (1998) Endothelium-independent relaxation and hyperpolarization to C-type natriuretic peptide in porcine coronary arteries *J. Cardiovasc. Pharmacol.* **31**: 377–383.

632. Chauhan, S.D., Nilsson, H., Ahluwalia, A., Hobbs, A.J. (2003) Release of C-type natriuretic peptide accounts for the biological activity of endothelium-derived hyperpolarizing factor. *Proc. Natl. Acad. Sci. USA* **100**: 1426–1431.

633. Koller, K.J., Lowe, D.G., Bennett, G.L., Minamino, N., Kangawa, K., Matsuo, H., Goeddel, D.V. (1991) Selective activation of the B natriuretic peptide receptor by C-type natriuretic peptide (CNP). *Science* **252**: 120–123.

634. Amin, J., Carretero, O.A., Ito, S. (1996) Mechanism of action of atrial natriuretic peptide and C-type natriuretic peptide. *Hypertension* **27**: 684–687.

635. Murthy, K.S., Makhlouf, G.M. (1999) Identification of the G protein-activating domain of the natriuretic peptide clearance receptor (NPR-C). *J. Biol. Chem.* **274**: 17587–17592.

636. Maack, T., Suzuki, M., Almeida, F.A., Nussenzveig, D., Scarborough, R.M., McEnroe, G.A., Lewicki, J.A. (1987) Physiological role of silent receptors of atrial natriuretic factors. *Science* **238**: 675–678.

637. Ahluwalia, A., MacAllister, R.J., Hobbs, A.J. (2004) Vascular actions of natriuretic peptides. Cyclic GMP-dependent and -independent mechanisms. *Basic Res. Cardiol.* **99**: 83–89.

638. Krapivinsky, G., Gordon, E., Wickman, K., Velimirovic, B., Krapivinsky, L., Clapham, D.E. (1995) The G-protein-gated atrial K+ channel, IKACh, is a heteromultimer of two inwardly rectifying K+ channel proteins. *Nature* **374**: 135–141.

639. Mark, M.D., Herlitze, S. (2000) G-protein mediated gating of inward-rectifier K+ channels. *Eur. J. Biochem.* **267**: 5830–5836.

640. Sadja, R., Alagem, N., Reuveny, E. (2003) Gating of GIRK channels: details of an intricate, membrane-delimited complex. *Neuron* **39**: 9–12.

641. Bradley, K.K., Hatton, W.J., Mason, H.S., Walker, R.L., Flynn, E.R., Kenyon, J.L., Horowitz, B. (2000) Kir3.1/3.2 encodes an I(KACh)-like current in gastrointestinal myocytes. *Am. J. Physiol. Gastrointest. Liver Physiol.* **278**: G289–G296.

642. Ren, Y.J., Xu, X.H., Zhong, C.B., Feng, N., Wang, X.L. (2001) Hypercholesterolemia alters vascular functions and gene expression of potassium channels in rat aortic smooth muscle cells. *Acta Pharmacol. Sin.* **22**: 274–278.

643. Ren, Y.J., Xu, X.H., Wang, X.L. (2003) Atered mRNA expression of ATP-sensitive and inward rectifier potassium channel subunits in streptozotocin-induced diabetic rat heart and aorta. *J. Pharmacol. Sci.* **93**: 478–483.

644. Rose, R.A., Lomax, A.E., Giles, W.R. (2003) Inhibition of L-type Ca^{2+} current by C-type natriuretic peptide in bullfrog atrial myocytes: an NPR-C mediated effect. *Am. J. Physiol.* **285**: H2454–H2462.

645. Kitamura, K., Kangawa, K., Kawamoto, M., Ichiki, Y., Nakamura, S., Matsuo, H., Eto, K. (1993) Adrenomedullin: a novel hypotensive peptide isolated from human pheochromocytoma. *Biochem. Biophys. Res. Comm.* **192**: 553–560.

646. Poyner, D.R., Sexton, P.M., Marshall, I., Smith, D.M., Quirion, R., Born, W., Muff, R., Fisher, J.A., Foord, S.M. (2002) International Union of Pharmacology. XXXII. The mammalian calcitonin gene-related peptides, adrenomedullin, amylin and calcitonin receptors. *Pharmacol. Rev.* **54**: 233–246.

647. Brain, S.D., Grant, A.D. (2004) Vascular actions of calcitonin-gene related peptide and adrenomedullin. *Physiol. Rev.* **84**: 903–934.
648. Kis, B., Abraham, C.S., Deli, M.A., Kobayashi, H., Niwa, M., Yamashita, H., Busija, D.W., Ueta, Y. (2003) Adrenomedullin, an autocrine mediator of blood brain barrier function. *Hypertens. Res.* **26**: S61–S70.
649. Ozaka, T., Doi, Y., Kayashima, K., Fujimoto, S. (1997) Weibel-Palade bodies as a storage site of calcitonin gene related peptide and endothelin-1 in blood vessels of the rat carotid body. *Anatom. Record* **247**: 388–394.
650. Wu, D., Doods, H., Arndt, K., Schindler, M. (2002) Development and potential of non-peptide antagonists for calcitonin-gene related peptide (CGRP) receptors: evidence for CGRP receptor heterogeneity. *Biochem. Soc. Trans.* **30**: 468–473.
651. Terata, K., Miura, H., Liu, Y., Loberiza, F., Gutterman, D.D. (2000) Human coronary arteriolar dilation to adrenomedullin: role of nitric oxide and K^+ channels. *Am. J. Physiol.* **279**: H2620–H2626.
652. Gumusel, B., Hao, Q., Hyman, A.L., Kadowitz, P.J., Champion, H.C., Chang, J.K., Mehta, J.L., Lippton, H. (1998) Analysis of responses to adrenomedullin-(13-52) in the pulmonary vascular bed of rats. *Am. J. Physiol.* **274**: H1255–H1263.
653. Champion, H.C., Pierce, R.L., Bivalacqa, T.J., Murphy, W.A., Coy, D.H., Kadowitz, P.J. (2001) Analysis of responses to hAmylin, hCGRP and hADM in isolated resistance arteries from the mesenteric vascular bed of the rat. *Peptides* **22**: 1427–1434.
654. Wangensteen, R., Quesada, A., Sainz, J., Duarte, J., Vargas, F., Osuna, A. (2002) Role of endothelium-derived relaxing factors in adrenomedullin-induced vasodilation in the rat kidney. *Eur. J. Pharmacol.* **444**: 97–102.
655. Dettmann, E.S., Vysniauskiene, I., Wu, R., Flammer, J., Haefliger, I.O. (2003) Adrenomedullin-induced endothelium-dependent relaxation in porcine ciliary artery. *Invest. Ophthalmol. Sci.* **44**: 3961–3966.
656. Boussery, K., Delaey, C., Van de Voorde, J. (2004) Influence of adrenomedullin on tone of isolated bovine retinal arteries. *Invest. Ophthalmol. Vis. Sci.* **45**: 552–559.
657. Nishimura, Y., Suzuki, A. (1997) Relaxant effects of vasodilator peptides on isolated basilar arteries from stroke-prone spontaneously hypertensive rats. *Clin. Exp. Pharmacol. Physiol.* **24**: 157–161.
658. Eguchi, S., Hirata, Y., Iwasaki, H., Sato, K., Watanabe, T.X., Inui, T., Nakajima, K., Sakakibara, S., Marumo, F. (1994) Structure-activity relationship of adrenomedullin, a novel vasodilatory peptide, in cultured vascular smooth muscle cells. *Endocrinology* **135**: 2454–2458.
659. Ishizaka, Y., Ishiyama, Y., Tanaka, M., Kitamura, K., Kangawa, K., Minamino, N., Matsuo, H., Eto, T. (1994) Adrenomedullin stimulates cyclic AMP formation in rat vascular smooth muscle cells. *Biochem. Biophys. Res. Comm.* **200**: 642–646.
660. Nelson, M.T., Huang, Y., Brayden, J.E., Hescheler, J., Standen, N.B. (1990) Arterial dilations in response to calcitonin gene related peptide involve activation of K^+ channels. *Nature* **344**: 770–773.
661. Zschauer, A., Uusitalo, H., Brayden, J.E. (1992) Role of endothelium and hyperpolarization in CGRP-induced vasodilatation of rabbit ophthalmic artery. *Am. J. Physiol.* **263**: H359–H365.
662. Bodelsson, G., Stjernquist, M. (1992) Smooth muscle dilation in the human uterine artery induced by substance P, vasoactive intestinal polypeptide, calcitonin gene-related peptide and atrial natriuretic peptide: relation to endothelium-derived substances. *Hum. Reprod.* **7**: 1185–1188.

663. Luu, T.N., Dashwood, M.R., Tadjkarimi, S., Chester, A.H., Yacoub, M.H. (1997) ATP-sensitive potassium channels mediate vasodilatation by calcitonin gene related peptide in human internal mammary but not gastroepiploic arteries. *Eur. J. Clin. Invest.* **27**: 960–966.

664. Kitazono, T., Heistad, D.D., Faraci, F.M. (1993) Role of ATP-sensitive K⁺ channels in CGRP-induced dilatation of basilar artery in vivo. *Am. J. Physiol.* **265**: H851–H855.

665. Lei, S., Mulvany, M.J., Nyborg, N.C. (1994) Characterization of the CGRP receptor and mechanisms of action in rat mesenteric small arteries. *Pharmacol. Toxicol.* **74**: 130–135.

666. Gao, Y.J., Nishimura, Y., Nakai, Y. (1995) Calcitonin gene-related peptide-induced relaxation in isolated small superior mesenteric arteries from adult stroke-prone spontaneously hypertensive rats. *Clin. Exp. Pharmacol. Physiol.* **22**: S109–S111.

667. Miyoshi, H., Nakaya, Y. (1995) Calcitonin gene related peptide activates the K⁺ channels of vascular smooth muscle cells via adenylate cyclase. *Basic Res. Cardiol.* **90**: 332–336.

668. Wellman, G.C., Quayle, J.M., Standen, N.B. (1998) ATP-sensitive K⁺ channel activation by calcitonin gene related peptide and protein kinase A in pig coronary arterial smooth muscle. *J. Physiol.* **507**: 117–129.

669. Kobayashi, D., Todoki, K., Ozono, S., Okabe, E. (1995) Calcitonin-gene-related peptide mediated neurogenic vasorelaxation in the isolated canine lingual artery. *Jpn. J. Pharmacol.* **67**: 329–339.

670. Lo, Y.C., Wu, J.R., Wu, S.N., Chen, I.J. (1997) Glyceryl nonivamide: a capsaicin derivative with cardiac calcitonin gene-related peptide releasing, K⁺ channel opening and vasorelaxant properties. *J. Pharmacol. Exp. Ther.* **28**: 253–260.

671. Dunn, W.R., Hardy, T.A., Brock, J.A. (2003) Electrophysiological effects of activating the peptidergic primary afferent innervation of rat mesenteric arteries. *Br. J. Pharmacol.* **140**: 231–238.

672. Hogestatt, E.D., Johansson, R., Andersson, D.A., Zygmunt, P.M. (2000) Involvement of sensory nerves in vasodilator response to acetylcholine and potassium ions in rat hepatic artery. *Br. J. Pharmacol.* **130**: 27–32.

673. Scott, T.M., Chafe, L. (1994) The involvement of CGRP in acetylcholine-induced vascular relaxation. *Artery* **21**: 38–50.

674. Harmar, A.J., Arimura, A., Gozes, I., Journot, L., Laburthe, M., Pisegna, J.R., Rawlings, S.R., Robberecht, P., Said, S.I., Sreedharan, S.P., Wank, S.A., Waschek, J.A. (1998) International Union of Pharmacology. XVIII. Nomenclature of receptors for vasoactive intestinal peptide and pituitary adenylate cyclase-activating polypeptide. *Pharmacol. Rev.* **50**: 265–270.

675. Ignarro, L.J., Byrns, R.E., Buga, G.M., Wood, K.S. (1987) Mechanism of endothelium-dependent vascular smooth muscle relaxation elicited by bradykinin and VIP. *Am. J. Physiol.* **253**: H1074–H1082.

676. Jovanovic, A., Jovanovic, S., Tulic, I., Grbovic, L. (1998) Predominant role for nitric oxide in the relaxation induced by vasoactive intestinal polypeptide in human uterine artery. *Mol. Hum. Reprod.* **4**: 71–76.

677. Chan, S.L., Fiscus, R.R. (2003) Vasorelaxations induced by calcitonin gene-related peptide, vasoactive intestinal peptide, and acetylcholine in aortic rings of endothelial and inducible nitric oxide synthase-knock-out mice. *J. Cardiovasc. Pharmacol.* **41**: 434–443.

678. Itoh, H., Lederis, K.P., Rorstad, O.P. (1990) Relaxation of isolated bovine coronary arteries by vasoactive intestinal peptide. *Eur. J. Pharmacol.* **181**: 199–205.

679. Ganz, P., Sandbrock, A.W., Landis, S.C., Leopold, J., Gimbrone, M.A., Alexander, R.W. (1986) Vasoactive intestinal peptide: vasodilatation and cyclic AMP generation. *Am. J. Physiol.* **250**: H755–H760.

680. Tanaka, Y., Mochizuki, Y., Hirano, H., Aida, M., Tanaka, H., Toro, L., Shigenobu, K. (1999) Role of maxiK channels in vasoactive intestinal peptide-induced relaxation of rat mesenteric artery. *Eur. J. Pharmacol.* **383**: 291–296.

681. Standen, N.B., Quayle, J.M., Davies, N.W., Brayden, J.E., Huang, Y., Nelson, M.T. (1989) Hyperpolarizing vasodilators activate ATP-sensitive K^+ channels in arterial smooth muscle. *Science* **245**: 177–180.

682. Kawasaki, J., Kobayashi, S., Miyagi, Y., Nishimura, J., Fujishima, M., Kanaide, H. (1997) The mechanisms of the relaxation induced by vasoactive intestinal peptide in the porcine coronary artery. *Br. J. Pharmacol.* **121**: 977–985.

683. Bruch, L., Rubel, S., Kastner, A., Gellen, K., Gollasch, M. (1998) Pituitary adenylate cyclase activating peptides relax human pulmonary arteries by opening K_{ATP} and K_{Ca} channels. *Thorax* **53**: 586–587.

684. Kastner, A., Bruch, L., Will-Sahab, L., Modersohn, D., Baumann, G. (1995) Pituitary adenylate cyclase peptides are endothelium-independent dilators of human and porcine coronary arteries. *Agents Actions Suppl.* **45**: 283–289.

685. Bruch, L., Bychkov, R., Kastner, A., Bulow, T., Ried, C., Gollasch, M., Baumann, G., Luft, F.C., Haller, H. (1997) Pituitary adenylate cyclase activating peptides relax human coronary arteries by activating K_{ATP} and K_{Ca} channels in smooth muscle cells. *J. Vasc. Res.* **34**: 11–18.

686. Lange, D., Funa, K., Ishisaki, A., Bauer, R., Wollina, U. (1999) Autocrine endothelial regulation in brain stem vessels of newborn piglets. *Histol. Histopathol.* **14**: 821–825.

687. Koh, P.O., Noh, H.S., Kim, Y.S., Cheon, E.W., Kim, H.J., Kang, S.S., Cho, G.J., Choi, W.S. (2003) Cellular localization of pituitary adenylate cyclase-activating polypeptide in the rat testis. *Mol. Cells* **15**: 271–276.

688. Wang, R. (2002) Two's company, three's a crowd: can H_2S be the third endogenous gaseous transmitter? *FASEB J.* **16**: 1792–1798.

689. Durante, W. (2002) Carbon monoxide and bile pigments: surprising mediators of vascular function. *Vasc. Med.* **7**: 195–202.

690. Wang, R. (1998) Resurgence of carbon monoxide: an endogenous gaseous vasorelaxing factor. *Can. J. Physiol. Pharmacol.* **76**: 1–15.

691. Chen, Y-H., Yet, S-F., Perrella, M.A. (2003) Role of heme-oxygenase-1 in the regulation of blood pressure and cardiac function. *Exp. Biol. Med.* **228**: 447–453.

692. Wang, R., Wang, Z.Z., Wu, L.Y. (1997) Carbon-monoxide-induced vasorelaxation and the underlying mechanisms. *Br. J. Pharmacol.* **121**: 927–934.

693. Nishikawa, Y., Stepp, D.W., Merkus, D., Jones, D., Chilian, W.M. (2004) In vivo role of heme oxygenase in ischemic coronary vasodilation. *Am. J. Physiol.* **286**: H2296–H2304.

694. Wang, R., Wu, L.Y. (1997) The chemical modification of KCa channels by carbon monoxide in vascular smooth muscle cells. *J. Biol. Chem.* **272**: 8222–8226.

695. Wang, R., Wu, L.Y., Wang, Z.Z. (1997) The direct effect of carbon monoxide on KCa channels in vascular smooth muscle cells. *Pflügers Arch.* **434**: 285–291.

696. Wu, L., Cao, K., Lu, Y., Wang, R. (2002) Different mechanisms underlying the stimulation of K(Ca) channels by nitric oxide and carbon monoxide. *J. Clin. Invest.* **110**: 691–700.

697. Jaggar, J.H., Leffler, C.W., Cheranov, S.Y., Tcheranova, D.E.S., Cheng, X. (2002) Carbon monoxide dilates cerebral arterioles by enhancing the coupling of Ca^{2+} sparks to Ca^{2+}-activated K^+ channels. *Circ. Res.* **91**: 610–617.

698. Xi, Q., Tcheranova, D., Parfenova, H., Horowitz, B., Leffler, C.W., Jaggar, J.H. (2004) Carbon monoxide activates K_{Ca} channels in newborn arteriole smooth muscle cells by increasing apparent Ca^{2+} sensitivity of alpha-subunits. *Am. J. Physiol.* **286**: H610–H618.

699. Naik, J.S., Walker, B.R. (2003) Heme oxygenase-mediated vasodilation involves vascular smooth muscle cell hyperpolarization. *Am. J. Physiol.* **285**: H220–H228.

700. Komuro, T., Borsody, M.K., Ono, S., Marton, L.S., Weir, B.K., Zhang, Z.D., Paik, E., Macdonald, R.L. (2001) The vasorelaxation of cerebral arteries by carbon monoxide. *Exp. Biol. Med.* **226**: 860–865.

701. Coceani, F., Kelsey, L., Seidlitz, E. (1996) Carbon monoxide-induced relaxation of the ductus arteriosus in the lamb: evidence against the prime role of guanylyl cyclase. *Br. J. Pharmacol.* **118**: 1689–1696.

702. Coceani, F., Kelsey, L., Seidlitz, E., Korzekwa, K. (1996) Inhibition of the contraction of the ductus arteriosus to oxygen by 1-aminobenzotriazole, a mechanism-based inactivation of cytochrome P-450. *Br. J. Pharmacol.* **117**: 1586–1592.

703. Roman, R.J. (2002) P450 metabolites of arachidonic acid in the control of cardiovascular function. *Physiol. Rev.* **82**: 131–185.

704. Leffler, C.W., Nasjletti, A., Yu, C., Johnson, R.A., Fedinec, A.L., Walker, N. (1999) Carbon monoxide and cerebral microvascular tone in newborn pigs. *Am. J. Physiol.* **276**: H1641–H1646.

705. Kaide, J.I., Zhang, F., Wei, Y., Jiang, H., Yu, C., Wang, W.H., Balazy, M., Abraham, N.G., Nasjletti, A. (2001) Carbon monoxide of vascular origin attenuates the sensitivity of renal arterial vessels to vasoconstrictors. *J. Clin. Invest.* **107**: 1163–1171.

706. Zhang, F., Kaide, J., Wei, Y., Jiang, H., Yu, C., Balazy, M., Abraham, N.G., Wang, W., Nasjletti, A. (2001) Carbon monoxide produced by isolated arterioles attenuates pressure-induced vasoconstriction. *Am. J. Physiol.* **281**: H350–H358.

707. Suematsu, M., Suganuma, K., Kashiwagi, S. (2003) Mechanistic probing of gaseous signal transduction in microcirculation. *Antioxid. Redox Signal.* **5**: 485–492.

708. Zygmunt, P.M., Högestätt, E.D., Grundemar, L. (1994) Light-dependent effects of zinc protoporphyrin IX on endothelium-dependent relaxation resistant to N-omega-nitroL-arginine. *Acta Physiol. Scand.* **152**: 137–143.

709. Ishikawa, K. (2003) Heme-oxygenase-1 against vascular insufficiency: role of atherosclerotic disorders. *Curr. Pharm. Des.* **9**: 2489–2497.

710. Hosoki, R., Matsuki, N., Kimura, H. (1997) The possible role of hydrogen sulfide as an endogenous smooth muscle relaxant in synergy with nitric oxide. *Biochem. Biophys. Res. Commun.* **237**: 527–531.

711. Zhao, W., Zhang, J., Lu, Y., Wang, R. (2001) The vasorelaxant effect of H_2S as a novel endogenous gaseous K_{ATP} channel opener. *EMBO J.* **20**: 6008–6016.

712. Zhao, W., Wang, R. (2002) H_2S-induced vasorelaxation and underlying cellular and molecular mechanisms. *Am. J. Physiol.* **283**: H474–H480.

713. Cheng, Y., Ndisang, J.F., Tang, G., Cao, K., Wang, R. (2004) Hydrogen sulfide induced relaxation of resistance mesenteric artery beds of rats. *Am. J. Physiol.* **287**: H2316–H2323.

714. Burnstock, G. (1997) The past, present and future of purine nucleotides as signaling molecules. *Neuropharmacology* **36**: 1127–1139.

715. Ralevic, V., Burnstock, G. (1998) Receptors for purines and pyrimidines. *Pharmacol. Rev.* **50**: 413–492.

716. Burnstock, G. (2002) Purinergic signaling and vascular cell proliferation and death. *Arterioscler. Thromb. Vasc. Biol.* **22**: 364–373.

717. Olsson, R.A., Pearson, J.D. (1990) Cardiovascular purinoceptors. *Physiol. Rev.* **70**: 761–845.

718. Ralevic, V., Burnstock, G. (1991) Roles of P_2-purinoceptors in the cardiovascular system. *Circulation* **84**: 2212.

719. Kunapuli, S.P., Daniel, J.L. (1998) P2 receptor subtypes in the cardiovascular system. *Biochem. J.* **336**: 513–523.

720. Berne, R.M. (1980) The role of adenosine in the regulation of coronary blood flow. *Circ. Res.* **47**: 807–813.
721. Clifford, P.S., Hellsten, Y. (2004) Vasodilatory mechanisms in contracting skeletal muscle. *J. Appl. Physiol.* **97**: 393–403.
722. Shryock, J.C., Rubio, R., Berne, R.M. (1988) Release of adenosine from pig aortic endothelial cells during hypoxia and metabolic inhibition. *Am. J. Physiol.* **254**: H223–H229.
723. Shinozuka, K., Hashimoto, M., Bjur, R.A., Westfall, W.P., Hattori, K. (1994) In vitro studies of release of adenine nucleotides and adenosine from rat vascular endothelium in response to alpha(1)-adrenoceptor stimulation. *Br. J. Pharmacol.* **113**: 1203–1208.
724. Nees, S., Gerbes, A.L., Willershausen-Zonnchen, B., Gerlach, E. (1980) Purine metabolism in cultured coronary endothelial cells. *Adv. Exp. Med. Biol.* **122**: 25–30.
725. Milner, P., Kirkpatrick, K.A., Ralevic, V., Toothill, V., Burnstock, G. (1990) Endothelial cells cultured from human umbilical vein release ATP and acetylcholine in response to increased flow. *Proc. R. Soc. Lond. B* **241**: 245–248.
726. Bodin, P., Bailey, D.J., Burnstock, G. (1991) Increased flow-induced ATP release from isolated vascular endothelial but not smooth muscle cells. *Br. J. Pharmacol.* **103**: 1203–1205.
727. Burnstock, G. (1999) Release of vasoactive substances from endothelial cells by shear stress and purinergic mechanosensory transduction. *J. Anat.* **194**: 335–342.
728. Sedaa, K.O., Bjur, R.A., Shinozuka, K., Westfall, D.P. (1990) Nerve and drugs-induced release of adenine nucleosides and nucleotides from rabbit aorta. *J. Pharmacol. Exp. Ther.* **252**: 1060–1067.
729. Bodin, P., Burnstock, G. (1996) ATP-stimulated release of ATP by human endothelial cells. *J. Cardiovasc. Pharmacol.* **27**: 872–875.
730. Grbovic, L., Radenkovic, M., Prostran, M., Pesic, S. (2000) Characterization of adenosine action in isolated renal artery. Possible role of adenosine A(2A) receptors. *Gen. Pharmacol.* **35**: 29–36.
731. Bryan, P.T., Marshall, J.M. (1999) Cellular mechanisms by which adenosine induces vasodilatation in rat skeletal muscle: significance for systemic hypoxia. *J. Physiol.* **514**: 163–175.
732. Hinschen, A.K., Rose'Meyer, R.B., Headrick, J.P. (2001) Age-related changes in adenosine-mediated relaxation of coronary and aortic smooth muscle. *Am. J. Physiol.* **280**: H2380–H2389.
733. Rump, L.C., Jabbari, T.J., von Kugelgen, I., Oberhauser, V. (1999) Adenosine mediates nitric oxide-independent renal vasodilation by activation of A_{2A} receptors. *J. Hypertens.* **17**: 1987–1993.
734. Prior, H.M., Yates, M.S., Beech, D.J. (1999) Role of K^+ channels in A_{2A} adenosine receptor-mediated dilation of the pressurized renal arcuate artery. *Br. J. Pharmacol.* **126**: 494–500.
735. Rubanyi, G.M., Vanhoutte, P.M. (1985) Endothelium removal decreases relaxations of canine coronary arteries caused by beta-adrenergic agonists and adenosine. *J. Cardiovasc. Pharmacol.* **7**: 139–144.
736. Kuo, L., Chancellor, J.D. (1995) Adenosine potentiates flow-induced dilation of coronary arterioles by activating K-ATP channels in endothelium. *Am. J. Physiol.* **38**: H541–H549.
737. Olanrewaju, H.A., Hargittai, P.T., Lieberman, E.A., Mustafa, S.J. (1995) Role of endothelium in hyperpolarisation of coronary smooth muscle by adenosine and its analogues. *J. Cardiovasc. Pharmacol.* **25**: 234–239.

738. Olanrewaju, H.A., Hargittai, P.T., Lieberman, E.M., Mustafa, S.J. (1997) Effect of ouabain on adenosine receptor-mediated hyperpolarization in porcine coronary artery smooth muscle. *Eur. J. Pharmacol.* **322**: 185–190.

739. Smits, P., Williams, S.B., Lipson, D.E., Banitt, P., Ronge, G.A., Creager, M.A. (1995) Endothelial release of nitric oxide contributes to the vasodilator effect of adenosine in humans. *Circulation* **92**: 2135–2141.

740. Olanrewaju, H.A., Gafurov, B.S., Mustafa, S.J. (2002) Involvement of K^+ channels in adenosine A_{2A} and A_{2B} receptor-mediated hyperpolarization of porcine coronary endothelial cells. *J. Cardiovasc. Pharmacol.* **40**: 43–49.

741. Imai, S., Takeda, K. (1967) Effects of vasodilators upon the isolated taenia coli of the guinea pig. *J. Pharmacol. Exp. Ther.* **156**: 557–564.

742. Kleppisch, T., Nelson, M.T. (1995) Adenosine activates ATP-sensitive potassium channels in arterial myocytes via A_2 receptor and cAMP-dependent protein kinase. *Proc. Natl. Acad. Sci. USA* **92**: 12441–12445.

743. Herlihy, J.T., Bockman, E.L., Berne, R.M., Rubio, R. (1976) Adenosine relaxation of isolated vascular smooth muscle. *Am. J. Physiol.* **239**: 1239–1243.

744. Akatsuka, Y., Egashira, K., Katsuda, Y., Narishige, T., Ueno, H., Shimokawa, H., Takeshita, A. (1994) ATP-sensitive potassium channels are involved in adenosine A(2) receptor mediated coronary vasodilatation. *Cardiovasc. Res.* **28**: 906–911.

745. Mustafova-Yambolieva, V.N., Keef, K.D. (1997) Adenosine-induced hyperpolarization in guinea pig coronary artery involves A(2B) receptors and K-ATP channels. *Am. J. Physiol.* **42**: H2687–H2695.

746. Kemp, B.K., Cocks, T.M. (1999) Adenosine mediates relaxation of human resistance-like coronary arteries via A_{2B} receptors. *Br. J. Pharmacol.* **126**: 1796–1800.

747. Ongini, E., Fredholm, B.B. (1996) Pharmacology of adenosine A_{2A} receptors. *Trends Pharmacol. Sci.* **17**: 364–372.

748. Gidday, J.M., Maceren, R.G., Shah, A.R., Meier, J.A., Zhu, Y. (1996) K-ATP channels mediate adenosine-induced hyperemia in retina. *Invest. Ophthalmol. Visual Sci.* **37**: 2624–2633.

749. Sheridan, B.C., McIntyre, R.C., Meldrum, D.R., Fullerton, D.A. (1997) K-ATP channels contribute to beta and adenosine receptor-mediated pulmonary vasorelaxation. *Am. J. Physiol.* **17**: L950–L956.

750. Dart, C., Standen, N.B. (1993) Adenosine-activated potassium current in smooth muscle cells isolated from the pig coronary artery. *J. Physiol.* **471**: 767–786.

751. Makujina, S.R., Olanrewaju, H.A., Mustafa, S.J. (1994) Evidence against K-ATP channel involvement in adenosine receptor-mediated dilation of epicardial vessels. *Am. J. Physiol.* **267**: H716–H724.

752. Cabell, F., Weiss, D.S., Price, J.M. (1994) Inhibition of adenosine-induced coronary vasodilatation by block of large-conductance Ca^{2+}-activated K^+ channels. *Am. J. Physiol.* **36**: H1455–H1460.

753. Ruiz, E., Tejerina, T. (1998) Relaxant effects of L-citrulline in rabbit vascular smooth muscle. *Br. J. Pharmacol.* **125**: 186–192.

754. Marx, S., Vedernikov, Y., Saade, G.R., Garfield, R.E. (2000) Citrulline does not relax isolated rat and rabbit vessels. *Br. J. Pharmacol.* **130**: 713–716.

755. He, G.W. (1997) Hyperkalemia exposure impairs EDHF-mediated endothelial function in the human coronary artery. *Ann. Thorac. Surg.* **63**: 84–87.

756. Dessy, C., Moniotte, S., Ghisdal, P., Havaux, X., Noirhomme, P., Balligand, J.L. (2004) Endothelial beta3-adrenoceptors mediate vasorelaxation of human coronary arteries through nitric oxide and endothelium-dependent hyperpolarization. *Circulation* **110**: 948–954.

757. Hamilton, C.A., Williams, R., Pathi, V., Berg, G., McArthur, K., McPhaden, A.R., Reid, J.L., Dominiczack, A.F. (1999) Pharmacological characterization of endothelium-dependent relaxation in human radial artery: comparison with internal thoracic artery. *Cardiovasc. Res.* **42**: 214–223.

758. Pascoal, I.F., Umans, J.G. (1996) Effect of pregnancy on mechanisms of relaxation in human omental microvessels. *Hypertension* **28**: 183–187.

759. Tottrup, A., Kraglund, K. (2004) Endothelium-dependent responses in small human mesenteric arteries. *Physiol. Res.* **52**: 255–263.

760. Martinez-Leon, J.B., Segarra, G., Medina, P., Vila, J.M., Lluch, P., Peiro, M., Otero, E. (2003) Ca^{2+}-activated K^+ channels mediate relaxation of forearm veins in chronic renal failure. *J. Hypertens.* **21**: 1927–1934.

761. Woolfson, R.G., Poston, L. (1990) Effect of NG-monomethyl-L-arginine on endothelium-dependent relaxation of human subcutaneous resistance arteries. *Clin. Sci. London* **79**: 273–278.

762. Miura, H., Wachtel, R.E., Liu, Y., Loberiza, F.R., Jr., Saito, T., Miura, M., Gutterman, D.D. (2001) Flow-induced dilation of human coronary arterioles: important role of Ca(2+)-activated K(+) channels. *Circulation* **103**: 1992–1998.

763. Tagawa, H., Shimokawa, H., Tagawa, T., Kuroiwa-Matsumoto, M., Hirooka, Y., Takeshita, A. (1997) Short-term estrogen augments both nitric oxide-mediated and non-nitric oxide-mediated endothelium-dependent vasodilation in postmenopausal women. *J. Cardiovasc. Pharmacol.* **30**: 481–488.

764. Honing, M.L., Smits, P., Morrison, P.J., Rabelink, T.J. (2000) Bradykinin-induced vasodilatation of human forearm resistance vessels is primarily mediated by endothelium-dependent hyperpolarization. *Hypertension* **35**: 1314–1318.

765. Brown, N.J., Gainer, J.V., Murphey, L.J., Vaughan, D.E. (2000) Bradykinin stimulates tissue plasminogen activator release from human forearm vasculature through B(2) receptor-dependent, NO synthase-independent and cyclooxygenase-independent pathway. *Circulation* **102**: 2190–2196.

766. Katz, S.D., Krum, H. (2001) Acetylcholine-mediated vasodilatation in the forearm circulation of patients with heart failure: indirect evidence for the role of endothelium-derived hyperpolarizing factor. *Am. J. Physiol.* **87**: 1089–1092.

767. Schrage, W.G., Dietz, N.M., Eisenach, J.H., Joyner, M.J. (2005) Agonist-dependent variability of contributions of nitric oxide and prostaglandins in human skeletal muscle. *J. Appl. Physiol.* **98**: 1251–1257.

768. Inokuchi, K., Hirooka, Y., Shimokawa, H., Sakai, K., Kishi, T., Ito, K., Kimura, Y., Takeshita, A. (2003) Role of endothelium-derived hyperpolarizing factor in human forearm circulation. *Hypertension* **42**: 919–924.

769. Galanakis, D., Ganellin, C.R., Malik, S., Dunn, P.M. (1996) Synthesis and pharmacological testing of dequalinium analogues as blockers of the apamin-sensitive Ca(2+)-activated K+ channel: variation of the length of the alkylene chain. *J. Med. Chem.* **39**: 3592–3595.

770. Liegeois, J.F., Mercier, F., Graulich, A., Graulich-Lorge, F., Scuvee-Moreau, J., Setin, V. (2003) Modulation of small conductance calcium-activated potassium (SK) channels: a new challenge in medicinal chemistry. *Curr. Med. Chem.* **10**: 625–647.

771. Corriu, C., Félétou, M., Puybasset, L., Bea, M-L., Berdeaux, A., Vanhoutte, P.M. (1998) Endothelium-dependent hyperpolarization in isolated arteries taken from animals treated with NO-synthase inhibitors. *J. Cardiovasc. Pharmacol.* **32**: 944–950.

772. Chen, G., Cheung, D.W. (1997) Effects of K^+ channel blockers on ACh-induced hyperpolarization and relaxation in mesenteric arteries. *Am. J. Physiol.* **41**: H2306–H2312.

773. Hashitani, H., Suzuki, H. (1997) K⁺ channels which contribute to the acetylcholine-induced hyperpolarization in smooth muscle of the guinea-pig submucosal arterioles. *J. Physiol.* **501**: 319–29.

774. Ayajiki, K., Ozaki, M., Shiomi, M., Okamura, T., Toda, N. (2000) Comparison of endothelium-dependent relaxation in carotid arteries from Japanese white and Watanabe heritable hyperlipidemic rabbits. *J. Cardiovasc. Pharmacol.* **36**: 622–630.

775. Wang, X., Loutzenhiser, R. (2002) Determinants of renal microvascular response to ACh: afferent and efferent arteriolar actions of EDHF. *Am. J. Physiol.* **282**: F124–F132.

776. You, J., Johnson, T.D., Marelli, S.P., Mombouli, J.V., Bryan, R.M., Jr. (1999) P2u receptor mediated release of endothelium-derived relaxing factor/nitric oxide and endothelium-derived hyperpolarizing factor from cerebrovascular endothelium in rats. *Stroke* **30**: 1125–1133.

777. Dong, H., Waldron, G.J., Galipeau, D., Cole W.C., Triggle, C.R. (1997) NO/PGI2-independent vasorelaxation and the cytochrome P450 pathway in rabbit carotid artery. *Br. J. Pharmacol.* **120**: 695–701.

778. Nakashima, Y., Toki, Y., Fukami, Y., Hibino, M., Okumura, K., Ito, T. (1997) Role of K⁺ channels in EDHF-mediated relaxation induced by acetylcholine in canine coronary artery. *Heart Vessels* **12**: 287–293.

779. Fujioka, H., Ayajiki, K., Shinozaki, K., Toda, N., Okamura, T. (2002) Mechanisms underlying endothelium-dependent, nitric oxide- and prostanoid-independent relaxation in monkey and dog coronary arteries. *Naunyn Schmiedebergs Arch. Pharmacol.* **366**: 488–495.

780. McNeish, A.J., Wilson, W.S., Martin, W. (2001) Dominant role of an endothelium-derived hyperpolarizing factor (EDHF)-like vasodilator in the ciliary vascular bed of the bovine isolated perfused eye. *Br. J. Pharmacol.* **134**: 912–920.

781. Tracey, A., Bunton, D., Irvine, J., MacDonald, A., Shaw, A.M. (2002) Relaxation to bradykinin in bovine pulmonary supernumerary arteries can be mediated by both nitric oxide-dependent and –independent mechanism. *Br. J. Pharmacol.* **137**: 538–544.

782. Prieto, D., Simonsen, U., Hernandez, M., Garcia-Sacristan, A. (1998) Contribution of K⁺ channels and oubain-sensitive mechanisms to the endothelium-dependent relaxations of horse penile small arteries. *Br. J. Pharmacol.* **123**: 1609–1620.

783. Ayajiki, K., Okamura, T., Fujioka, H., Imaoka, S., Funae, Y., Toda, N. (1999) Involvement of CYP3A-derived arachidonic acid metabolite(s) in response to endothelium-derived K⁺ channel opening substance in monkey lingual artery. *Br. J. Pharmacol.* **128**: 802–808.

CHAPTER 4

1. Chataigneau, T., Félétou, M., Huang, P.L., Fishman, M.C., Duhault, J., Vanhoutte, P.M. (1999) Acetylcholine-induced relaxation in blood vessels from endothelial nitric oxide synthase knockout mice. *Br. J. Pharmacol.* **126**: 219–226.

2. Waldron, G.J., Ding, H., Lovren, F., Kubes, P., Triggle, C.R. (1999) Acetylcholine-induced relaxation of peripheral arteries isolated from mice lacking endothelial nitric oxide synthase. *Br. J. Pharmacol.* **128**: 653–658.

3. Huang, A., Sun, D., Smith, C.J., Connetta, J.A., Shesely, E.G., Koller, A., Kaley, G. (2000) In eNOS knockout mice skeletal muscle, arteriolar dilation is mediated by EDHF. *Am. J. Physiol.* **278**: H762–H768.

4. Huang, A., Sun, D., Carroll, M.A., Jiang, H., Smith, C.J., Connetta, J.A., Falck, J.R., Shesely, E.G., Koller, A., Kaley, G. (2001) EDHF mediates flow-induced dilation in skeletal muscle arterioles of female eNOS-KO mice. *Am. J. Physiol.* **280**: H2462–H2469.

5. Ding, H., Kubes, P., Triggle, C. (2000) Potassium and acetylcholine-induced vasore-laxation in mice lacking endothelial nitric oxide synthase. *Br. J. Pharmacol.* **129**: 1194–1200.

6. Brandes, R.P., Schmitz-Winnenthal, F-H., Félétou, M., Gödecke, A., Huang, P-L., Vanhoutte, P.M., Fleming, I., Busse, R. (2000) An endothelium-derived hyperpolarizing factor distinct from NO and prostacyclin is a major endothelium-dependent vasodi-lator in resistance vessels of wild type and endothelial NO synthase knock-out mice. *Proc. Natl. Acad. Sci. USA* **97**: 9747–9752.

7. Scotland, R.S., Chauhan, S., Vallance, P.J., Ahluwalia, A. (2001) An endothelium-derived hyperpolarizing factor moderates myogenic constriction of mesenteric resis-tance arteries in the absence of endothelial nitric oxide synthase derived nitric oxide. *Hypertension* **38**: 833–839.

8. Scotland, R.S., Madhani, M., Chauhan, S., Moncada, S., Andresen, J., Nilsson, H., Hobbs, A.J., Ahluwalia, A. (2005) Investigation of vascular responses in endothelial nitric oxide synthase/cyclooxygenase-1 double knock-out mice. Key role for endo-thelium-derived hyperpolarizing factor in the regulation of blood pressure in vivo. *Circulation* **111**: 796–803.

9. Nagao, T., Illiano, S.C., Vanhoutte, P.M. (1992) Heterogeneous distribution of endothelium-dependent relaxations resistant to N^G-nitro-L-arginine in rats. *Am. J. Physiol.* **263**: H1090–H1094.

10. Shimokawa, H., Yasutake, H., Fujii, K., Owada, M.K., Nakaike, R., Fukumoto, Y., Takayanagani, T., Nagao, T., Egashira, K., Fujishima, M., Takeshita, A. (1996) The importance of the hyperpolarizing mechanism increases as the vessel size decreases in endothelium-dependent relaxations in rat mesenteric circulation. *J. Cardiovasc. Pharmacol.* **28**: 703–711.

11. Félétou, M., Hoeffner, U., Vanhoutte, P.M. (1989) Endothelium-dependent relaxing factors do not affect the smooth muscle of portal mesenteric veins. *Blood Vessels* **26**: 21–32.

12. Zhang, R.Z., Yang, Q., Yim, A.P., Huang, Y., He, G.W. (2004) Different role of nitric oxide and endothelium-derived hyperpolarizing factor in endothelium-dependent hyperpolarization and relaxation in porcine coronary arterial and venous system. *J. Cardiovasc. Pharmacol.* **43**: 839–850.

13. Parkington, H.C., Chow, J.A., Evans, R.G., Coleman, H.A., Tare, M. (2002) Role for endothelium-derived hyperpolarizing factor in vascular tone in rat mesenteric and hindlimb circulations in vivo. *J. Physiol.* **542**: 929–937.

14. Chaytor, A.Y., Evans, W.H., Griffith, T.M. (1998) Central role of heterocellular gap junction communication in endothelium-dependent relaxations of rabbit arteries. *J. Physiol. (Lond.)* **508**: 561–573.

15. De Vriese, A.S., Van de Voorde, J., Lameire, N.H. (2002) Effects of connexin-mimetic peptides on nitric oxide synthase- and cyclooxygenase-independent renal vasodilata-tion. *Kidney Int.* **61**: 177–185.

16. Taylor, M.S., Bonev, A.D., Gross, T.P., Eckman, D.M., Brayden, J.E., Bond, C.T., Adel-man, J.P., Nelson, M.T. (2003) Altered expression of small-conductance Ca^{2+}-activated K^+ (SK3) channels modulate arterial tone and blood pressure. *Circ. Res.* **93**: 124–131.

17. Kawabata, A., Nakaya, Y., Kuroda, R., Wakisaka, R., Masuko, T., Nishikawa, H., Kawai, K. (2003) Involvement of EDHF in the hypotension and increased gastric mucosal blood flow caused by PAR-2 activation in rats. *Br. J. Pharmacol.* **140**: 247–254.

18. Hoepfl, B., Rodenwaldt, B., Pohl, U., De Wit, C. (2002) EDHF, but not NO or prostaglandins, is critical to evoke a conducted dilation upon ACh in hamster arterioles. *Am. J. Physiol.* **283**: H996–H1004.

19. Wulff, H., Miller, M.J., Haensel, W., Grissner, S., Cahalan, M.D., Chandy, K.G. (2000) Design of potent and selective inhibitor of the intermediate-conductance Ca^{2+}-activated K^+ channel, IKCa1: a potential immunosuppressant. *Proc. Natl. Acad. Sci. USA* **97**: 8151–8156.

20. Liegeois, J.F., Mercier, F., Graulich, A., Graulich-Lorge, F., Scuvee-Moreau, J., Setin, V. (2003) Modulation of small conductance calcium-activated potassium (SK) channels: a new challenge in medicinal chemistry. *Curr. Med. Chem.* **10**: 625–647.

21. Gluais, P., Edwards, G., Weston, A.H., Falck, J.R., Vanhoutte, P.M., Félétou, M. (2005) SK_{Ca} and IK_{Ca} in the endothelium-dependent hyperpolarization of the guinea-pig isolated carotid artery. *Br. J. Pharmacol.* **144**: 477–485.

22. Hoyer, J., Kohler, R., Distler, A. (1998) Mechanosensitive Ca^{2+} oscillations and STOC activation in endothelial cells. *FASEB J.* **12**: 359–366.

23. Brakemier, S., Eichler, I., Hopp, H., Kohler, R., Hoyer, J. (2002) Up-regulation of endothelial stretch-activated cation channels by fluid shear stress. *Cardiovasc. Res.* **53**: 209–218.

24. Qiu, W.P., Hu, Q., Paolocci, N., Ziegelstein, R.C., Kass, D.A. (2003) Differential effects of pulsatile versus steady flow on coronary endothelial membrane potential. *Am. J. Physiol.* **285**: H341–H346.

25. Busse, R., Edwards, G., Félétou, M., Fleming, I., Vanhoutte, P.M., Weston, A.H. (2002) Endothelium-dependent hyperpolarization, bringing the concepts together. *Trends Pharmacol. Sci.* **23**: 374–380.

26. Brakemier, S., Kersten, A., Eichler, I., Grgic, I., Zakrzewicz, A., Hopp, H., Kohler, R., Hoyer, J. (2003) Shear stress-induced up-regulation of the intermediate-conductance Ca(2+) activated K(+) channel in human endothelium. *Cardiovasc. Res.* **60**: 488–496.

27. Takamura, Y., Shimokawa, H., Zhao, H., Igarashi, H., Egashira, K., Takeshita, A. (1999) Important role of endothelium-derived hyperpolarizing factor in shear stress-induced endothelium-dependent relaxations in the rat mesenteric artery. *J. Cardiovasc. Pharmacol.* **34**: 381–387.

28. Dube, S., Canty, J.M., Jr. (2001) Shear stress-induced vasodilation in porcine coronary conduit arteries is independent of nitric oxide release. *Am. J. Physiol.* **280**: H2581–H2590.

29. Miura, H., Wachtel, R.E., Liu, Y., Loberiza, F.R., Jr., Saito, T., Miura, M., Gutterman, D.D. (2001) Flow-induced dilation of human coronary arterioles: important role of Ca(2+)-activated K(+) channels. *Circulation* **103**: 1992–1998.

30. Miura, H., Bosnjak, J.J., Ning, G., Salto, T., Miura, M., Gutterman, D.D. (2003) Role for hydrogen peroxide in flow-induced dilation of human coronary arterioles. *Circ. Res.* **92**: e31–e40.

31. Matsumoto, T., Oda, S.I., Kobayashi, T., Kamata, K. (2004) Flow-induced endothelium-dependent vasoreactivity in rat mesenteric arterial bed. *J. Smooth Muscle Res.* **40**: 1–14.

32. Watanabe, S., Yashiro, Y., Mizuno, R., Ohhashi, T. (2005) Involvement of NO and EDHF in flow-induced vasodilatation in isolated hamster cremaster arterioles. *J. Vasc. Res.* **42**: 137–147.

33. Huang, A., Sun, D., Jacobson, A., Carroll, M.A., Falck, J.R., Kaley, G. (2005) Epoxyeicosatrienoic acids are released to mediate shear stress-dependent hyperpolarization of arteriolar smooth muscle. *Circ. Res.* **96**: 376–383.

34. Popp, R., Fleming, I., Busse, R. (1998) Pulsatile stretch in coronary arteries elicits release of endothelium-derived hyperpolarizing factor: a modulator of arterial compliance. *Circ. Res.* **82**: 696–703.
35. Fisslthaler, B., Popp, R., Michaelis, U.R., Kiss, L., Fleming, I., Busse, R. (2001) Cyclic stretch enhances the expression and activity of coronary endothelium-derived hyperpolarizing factor synthase. *Hypertension* **38**: 1427–1432.
36. Paolocci, N., Pagliaro, P., Isoda, T., Saavedra, F.W., Kass, D.A. (2001) Role of calcium-sensitive K(+) channels and nitric oxide in *in vivo* coronary vasodilatation from enhanced perfusion pulsatility. *Circulation* **103**: 119–124.
37. Duling, B.R., Berne, R.M. (1970) Propagated vasodilation in the microcirculation of the hamster cheek pouch. *Circ. Res.* **26**: 163–170.
38. Segal, S.S., Duling, B.R. (1986) Flow control among microvessels coordinated by intracellular conduction. *Science* **234**: 868–870.
39. Budel, S., Bartlett, I.S., Segal, S.S. (2003) Homocellular conduction along endothelium and smooth muscle of arterioles in hamster cheek pouch. Unmasking an NO wave. *Circ. Res.* **93**: 61–68.
40. Payne, G.W., Madri, J.A., Sessa, W.C., Segal, S.S. (2004) Histamine inhibits conducted vasodilation through endothelium-derived NO production in arterioles of mouse skeletal muscle. *FASEB J.* **18**: 280–286.
41. Dora, K., Xia, J., Duling, B.R. (2003) Endothelial cell signaling during conducted vasomotor responses. *Am. J. Physiol.* **285**: H119–H126.
42. Duza, T., Sarelius, I.H. (2003) Conducted dilations initiated by purines in arterioles are endothelium dependent and require endothelial Ca^{2+}. *Am. J. Physiol.* **285**: H26–H37.
43. Yashiro, Y., Duling, B.R. (2003) Participation of intracellular Ca^{2+} stores in arteriolar conducted responses. *Am. J. Physiol.* **285**: H54–H73.
44. Takano, H., Dora, K.A., Spitaler, M.M., Garland, C.J. (2004) Spreading vasodilation in rat mesenteric arteries associated with calcium-independent endothelial cell hyperpolarization. *J. Physiol.* **556**: 887–903.
45. Yashiro, Y., Duling, B.R. (2000) Integrated Ca(2+) signaling between smooth muscle and endothelium of resistance vessels. *Circ. Res.* **87**: 1048–1054.
46. Ungvari, Z., Csiszar, A., Koller, A. (2002) Increases in endothelial Ca(2+) activate K(Ca) channels and elicit EDHF-type arteriolar dilation via gap junctions. *Am. J. Physiol.* **282**: H1760–1767.
47. De Wit, C., Roos, F., Bolz, S.S., Kirchhoff, S., Kruger, O., Willecke, K., Pohl, U. (2000) Impaired conduction of vasodilatation along arterioles in connexin 40-deficient mice. *Circ. Res.* **86**: 649–655.
48. Yamamoto, Y., Fukuta, H., Nakahira, Y., Suzuki, H. (1998) Blockade by 18β-glycyrrhetinic acid of intercellular electrical coupling in guinea-pig arterioles. *J. Physiol.* **511**: 501–518.
49. Yamamoto, Y., Imaeda, K., Suzuki, H. (1999) Endothelium-dependent hyperpolarization and intercellular electrical coupling in guinea-pig mesenteric arterioles. *J. Physiol.* **514**: 505–513.
50. Emerson, G.G., Segal, S.S. (2000) Electrical coupling between endothelial cells and smooth muscle cells in hamster feed arteries: role in vasomotor control. *Circ. Res.* **87**: 474–479.
51. Emerson, G.G., Segal, S.S. (2000) Endothelial cell pathway for conduction of hyperpolarization and vasodilation along hamster feed artery. *Circ. Res.* **86**: 94–100.
52. Emerson, G.G., Segal, S.S. (2001) Electrical activation of endothelium evokes vasodilatation and hyperpolarization along hamster feed arteries. *Am. J. Physiol.* **280**: H160–H167.

53. Coleman, H.A., Tare, M., Parkington, H.C. (2001) K$^+$ currents underlying the action of endothelium-derived hyperpolarizing factor in guinea-pig, rat and human blood vessels. *J. Physiol.* **531**: 359–373.

54. Coleman, H.A., Tare, M., Parkington, H.C. (2001) EDHF is not K$^+$ but may be due to spread of current from the endothelium in guinea pig arterioles. *Am. J. Physiol.* **280**: H2478–H2483.

55. Bartlett, I.S., Segal, S.S. (2000) Resolution of smooth muscle and endothelial pathway for conduction along hamster cheek pouch arterioles. *Am. J. Physiol.* **278**: H604–H612.

56. Welsh, D.G., Segal, S.S. (1998) Endothelial and smooth muscle cell conduction in arterioles controlling blood flow. *Am. J. Physiol.* **274**: H178–H186.

57. Looft-Wilson, R.C., Payne, G.W., Segal, S.S. (2004) Connexin expression and conducted vasodilation along arteriolar endothelium in mouse skeletal muscle. *J. Appl. Physiol.* **97**: 1152–1158.

58. Crane, G.J., Neild, T.O., Segal, S.S. (2004) Contribution of active membrane processes to conducted hyperpolarization in arterioles of hamster cheek pouch. *Microcirculation* **11**: 425–433.

59. Povstyan, O.V., Gordienko, D.V., Harhun, M.I., Bolton, T.B. (2003) Identification of interstitial cells of Cajal in the rabbit portal vein. *Cell Calcium* **33**: 223–239.

60. Harhun, M.I., Gordienko, D.V., Povstyan, O.V., Moss, R.F., Bolton, T.B. (2004) Function of interstitial cells of Cajal in the rabbit portal vein. *Circ. Res.* **17**: 619–626.

61. Pucovsky, V., Moss, R.F., Bolton, T.B. (2003) Non-contractile cells with thin processes resembling cells of Cajal found in the wall of guinea-pig mesenteric arteries. *J. Physiol.* **552**: 119–133.

62. Daniel, E.E. (2004) Communication between interstitial cells of Cajal and gastrointestinal muscle. *Neurogastroenterol. Motil.* **16**: S118–S122.

63. Welsh, D.G., Segal, S.S. (2000) Role of EDHF in conduction of vasodilation along hamster cheek pouch arterioles in vivo. *Am. J. Physiol.* **278**: 1832–1839.

64. Rivers, R.J., Hein, T.W., Zhang, C., Kuo, L. (2001) Activation of barium-sensitive inward rectifier potassium channels mediates remote dilation of coronary arterioles. *Circulation* **104**: 1749–1753.

65. Nilsson, H., Aalkjaer, C. (2003) Vasomotion—mechanisms and physiological importance. *Mol. Interv.* **3**: 79–89.

66. Peng, H., Matchkov, V., Ivarsen, A., Aalkjaer, C., Nilsson, H. (2001) Hypothesis for the initiation of vasomotion. *Circ. Res.* **88**: 810–815.

67. Okazaki, K., Seki, S., Kanaya, N., Hattori, J., Toshe, N., Namiki, A. (2003) Role of endothelium-derived hyperpolarizing factor in phenylephrine-induced oscillatory vasomotion in rat small mesenteric artery. *Anesthesiology* **98**: 1164–1171.

68. Mauban, J.R.H., Weir, W.G. (2004) Essential role of EDHF in the initiation and maintenance of adrenergic vasomotion in rat mesenteric arteries. *Am. J. Physiol.* **287**: H608–H616.

69. De Wit, C., Roos, F., Bolz, S.S., Pohl, U. (2003) Lack of vascular connexin 40 is associated with hypertension and irregular arteriolar vasomotion. *Physiol. Genomics* **13**: 169–177.

70. Figueroa, X.F., Paul, D.L., Simon, A.M., Goodenough, D.A., Day, K.H., Damon, D.N., Duling, B.R. (2003) Central role of connexin 40 in the propagation of electrically activated vasodilation in mouse cremasteric arterioles in vivo. *Circ. Res.* **92**: 793–800.

71. Looft-Wilson, R.C., Haug, S.J., Neufer, P.D., Segal, S.S. (2004) Independence of connexin expression and vasomotor conduction from sympathetic innervation in hamster feed arteries. *Microcirculation* **11**: 397–408.

72. Furchgott, R.F., Vanhoutte, P.M. (1989) Endothelium-derived relaxing and contracting factors. *FASEB J.* **3**: 2007–2018.

73. Wadsworth, R.M. (1994) Vasoconstrictor and vasodilator effects of hypoxia. *Trends Pharmacol. Sci.* **15**: 47–53.

74. Furchgott, R.F., Zawadzki, J.V. (1980) The obligatory role of the endothelial cells in the relaxation of arterial smooth muscle by acetylcholine. *Nature* **288**: 373–376.

75. Palmer, R.M.J., Ashton, D.S., Moncada, S. (1988) Vascular endothelial cells synthesize nitric oxide from L-arginine. *Nature* **333**: 664–666.

76. Félétou, M., Girard, V., Canet, E. (1995) Different involvement of nitric oxide in endothelium-dependent relaxation of porcine pulmonary artery and vein: influence of hypoxia. *J. Cardiovasc. Pharmacol.* **25**: 665–673.

77. Petersson, J., Zygmunt, P.M., Jonsson, P., Högestätt, E.D. (1998) Characterization of endothelium-dependent relaxation in guinea pig basilar artery—effect of hypoxia and role of cytochrome P450 monooxygenase. *J. Vasc. Res.* **35**: 285–294.

78. Shimizu, S., Paul, R.J. (1999) Hypoxia and alkalinization inhibit endothelium-derived nitric oxide but not endothelium-derived hyperpolarizing factor responses in porcine coronary artery. *J. Pharmacol. Exp. Ther.* **291**: 335–344.

79. Earley, S., Walker, B.R. (2003) Increased nitric oxide production following chronic hypoxia contributes to attenuated systemic vasoconstriction. *Am. J. Physiol.* **284**: H1655–H1661.

80. Earley, S., Naik, J.S., Walker, B.R. (2002) 48-h hypoxic exposure results in endothelium-dependent systemic vascular smooth muscle hyperpolarization. *Am. J. Physiol.* **283**: R79–R85.

81. Earley, S., Pastuszyn, A., Walker, B.R. (2003) Cytochrome p-450 epoxygenase products contribute to attenuated vasoconstriction after chronic hypoxia. *Am. J. Physiol.* **285**: H127–H136.

82. Earley, S., Walker, B.R. (2002) Endothelium-dependent blunting of myogenic responsiveness after chronic hypoxia. *Am. J. Physiol.* **283**: H2202–H2209.

83. Hasunuma, K., Yamaguchi, T., Rodman, D., O'Brien, R., McMurtry, I. (1991) Effects of inhibitors of EDRF and EDHF on vasoreactivity of perfused rat lungs. *Am. J. Physiol.* **260**: L97–L104.

84. Carter, E.P., Sato, K., Morio, Y., McMurtry, I.F. (2000) Inhibition of K_{Ca} channels restores blunted hypoxic pulmonary vasoconstriction with cirrhosis. *Am. J. Physiol.* **279**: L903–L910.

85. Morio, Y., Carter, E.P., Oka, M., McMurtry, I.F. (2003) EDHF-mediated vasodilation involves different mechanisms in normotensive and hypertensive rat lungs. *Am. J. Physiol.* **284**: H1762–H1770.

86. Smani, T., Hernandez, A., Urena, J., Castellano, A.G., Franco-Obregon, A., Ordonez, A., Lopez-Barneo, J. (2002) Reduction of Ca^{2+} channel activity by hypoxia in human and porcine coronary myocytes. *Cardiovasc. Res.* **53**: 97–104.

87. Shepherd, J.T., Vanhoutte, P.M. (1979) *The Human Cardiovascular System. Facts and Concepts.* Raven Press, New York, pp. 1–351.

88. Clifford, P.S., Hellsten, Y. (2004) Vasodilatory mechanisms in contracting skeletal muscle. *J. Appl. Physiol.* **97**: 393–403.

89. Segal, S.S., Jacobs, T.L. (2001) Role for endothelial cell conduction in ascending vasodilatation and exercise hyperaemia in hamster skeletal muscle. *J. Physiol.* **536**: 937–946.

90. Welsh, D.G., Segal, S.S. (1997) Coactivation of resistance vessels and muscle fibers with acetylcholine release from motor nerves. *Am. J. Physiol.* **273**: H156–H163.

91. Orshal, J.M., Khalil, R.A. (2004) Gender, sex hormones, and vascular tone. *Am. J. Physiol.* **286**: R233–R249.

92. Santos, R.L., Abreu, G.R., Bissoli, N.S., Moyses, M.R. (2004) Endothelial mediators of 17 beta-estradiol-induced coronary vasodilation in the isolated heart. *Braz. J. Med. Biol. Res.* **37**: 569–575.

93. Mendelsohn, M.E. (2002) Genomic and non-genomic effects of estrogen in the vasculature. *Am. J. Cardiol.* **90**: 3F–6F.

94. McCulloch, A.I., Randall, M.D. (1998) Sex differences in the relative contributions of nitric oxide and EDHF to agonist-stimulated endothelium-dependent relaxations in the rat isolated mesenteric arterial bed. *Br. J. Pharmacol.* **123**: 1700–1706.

95. White, R.M., Rivera, C.O., Davison, C.A. (2000) Nitric oxide-dependent and -independent mechanisms account for gender differences in vasodilatation to acetylcholine. *J. Pharmacol. Exp. Ther.* **292**: 375–380.

96. Pak, K.J., Geary, G.G., Duckles, S.P., Krause, D.N. (2002) Male-female differences in the relative contribution of endothelial vasodilators released by rat tail artery. *Life Sci.* **71**: 1633–1642.

97. Wangensteen, R., Moreno, J.M., Sainz, J., Rodriguez-Gomez, I., Chamorro, V., Luna, J. de D., Osuna, A., Vargas, F. (2004) Gender differences in the role of endothelium-derived relaxing factors modulating renal vascular reactivity. *Eur. J. Pharmacol.* **486**: 281–288.

98. Liu, M.Y., Hattori, Y., Fukao, M., Sato, A., Sakuma, I., Kanno, M. (2001) Alterations of EDHF-mediated hyperpolarization and relaxation in mesenteric arteries of female rats in long term deficiency of oestrogen and during oestrus cycle. *Br. J. Pharmacol.* **132**: 1035–1046.

99. Liu, M.Y., Hattori, Y., Sato, A., Ichikawa, R., Zhang, X.H., Sakuma, I. (2002) Ovariectomy attenuates hyperpolarization and relaxation mediated by endothelium-derived hyperpolarizing factor in female rat mesenteric artery: a concomitant decrease in connexin-43 expression. *J. Cardiovasc. Pharmacol.* **40**: 938–948.

100. Sakuma, I., Liu, M.Y., Sato, A., Hayashi, T., Iguchi, A., Kitabatake, A., Hattori, Y. (2002) Endothelium-dependent hyperpolarization and relaxation in mesenteric arteries of middle-aged rats: influence of oestrogen. *Br. J. Pharmacol.* **135**: 48–54.

101. Nawate, S., Fukao, M., Sakuma, I., Soma, T., Nagai, K., Takikawa, O., Miwa, S., Kitabatake, A. (2005) Reciprocal changes in the endothelium-derived hyperpolarizing factor- and nitric oxide-system in the mesenteric artery of adult female rats following ovariectomy. *Br. J. Pharmacol.* **144**: 178–179.

102. Burnham, M.P., Bychkov, R., Félétou, M., Richards, G.R., Vanhoutte, P.M., Weston, A.H., Edwards, G. (2002) Characterization of an apamin-sensitive small conductance Ca^{2+}-activated K^+ channel in porcine coronary artery endothelium: relevance to EDHF. *Br. J. Pharmacol.* **135**: 1133–1143.

103. Jacobson, D., Pribnow, D., Herson, P.S., Maylie, J., Adelman, J.P. (2003) Determinants contributing to estrogen-regulated expression of SK3. *Biochem. Biophys. Res. Commun.* **303**: 660–668.

104. Golding, E.M., Kepler, T.E. (2001) Role of estrogen in modulating EDHF-mediated dilations in the female rat middle cerebral artery. *Am. J. Physiol.* **280**: H2417–H2423.

105. Xu, H.L., Santizo, R.A., Koenig, H.M., Pelligrino, D.A. (2001) Chronic estrogen depletion alters adenosine diphosphate-induced pial arteriolar dilation in female rats. *Am. J. Physiol.* **281**: H2105–H2112.

106. Golding, E.M., You, J., Robertson, C.S., Bryan, R.M., Jr. (2001) Potentiated endothelium-derived hyperpolarizing factor-mediated dilations in cerebral arteries following mild head injury. *J. Neurotrauma* **18**: 691–697.

107. Xu, H.L., Santizo, R.A., Baughman, V.L., Pelligrino, D.A. (2002) ADP-induced pial arteriolar dilation in ovariectomized rats involved gap junctional communication. *Am. J. Physiol.* **283**: H1082–H1091.

108. Schildmeyer, L.A., Bryan, R.M., Jr. (2002) Effect of NO on EDHF response in rat middle cerebral arteries. *Am. J. Physiol.* **282**: H734–H738.

109. Gonzales, R.J., Krause, D.N., Duckles, S.P. (2004) Testosterone suppresses endothelium-dependent dilation of rat middle cerebral arteries. *Am. J. Physiol.* **286**: H552–H560.

110. Chataigneau, T., Zerr, M., Chataigneau, M., Hudlett, F., Hirn, C., Pernot, F., Schini-Kerth, V.B. (2004) Chronic treatment with progesterone but not medroxyprogesterone acetate restores the endothelial control of vascular tone in the mesenteric artery of ovariectomized rats. *Menopause* **11**: 255–263.

111. Tagawa, H., Shimokawa, H., Tagawa, T., Kuroiwa-Matsumoto, M., Hirooka, Y., Takeshita, A. (1997) Short-term estrogen augments both nitric oxide-mediated and non-nitric oxide-mediated endothelium-dependent vasodilation in postmenopausal women. *J. Cardiovasc. Pharmacol.* **30**: 481–488.

112. Gerber, R.T., Anwar, M.A., Poston, L. (1998) Enhanced acetylcholine induced relaxation in small mesenteric arteries from pregnant rats: an important role for endothelium-derived hyperpolarizing factor (EDHF). *Br. J. Pharmacol.* **125**: 455–460.

113. Dalle Lucca, J.J., Adeagbo, A.S., Alsip, N.L. (2000) Influence of the oestrous cycle and pregnancy on the reactivity of the rat mesenteric artery. *Hum. Reprod.* **15**: 961–968.

114. Dalle Lucca, J.J., Adeagbo, A.S., Alsip, N.L. (2000) Oestrous cycle and pregnancy alter the reactivity of the rat uterine vasculature. *Hum. Reprod.* **15**: 2496–2503.

115. Fulep, E.E., Vedernikov, Y.P., Saade, G.R., Garfield, R.E. (2001) The role of endothelium-derived hyperpolarizing factor in the regulation of the uterine circulation in pregnant rats. *Am. J. Obstet. Gynecol.* **185**: 638–642.

116. Ang, C., Hillier, C., Johnston, F., Cameron, A., Greer, I., Lumsden, M.A. (2002) Endothelial function is preserved in pregnant women with well controlled type I diabetes. *BJOG* **109**: 699–707.

117. Pascoal, I.F., Umans, J.G. (1996) Effect of pregnancy on mechanisms of relaxation in human omental microvessels. *Hypertension* **28**: 183–187.

118. Kenny, L.C., Baker, P.N., Kendall, D.A., Randall, M.D., Dunn, W.R. (2002) The role of gap junctions in mediating endothelium-dependent responses to bradykinin in myometrial small arteries isolated from pregnant women. *Br. J. Pharmacol.* **136**: 1085–1088.

119. Luksha, L., Nisell, H., Kublickiene, K. (2004) The mechanism of EDHF-mediated responses in subcutaneous small arteries of healthy pregnant women. *Am. J. Physiol.* **286**: R1102–R1109.

CHAPTER 5

1. Cai, H., Harrison, D.G. (2000) Endothelial dysfunction in cardiovascular diseases: the role of oxidant stress. *Circ. Res.* **10**: 840–844.

2. Félétou, M., Vanhoutte, P.M. (2004) EDHF: new therapeutic targets? *Pharmacol. Res,* **49**: 565–580.

3. Elliott, H.L. (1998) Endothelial dysfunction in cardiovascular disease: risk factor, risk marker or surrogate end point. *J. Cardiovasc. Pharmacol.* **32**: S74–S77.

4. Sarkis, A., Roman, R.J. (2004) Role of cytochrome P450 metabolites of arachidonic acid in hypertension. *Curr. Drug Metab.* **5**: 245–256.

5. Lüscher, T.F., Vanhoutte, P.M. (1986) Endothelium-dependent contractions to acetylcholine in the aorta of spontaneously hypertensive rat. *Hypertension* **8**: 344–348.

6. Fujii, K., Tominaga, M., Ohmori, S., Kobayashi, K., Koga, T., Takata, Y., Fujishima, M. (1992) Decreased endothelium-dependent hyperpolarization to acetylcholine in smooth muscle of the mesenteric artery of spontaneously hypertensive rats. *Circ. Res.* **70**: 660–669.

7. Fujii, K., Ohmori, S., Tominaga, M., Abe, I., Takata, Y., Ohya, Y., Kobayashi, K., Fujishima, M. (1993) Age-related changes in endothelium-dependent hyperpolarization in the rat mesenteric artery. *Am. J. Physiol.* **265**: H509–H516.

8. Mantelli, L., Amerini, S., Ledda, F. (1995) Role of nitric oxide and endothelium-derived hyperpolarizing factor in vasorelaxant effect of acetylcholine as influenced by aging and hypertension. *J. Cardiovasc. Pharmacol.* **25**: 595–602.

9. Hutri-Kahonen, N., Kahonen, M., Tolvanen, J.P., Wu, X., Sallinen, K., Porsti, I. (1997) Ramipril therapy improves arterial dilation in experimental hypertension. *Cardiovasc. Res.* **33**: 188–195.

10. Vanhoutte, P.M., Félétou, M., Taddei, S. (2005) Endothelium-dependent contractions in hypertension. *Br. J. Pharmacol.* **144**: 449–458.

11. Kansui, Y., Fujii, K., Nakamura, K., Goto, K., Oniki, H., Abe, I., Shibata, Y., Iida, M. (2004) Angiotensin II receptor blockade corrects altered expression of gap junctions in vascular endothelial cells from hypertensive rats. *Am. J. Physiol.* **287**: H216–H224.

12. Rummery, N.M., Hill, C.E. (2004) Vascular gap junctions and implications for hypertension. *Clin. Exp. Pharmacol. Physiol.* **31**: 659–667.

13. Goto, K., Rummery, N.M., Grayson, T.H., Hill, C.E. (2004) Attenuation of conducted vasodilation in rat mesenteric arteries during hypertension: role of inwardly rectifying potassium channels. *J. Physiol.* **561**: 215–231.

14. Kurjiaka, D.T., Bender, S.B., Nye, D.D., Wiehler, W.B., Welsh, D.G. (2005) Hypertension attenuates cell-to-cell communication in hamster retractor muscle feed arteries. *Am. J. Physiol.* **288**: H861–H870.

15. Hayakawa, H., Hirata, Y., Suzuki, E., Sugimoto, T., Matsuoka, H., Kihuchi, K., Nagano, T., Hirobe, M., Sugimoto, T. (1993) Mechanisms for altered endothelium-dependent relaxation in isolated kidneys from experimental hypertensive rats. *Am. J. Physiol.* **264**: H1535–H1541.

16. Dohi, Y., Kojima, M., Sato, K. (1996) Benidipine improves endothelial function in renal resistance arteries of hypertensive rats. *Hypertension* **28**: 58–63.

17. Büssemaker, E., Popp, R., Fisslthaler, B., Larson, C.M., Fleming, I., Busse, R., Brandes, R.P. (2003) Aged spontaneously hypertensive rats exhibit a selective loss of EDHF-mediated relaxation in the renal artery. *Hypertension* **42**: 562–568.

18. Sandow, S.L., Branich, N.J., Bandy, H.P., Rummery, N.M., Hill, C.E. (2003) Structure, function, and endothelium-derived hyperpolarizing factor in the caudal artery of the SHR and WKY rat. *Arterioscler. Thromb. Vasc. Biol.* **23**: 822–828.

19. Pourageaud, F., Freslon, J-L. (1995) Impaired endothelial relaxations induced by agonists and flow in spontaneously hypertensive rat compared to Wistar-Kyoto rat perfused coronary arteries. *J. Vasc. Res.* **32**: 190–199.

20. Vasquez-Perez, S., Navarro-Cid, J., de las Heras, N., Cediel, E., Sanz-Roza, D., Ruilope, L.M., Cachofeiro, V., Lahera, V. (2001) Relevance of endothelium-derived hyperpolarizing factor in the effects of hypertension on rat coronary relaxations. *J. Hypertens.* **19**: 539–545.

21. Tschudi, M.R., Criscione, L., Novosel, D., Pfeiffer, K., Luscher, T.F. (1994) Antihypertensive therapy augments endothelium-dependent relaxation in coronary arteries from normotensive and spontaneously hypertensive rats. *Circulation* **89**: 2212–2218.

22. Bund, S.J. (1998) Influence of mode of contraction on the mechanism of acetylcho-line-mediated relaxation of coronary arteries from normotensive and spontaneously hypertensive rats. *Clin. Sci.* **94**: 231–238.

23. Fuchs, L.C., Nuno, D., Lamping, K.G., Johnson, A.K. (1996) Characterization of endothelium-dependent vasodilation and vasoconstriction in coronary arteries from spontaneously hypertensive rats. *Am. J. Hypertens.* **9**: 475–486.

24. Gschwend, S., Pinto-Sietsma, S.J., Buikema, H., Pinto, Y.M., van Gilst, V.H., Schultz, A., de Zeeuw, D., Kreutz, R. (2002) Impaired coronary endothelial function in a rat model of spontaneous albuminuria. *Kidney Int.* **62**: 181–191.

25. Kahonen, M., Tolvanen, J.P., Sallinen, K., Wu, X., Porsti, I. (1998) Influence of gender on control of arterial tone in experimental hypertension. *Am. J. Physiol.* **275**: H15–H22.

26. Kagota, S., Tamashiro, A., Yamaguchi, Y., Nakamura, K., Kunimoto, M. (1999) Excessive salt or cholesterol intake alters the balance among endothelium-derived factors released from renal arteries in spontaneously hypertensive rats. *J. Cardiovasc. Pharmacol.* **34**: 533–539.

27. Sunano, S., Watanabe, H., Tanaka, S., Sekiguchi, F., Shimamura, K. (1999) Endo-thelium-derived relaxing, contracting and hyperpolarizing factors of mesenteric arter-ies of hypertensive and normotensive rats. *Br. J. Pharmacol.* **126**: 709–716.

28. Freitas, M.R., Schott, C., Corriu, C., Sassard, J., Stoclet, J-C., Andriantsitohaina, R. (2003) Heterogeneity of endothelium-dependent vasorelaxation in conductance and resistance arteries from Lyon normotensive and hypertensive rats. *J. Hypertens.* **21**: 1505–1512.

29. Ozawa, Y., Hayashi, K., Kanda, T., Homma, K., Takamatsu, I., Tatematsu, S., Yoshioka, K., Kumagai, H., Wakino, S., Saruta, T. (2004) Impaired nitric oxide- and endothelium-derived hyperpolarizing factor-dependent dilation of renal afferent arte-riole in Dahl salt-sensitive rats. *Nephrology (Carlton)* **9**: 272–277.

30. Onaka, U., Fujii, K., Abe, I., Fujishima, M. (1999) Antihypertensive therapy improves endothelium-dependent hyperpolarization. In *Endothelium-Dependent Hyperpolar-izations,* P.M. Vanhoutte, ed., Harwood Academic Publishers, Amsterdam, The Neth-erlands, pp. 305–312.

31. Sofola, O.A., Knill, A., Hainsworth, R., Drinkhill, M. (2002) Changes in endothelial function in mesenteric arteries of Sprague-Dawley rats fed a high salt diet. *J. Physiol.* **543**: 255–260.

32. Makynen, H., Kahonen, M., Wu, X., Arvola, P., Porsti, I. (1996) Endothelial function in deoxycorticosterone-NaCl hypertension: effect of calcium supplementation. *Cir-culation* **93**: 1000–1008.

33. Makynen, H., Kahonen, M., Wu, X., Wuorela, H., Porsti, I. (1996) Reversal of hypertension and endothelial dysfunction in deoxycorticosterone-NaCl-treated rats by high-Ca^{2+} diet. *Am. J. Physiol.* **270**: H1250–H1257.

34. Wu, X., Makynen, H., Kahonen, M., Arvola, P., Porsti, I. (1996) Mesenteric arterial function in vitro in three models of experimental hypertension. *J. Hypertens.* **14**: 365–372.

35. Adeagbo, A.S., Joshua, I.G., Falkner, C., Matheson, P.J. (2003) Tempol, an antioxi-dant, restores endothelium-derived hyperpolarizing factor-mediated vasodilation dur-ing hypertension. *Eur. J. Pharmacol.* **481**: 91–100.

36. Sendao Oliveira, A.P., Bendhack, L.M. (2004) Relaxation induced by acetylcholine involves endothelium-derived hyperpolarizing factor in 2-kidney 1-clip hypertensive rat carotid arteries. *Pharmacology* **72**: 231–239.

37. Benchetrit, S., Green, J., Katz, D., Berheim, J., Rathaus, M. (2003) Early endothelial dysfunction following renal mass reduction in rats. *Eur. J. Clin. Invest.* **33**: 26–33.

38. Kimura, K., Nishio, I. (1999) Impaired endothelium-dependent relaxation in mesenteric arteries of reduced renal mass hypertensive rats. *Scand. J. Clin. Lab. Invest.* **59**: 199–204.
39. Kalliovalkama, J., Jolma, P., Tolvanen, J.P., Kahonen, M., Hutri-Kahonen, N., Saha, H., Tuorila, S., Moilanem, E., Porsti, I. (1999) Potassium channel-mediated vasorelaxation is impaired in experimental renal failure. *Am. J. Physiol.* **277**: H1622–H1629.
40. Vargas, F., Osuna, A., Fernandez-Rivas, A. (1996) Vascular reactivity and flow-pressure curve in isolated kidneys from rats with N-nitro-L-arginine methyl ester-induced hypertension. *J. Hypertens.* **14**: 373–379.
41. Maeso, R., Navarro-Cid, J., Rodrigo, E., Ruilope, L.M., Cachofeiro, V., Lahera, V. (1999) Effects of antihypertensive therapy on factors mediating endothelium-dependent relaxations in rats treated chronically with L-NAME. *J. Hypertens.* **17**: 221–227.
42. Ruiz-Marcos, F.M., Ortiz, M.C., Fortepiani, L.A., Nadal, F.J., Atucha, N.M., Garcia-Estan, J. (2001) Mechanism of the increase pressor response to vasopressors in the mesenteric bed of nitric oxide-deficient hypertensive rats. *Eur. J. Pharmacol.* **412**: 273–279.
43. Zhao, H., Shimokawa, H., Uragami-Harasawa, L., Igarashi, H., Takeshita, A. (1999) Long-term vascular effects of N-omega-nitro-L-arginine methyl ester are not solely mediated by inhibition of endothelial nitric oxide synthesis in the rat mesenteric artery. *J. Cardiovasc. Pharmacol.* **33**: 554–566.
44. Kimura, K., Tsuda, K., Sasajima, H., Shiotani, M., Baba, A., Hano, T., Nishio, I. (1999) Arterial relaxation mediated by endothelium-derived hyperpolarizing factor in hypertension induced by chronic inhibition of nitric oxide synthesis. *Clin. Exp. Hypertens.* **21**: 1203–1221.
45. Kalliovalkama, J., Jolma, P., Tolvanen, J.P., Kahonen, M., Hutri-Kahonen, N., Wu, X., Holm, P., Porsti, I. (1999) Arterial function in nitric oxide-deficient hypertension: influence of long-term angiotensin II receptor antagonism. *Cardiovasc. Res.* **42**: 772–782.
46. Tolvanen, J.P., Sallinen, K., Wu, X., Kahonen, M., Arvola, P., Porsti, I. (1998) Variations of arterial responses in vitro in different sections of rat main superior mesenteric artery. *Pharmacol. Toxicol.* **83**: 75–82.
47. Chataigneau, T., Félétou, M., Huang, P.L., Fishman, M.C., Duhault, J., Vanhoutte, P.M. (1999) Acetylcholine-induced relaxation in blood vessels from endothelial nitric oxide synthase knockout mice. *Br. J. Pharmacol.* **126**: 219–226.
48. Waldron, G.J., Ding, H., Lovren, F., Kubes, P., Triggle, C.R. (1999) Acetylcholine-induced relaxation of peripheral arteries isolated from mice lacking endothelial nitric oxide synthase. *Br. J. Pharmacol.* **128**: 653–658.
49. Aisaka, K., Gros, S.S., Griffith, O.W., Levi, R. (1989) N^G-methylarginine, an inhibitor of endothelium-derived nitric oxide synthesis, is a potent pressor agent in the guinea pig: does nitric oxide regulate blood pressure in vivo? *Biochem. Biophys. Res. Commun.* **160**: 881–886.
50. Corriu, C., Félétou, M., Puybasset, L., Bea, M-L., Berdeaux, A., Vanhoutte, P.M. (1998) Endothelium-dependent hyperpolarization in isolated arteries taken from animals treated with NO-synthase inhibitors. *J. Cardiovasc. Pharmacol.* **32**: 944–950.
51. Puybasset, L., Bea, M.L., Ghaleb, B., Guidicelli, J-F., Berdeaux, A. (1996) Coronary and systemic hemodynamic effects of sustained inhibition of nitric oxide synthesis in conscious dogs. Evidence for cross-talk between nitric oxide and cyclooxygenase in coronary vessels. *Circ. Res.* **79**: 343–357.
52. Huang, P.L., Huang, Z., Mashimo, K.D., Bloch, M.A., Moskowitz, J.A., Bevan, J.A., Fishman, M.C. (1995) Hypertension in mice lacking the gene for endothelial nitric oxide synthase. *Nature* **377**: 239–242.

53. Huang, A., Sun, D., Smith, C.J., Connetta, J.A., Shesely, E.G., Koller, A., Kaley, G. (2000) In eNOS knockout mice skeletal muscle, arteriolar dilation is mediated by EDHF. *Am. J. Physiol.* **278**: H762–H768.

54. Huang, A., Sun, D., Carroll, M.A., Jiang, H., Smith, C.J., Connetta, J.A., Falck, J.R., Shesely, E.G., Koller, A., Kaley, G. (2001) EDHF mediates flow-induced dilation in skeletal muscle arterioles of female eNOS-KO mice. *Am. J. Physiol.* **280**: H2462–H2469.

55. Ding, H., Kubes, P., Triggle, C. (2000) Potassium and acetylcholine-induced vasorelaxation in mice lacking endothelial nitric oxide synthase. *Br. J. Pharmacol.* **129**: 1194–1200.

56. Brandes, R.P., Schmitz-Winnenthal, F-H., Félétou, M., Gödecke, A., Huang, P-L., Vanhoutte, P.M., Fleming, I., Busse, R. (2000) An endothelium-derived hyperpolarizing factor distinct from NO and prostacyclin is a major endothelium-dependent vasodilator in resistance vessels of wild type and endothelial NO synthase knock-out mice. *Proc. Natl. Acad. Sci. USA* **97**: 9747–9752.

57. Scotland, R.S., Chauhan, S., Vallance, P.J., Ahluwalia, A. (2001) An endothelium-derived hyperpolarizing factor moderates myogenic constriction of mesenteric resistance arteries in the absence of endothelial nitric oxide synthase derived nitric oxide. *Hypertension* **38**: 833–839.

58. Hasunuma, K., Yamaguchi, T., Rodman, D., O'Brien, R., McMurtry, I. (1991) Effects of inhibitors of EDRF and EDHF on vasoreactivity of perfused rat lungs. *Am. J. Physiol.* **260**: L97–L104.

59. Resta, T.C., Walker, B.R. (1996) Chronic hypoxia selectively augments endothelium-dependent pulmonary arterial vasodilatation. *Am. J. Physiol.* **270**: H888–H996.

60. Carter, E.P., Sato, K., Morio, Y., McMurtry, I.F. (2000) Inhibition of K_{Ca} channels restores blunted hypoxic pulmonary vasoconstriction with cirrhosis. *Am. J. Physiol.* **279**: L903–L910.

61. Morio, Y., Carter, E.P., Oka, M., McMurtry, I.F. (2003) EDHF-mediated vasodilation involves different mechanisms in normotensive and hypertensive rat lungs. *Am. J. Physiol.* **284**: H1762–H1770.

62. Taddei, S., Virdis, A., Mattei, P., Natali, A., Ferrannini, E., Salvetti, A. (1995) Effect of insulin on acetylcholine-induced vasodilatation in normotensive subjects and patients with essential hypertension. *Circulation* **92**: 2911–2918.

63. Taddei, S., Ghiadoni, L., Virdis, A., Buralli, S., Salvetti, A. (1999) Vasodilatation to bradykinin is mediated by an ouabain-sensitive pathway as a compensatory mechanism for impaired nitric oxide availability in essential hypertensive patients. *Circulation* **100**: 1400–1405.

64. Taddei, S., Virdis, A., Ghiadoni, L., Sudano, I., Salvetti, A. (2001) Endothelial dysfunction in hypertension. *J. Cardiovasc. Pharmacol.* **38**: S11–S14.

65. De Vriese, A.S., Blom, H.J., Heil, S.G., Mortier, S., Kluijtmans, L.A., Van de Voorde, J., Lameire, N.H. (2004) Endothelium-derived hyperpolarizing factor-mediated renal vasodilatory response is impaired during acute and chronic hyperhomocysteinemia. *Circulation* **109**: 2331–2336.

66. Heil, S.G., De Vriese, A.S., Kluijtmans, L.A., Mortier, S., Den Heijer, M., Blom, H.J. (2004) The role of hyperhomocysteinemia in nitric oxide (NO) and endothelium-derived hyperpolarizing factor (EDHF)-mediated vasodilatation. *Cell. Mol. Biol. (Noisy-le-grand)* **50**: 911–916.

67. Nakashima, M., Vanhoutte, P.M. (1993) Age-dependent decrease in endothelium-dependent hyperpolarizations to endothelin-3 in the rat mesenteric artery. *J. Cardiovasc. Pharmacol.* **22**: S352–S354.

68. Imaoka, Y., Osanai, T., Kamada, T., Moi, Y., Satoh, K., Okumura, K. (1999) Nitric-oxide dependent vasodilator mechanism is not impaired by hypertension but is diminished with aging in the rat aorta. *J. Cardiovasc. Pharmacol.* **33**: 756–761.

69. Nakashima, M., Mombouli, J-V., Taylor, A.A., Vanhoutte, P.M. (1993) Endothelium-dependent hyperpolarisation caused by bradykinin in human coronary arteries. *J. Clin. Invest.* **92**: 2867–2871.

70. Urakami-Harasawa, L., Shimokawa, H., Nakashima, M., Egashira, K., Takeshita, A. (1997) Importance of endothelium-derived hyperpolarizing factor in human arteries. *J. Clin. Invest.* **100**: 2793–2799.

71. Taddei, S., Virdis, A., Mattei, P., Ghiadoni, L., Fasolo, C.B., Sudano, I., Salvetti, A. (1997) Hypertension causes premature aging of endothelial functions in humans. *Hypertension* **29**: 736–743.

72. Taddei, S., Virdis, A., Ghiadoni, L., Salvetti, G., Berninin, G., Magagna, A., Salvetti, A. (2001) Age-related reduction of NO availability and oxidative stress in humans. *Hypertension* **39**: 274–279.

73. Suzuki, H., Hattori, T., Kajikuri, J., Yamamoto, T., Suzumori, K., Itoh, T. (2002) Reduced function of endothelial prostacyclin in human omental resistance arteries in pre-eclampsia. *J. Physiol.* **545**: 269–277.

74. Kenny, L.C., Baker, P.N., Kendall, D.A., Randall, M.D., Dunn, W.R. (2002) The role of gap junctions in mediating endothelium-dependent responses to bradykinin in myometrial small arteries isolated from pregnant women. *Br. J. Pharmacol.* **136**: 1085–1088.

75. Kenny, L.C., Baker, P.N., Kendall, D.A., Randall, M.D., Dunn, W.R. (2002) Different mechanisms of endothelial-dependent vasodilator responses in human myometrial small arteries in normal pregnancy and pre-eclampsia. *Clin. Sci.* **103**: 67–73.

76. Najibi, S., Cowan, C.L., Palacino, J.J., Cohen, R.A. (1994) Enhanced role of K^+ channels in relaxations of hypercholesterolemic rabbit carotid artery to NO. *Am. J. Physiol.* **266**: H2061–H2067.

77. Cohen, R.A., Plane, F., Najibi, S., Huk, I., Malinski, T., Garland, C.J. (1997) Nitric oxide is the mediator of both endothelium-dependent relaxation and hyperpolarisation of the rabbit carotid artery. *Proc. Natl. Acad. Sci. USA* **94**: 4193–4198.

78. Brandes, R.P., Behra, A., Lebherz, C., Boger, R.H., Bode-Boger, S.M., Mugge, A. (1999) Lovastatin maintains nitric oxide—but not EDHF-mediated endothelium-dependent relaxation in the hypercholesterolemic rabbit carotid artery. *Atherosclerosis* **142**: 97–104.

79. Ayajiki, K., Ozaki, M., Shiomi, M., Okamura, T., Toda, N. (2000) Comparison of endothelium-dependent relaxation in carotid arteries from Japanese white and Watanabe heritable hyperlipidemic rabbits. *J. Cardiovasc. Pharmacol.* **36**: 622–630.

80. Brandes, R.P., Behra, A., Lebherz, C., Boger, R.H., Bode-Boger, S.M., Phivthong-Ngam, L., Mugge, A. (1997) N(G)-nitro-L-arginine- and indomethacin-resistant endothelium-dependent relaxation in the rabbit renal artery: effect of hypercholesterolemia. *Atherosclerosis* **135**: 49–55.

81. Honda, H., Moroe, H., Fujii, H., Arai, K., Notoya, Y., Kogo, H. (2001) Short term hypercholesterolemia alters N(G)-nitro-L-arginine and indomethacin-resistant endothelium relaxation by acetylcholine in rabbit renal artery. *Jpn. J. Pharmacol.* **85**: 203–206.

82. Moroe, H., Fujii, H., Honda, H., Arai, K., Kanazawa, M., Notoya, Y., Kogo, H. (2004) Characterization of endothelium-dependent relaxation and modulation by treatment with pioglitazone in the hypercholesterolemic rabbit renal artery. *Eur. J. Pharmacol.* **497**: 317–325.

83. Kagota, S., Yamaguchi, Y., Nakamura, K., Shinozuka, K., Kunitomo, M. (2004) Chronic nitric oxide exposure alters the balance between endothelium-derived relaxing factors from renal arteries: prevention with NOX-100, a NO scavenger. *Life Sci.* **74**: 2757–2767.

84. Yang, A.L., Jen, C.J., Chen, H.I. (2003) Effects of high cholesterol diet and parallel exercise training on the vascular function of rabbit aortas: a time course study. *J. Appl. Physiol.* **95**: 1194–1200.

85. Yang, A.L., Chen, H.I. (2003) Chronic exercise reduces adhesion molecules/iNOS expression and partially reverses vascular responsiveness in hypercholesterolemic rabbit aortae. *Atherosclerosis* **169**: 11–17.

86. Krummen, S., Falck, J.R., Thorin, E. (2005) Two distinct pathways account for EDHF-dependent dilatation in the gracilis artery of dyslipidaemic hApoB$^{+/+}$ mice. *Br. J. Pharmacol.* **145**: 269–270.

87. Hein, T.W., Liao, J.C., Kuo, L. (2000) oxLDL specifically impairs endothelium-dependent, NO-mediated dilation of coronary arterioles. *Am. J. Physiol.* **278**: H175–H183.

88. Thum, T., Borlak, J. (2004) Mechanistic role of cytochrome P450 monooxygenases in oxidized low-density lipoprotein-induced vascular injury: therapy through LOX-1 receptor antagonism. *Circ. Res.* **94**: e1–e13.

89. Kaw, S., Hecker, M. (2002) Endothelium-derived hyperpolarizing factor, but not nitric oxide or prostacyclin release is resistant to menadione-induced oxidative stress in the bovine coronary artery. *Naunyn -Schmiedebergs Arch. Pharmacol.* **359**: 133–139.

90. Selemidis, S., Cocks, T.M. (2002) Endothelium-dependent hyperpolarization as a remote anti-atherogenic mechanism. *Trends Pharmacol. Sci.* **23**: 213–220.

91. Goulter, A.B., Avella, M.A., Elliot, J., Botham, K.M. (2002) Chylomicron-like particles inhibit receptor-mediated endothelium-dependent vasorelaxation in pig coronary arteries. *Clin. Sci.* **103**: 451–460.

92. Hamilton, C.A., McPhaden, A.R., Berg, G., Pathi, V., Dominiczak, A.F. (2001) Is hydrogen peroxide an EDHF in human radial arteries? *Am. J. Physiol.* **280**: H2451–H2455.

93. Cowan, C.L., Steffen, R.P. (1995) Lysophosphatidylcholine inhibits relaxation of rabbit abdominal aorta mediated by endothelium-derived nitric oxide and endothelium-derived hyperpolarizing factor independent of protein kinase C activation. *Arterioscler. Thromb. Vasc. Biol.* **15**: 2290–2297.

94. Fukao, M., Hattori, Y., Kanno, M., Sakuma, I., Kitabatake, A. (1995) Evidence for selective inhibition by lysophosphatidylcholine of acetylcholine-induced endothelium-dependent hyperpolarization and relaxation in rat mesenteric artery. *Br. J. Pharmacol.* **116**: 1541–1543.

95. Eizawa, H., Yui, Y., Inoue, R., Kosuga, K., Hattori, R., Aoyama, T., Sasayama, S. (1995) Lysophosphatidylcholine inhibits endothelium-dependent hyperpolarization and Nomega-nitro-L-arginine/indomethacin-resistant endothelium-dependent relaxation in the porcine coronary artery. *Circulation* **92**: 3520–3526.

96. Leung, S.W., Teoh, H., Quan, A., Man, R.Y. (1997) Endothelial dysfunction exacerbates the impairment of relaxation by lysophosphatidylcholine in porcine coronary artery. *Clin. Exp. Pharmacol. Physiol.* **24**: 984–986.

97. Drexler, H., Hayoz, D., Munzel, T., Just, H., Zelis, R., Brummer, H.R. (1994) Endothelial dysfunction in chronic heart failure. Experimental and clinical studies. *Arzneimittelforschung* **44**: 455–458.

98. Linke, A., Recchia, F., Zhang, X., Hintze, T.H. (2003) Acute and chronic endothelial dysfunction: implications for the development of heart failure. *Heart Fail. Rev.* **8**: 87–97.

99. Malmsjo, M., Bergdahl, A., Zhao, X.H., Sun, X.Y., Hedner, T., Edvinsson, L., Erlinge, D. (1999) Enhanced acetylcholine and P2Y-receptor stimulated vascular EDHF-dilatation in congestive heart failure. *Cardiovasc. Res.* **43**: 200–209.

100. Gschwend, S., Buikema, H., Henning, R.H., Pinto, Y.M., de Zeeuw, D., van Gilst, V.H. (2003) Endothelial dysfunction and infarct-size relate to impaired EDHF-response in rat experimental chronic heart failure. *Eur. J. Heart Fail.* **5**: 147–154.

101. Rees, D.D., Palmer, R.M.J., Schulz, R., Hodson, H.F., Moncada, S. (1990) Characterization of three inhibitors of endothelial nitric oxide synthase in vitro and in vivo. *Br. J. Pharmacol.* **101**: 746–752.

102. Schini, V.B., Vanhoutte, P.M. (1992) Inhibitors of calmodulin impair the constitutive but not the inducible nitric oxide synthase activity in the rat aorta. *J. Pharmacol. Exp. Ther.* **261**: 553–559.

103. Katz, S.D., Krum, H. (2001) Acetylcholine-mediated vasodilatation in the forearm circulation of patients with heart failure: indirect evidence for the role of endothelium-derived hyperpolarizing factor. *Am. J. Physiol.* **87**: 1089–1092.

104. Lefer, A.M., Tsao, P.S., Lefer, D.J., Ma, X.L. (1991) Role of endothelial dysfunction in the pathogenesis of reperfusion injury after myocardial ischemia. *FASEB J.* **5**: 2029–2034.

105. Jugdutt, B.L. (2002) Nitric oxide and cardioprotection during ischemia-reperfusion. *Heart Fail. Rev.* **7**: 391–405.

106. Chan, E.C., Woodman, O.L. (1999) Enhanced role for the opening of potassium channels in relaxant responses to acetylcholine after myocardial ischemia and reperfusion in dog coronary arteries. *Br. J. Pharmacol.* **126**: 925–932.

107. McLean, P.G., Aston, D., Sarkar, D., Ahluwalia, A. (2002) Protease-activated receptor-2 activation causes EDHF-like coronary vasodilation: selective preservation in ischemia/reperfusion injury: involvement of lipoxygenase products, VR1 receptors and C-fibers. *Circ. Res.* **90**: 465–472.

108. Hobbs, A., Foster, P., Prescott, C., Scotland, R., Ahluwalia, A. (2004) Natriuretic peptide receptor-C regulates coronary blood flow and prevents myocardial ischemia/reperfusion injury: novel cardioprotective role for endothelium-derived C-type natriuretic peptide. *Circulation* **110**: 1231–1235.

109. Nishikawa, Y., Stepp, D.W., Merkus, D., Jones, D., Chilian, W.M. (2004) In vivo role of heme oxygenase in ischemic coronary vasodilation. *Am. J. Physiol.* **286**: H2296–H2304.

110. Lagneux, C., Adam, A., Lamontagne, D. (2003) A study of the mediators involved in the protection induced by exogenous kinins in the isolated rat heart. *Int. Immunopharmacol.* **3**: 1511–1518.

111. Liu, Y., Terata, K., Chai, Q., Li, H., Kleinman, L.H., Gutterman, D.D. (2002) Peroxynitrite inhibits Ca^{2+}-activated K^+ channel activity in smooth muscle of human coronary arterioles. *Circ. Res.* **91**: 1070–1076.

112. Marrelli, S.P., Khorovets, A., Johnson, T.D., Childres, W.F., Bryan, R.M., Jr. (1999) P2 purinoceptor-mediated dilations in the rat middle cerebral artery after ischemia-reperfusion. *Am. J. Physiol.* **276**: H33–H41.

113. Golding, E.M., You, J., Robertson, C.S., Bryan, R.M., Jr. (2001) Potentiated endothelium-derived hyperpolarizing factor-mediated dilations in cerebral arteries following mild head injury. *J. Neurotrauma* **18**: 691–697.

114. Schildmeyer, L.A., Bryan, R.M., Jr. (2002) Effect of NO on EDHF response in rat middle cerebral arteries. *Am. J. Physiol.* **282**: H734–H738.

115. Olmos, L., Mombouli, J.V., Iliano, S., Vanhoutte, P.M. (1995) cGMP mediates the desensitization to bradykinin in isolated canine coronary arteries. *Am. J. Physiol.* **265**: H865–H870.

116. Kessler, P., Popp, R., Busse, R., Schini-Kerth, V.B. (1999) Proinflammatory mediators chronically down-regulate the formation of the endothelium-derived hyperpolarizing factor in arteries via a nitric oxide/cyclic GMP-dependent mechanism. *Circulation* **99**: 1878–1884.

117. Marrelli, S.P. (2002) Altered endothelial Ca^{2+} regulation after ischemia/reperfusion produces potentiated endothelium-derived hyperpolarizing factor-mediated dilations. *Stroke* **33**: 2285–2291.

118. Borg-Capra, C., Fournet-Bourguignon, M.P., Janiak, P., Villeneuve, N., Bidouard, J.P., Vilaine, J.P., Vanhoutte, P.M. (1997) Morphological heterogeneity with normal expression but altered function of G-proteins in porcine cultured regenerated coronary endothelial cells. *Br. J. Pharmacol.* **122**: 999–1008.

119. Shimokawa, H., Flavahan, N.A., Vanhoutte, P.M. (1989) Natural course of the impairment of endothelium-dependent relaxations after balloon endothelium removal in porcine coronary arteries. Possible dysfunction of a pertussis toxin-sensitive G protein. *Circ. Res.* **65**: 740–753.

120. Thollon, C., Bidouard, J-P., Cambarrat, C., Delescluse, I., Villeneuve, N., Vanhoutte, P.M., Vilaine, J-P. (1999) Alteration of endothelium-dependent hyperpolarizations in porcine coronary arteries with regenerated endothelium. *Circ. Res.* **84**: 371–377.

121. Thollon, C., Fournet-Bourguignon, M-P., Saboureau, D., Lesage, L., Reure, H., Vanhoutte, P.M., Vilaine, J-P. (2002) Consequences of reduced production of NO on vascular reactivity of porcine coronary arteries after angioplasty: importance of EDHF. *Br. J. Pharmacol.* **136**: 1153–1161.

122. Thorin, E., Meerkin, D., Bertrand, O.F., Paiement, P., Joyal, M., Bonan, R. (2000) Influence of postangioplasty beta-irradiation on endothelial function in porcine coronary arteries. *Circulation* **101**: 1430–1435.

123. Kohler, R., Brakemier, S., Kuhn, M., Behrens, C., Real, R., Degenhardt, C., Orzechowski, H.D., Prie, A.R., Paul, M., Hoyer, J. (2001) Impaired hyperpolarization in regenerated endothelium after balloon catheter injury. *Circ. Res.* **89**: 174–179.

124. Lippolis, L., Sorrentino, R., Popolo, A., Maffia, P., Nasti, C., D'Emmanuele di Villa Bianca, R., Marzocco, S., Autore, G., Pinto, A. (2003) Time course of vascular reactivity to contracting and relaxing factors after endothelial denudation by balloon angioplasty in rat carotid arteries. *Atherosclerosis* **171**: 171–179.

125. Dina, J.P., Feres, T., Paiva, A.C., Paiva, T.B. (2004) Role of membrane potential and expression of endothelial factors in restenosis after angioplasty in SHR. *Hypertension* **43**: 132–135.

126. Edwards, G., Thollon, C., Gardener, M.J., Félétou, M., Vilaine, J-P., Vanhoutte, P.M., Weston, A.H. (2000) Role of gap junctions and EETs in endothelium-dependent hyperpolarization of porcine coronary artery. *Br. J. Pharmacol.* **129**: 1145–1162.

127. Weston, A.H., Félétou, M., Vanhoutte, P.M., Falck, J.R., Campbell, W.B., Edwards, G. (2005). Endothelium-dependent hyperpolarizations induced by bradykinin in the vasculature; clarification of the role of epoxyeicosatrienoic acids. *Br. J. Pharmacol.* Epub ahead of print.

128. Perrault, L.P., Bidouard, J.P., Janiak, P., Villeneuve, N., Bruneval, P., Vilaine, J-P., Vanhoutte, P.M. (1999) Impairment of G-protein-mediated signal transduction in the porcine coronary endothelium during rejection after heart transplantation. *Cardiovasc. Res.* **43**: 457–470.

129. Perrault, L.P., Thollon, C., Villeneuve, N., Bidouard, J-P., Vilaine, J-P., Vanhoutte, P.M. (1999) Endothelium-dependent responses to bradykinin in porcine coronary arteries after heart transplantation: reduced relaxation resistant to Nω-nitro-L-arginine despite preserved hyperpolarization. In *Endothelium-Dependent Hyperpolarizations*, P.M. Vanhoutte, ed., Harwood Academic Publishers, Amsterdam, The Netherlands, pp. 343–351.

130. He, G.W., Yang, C.Q. (1996) Hyperkalemia alters endothelium-dependent relaxation through non-nitric oxide non-cyclooxygenase pathway: a mechanism for coronary dysfunction due to cardioplegia. *Ann. Thorac. Surg.* **61**: 1394–1399.

131. He, G.W., Yang, C.Q. (1998) Impaired endothelium-derived hyperpolarizing factor-mediated relaxation in coronary arteries by cold storage with University of Wisconsin solution. *J. Thorac. Cardiovasc. Surg.* **116**: 122–130.

132. Ge, Z.D., He, G.W. (2000) Comparison of University of Wisconsin and St Thomas' Hospital solutions on endothelium-derived hyperpolarizing factor mediated function in coronary micro-arteries. *Transplantation* **70**: 22–31.

133. Zou, W., Yang, Q., Yim, A.P., He, G.W. (2003) Impaired endothelium-derived hyper-polarizing factor-mediated relaxation in porcine pulmonary microarteries after cold storage with Euro-Collins and University of Wisconsin solutions. *J. Thorac. Cardio-vasc. Surg.* **126**: 208–215.

134. Zhang, R.Z., Yang, Q., Yim, A.P., He, G.W. (2004) Alteration of cellular electrophysi-ology properties in pulmonary microcirculation after preservation with University of Wisconsin and Euro-Collins solutions. *Ann. Thorac. Surg.* **77**: 1944–1950.

135. Yang, Q., Zhang, R.Z., Yim, A.P., He, G.W. (2004) Histidin-tryptophan-ketoglutarate solution maximally preserves endothelium-derived hyperpolarizing factor-mediated function during heart preservation: comparison with University of Wisconsin solution. *J. Heart Lung Transplant.* **23**: 352–359.

136. Zou, W., Yang, Q., Yim, A.P., He, G.W. (2001) Epoxyeicosatrienoic acids (EET(11,12)) may partially restore endothelium-derived hyperpolarizing factor-mediated function in coronary microarteries. *Ann. Thorac. Surg.* **72**: 1970–1976.

137. Long, C., Li, W., Lin, D.M., Yang, J.G. (2002) Effects of potassium-channel openers on the release of endothelium-derived hyperpolarizing factor in porcine coronary arteries stored in cold hyperkalemic solution. *J. Extra Corpor. Technol.* **34**: 125–129.

138. Yang, Q., Zhang, R.Z., Yim, A.P., He, G.W. (2003) Effect of 11,12-epoxyeicosatrienoic acid as an additive to St. Thomas' cardioplegia and University of Wisconsin solutions on endothelium-derived hyperpolarizing factor-mediated function in coronary microar-teries: influence of temperature and time. *Ann. Thorac. Surg.* **76**: 1623–1630.

139. Fukao, M., Hattori, Y., Kanno, M., Sakuma, I., Kitabatake, A. (1997) Alterations in endothelium-dependent hyperpolarization and relaxation in mesenteric arteries from streptozotocin-induced diabetic rats. *Br. J. Pharmacol.* **121**: 1383–1391.

140. Terata, K., Coppey, L.J., Davidson, E.P., Dunlap, J.A., Gutterman, D.D., Yorek, M.A. (1999) Acetylcholine-induced arteriolar dilation is reduced in streptozotocin-induced diabetic rats with motor nerve dysfunction. *Br. J. Pharmacol.* **128**: 837–843.

141. De Vriese, A.S., Van de Voorde, J., Blom, H.J., Vanhoutte, P.M., Verbeke, M., Lameire, N.H. (2000) The impaired renal vasodilatator response attributed to endothelium-derived hyperpolarizing factor in streptozotocin-induced diabetic rats is restored by 5-methyltetrahydrofolate. *Diabetes* **43**: 1116–1125.

142. Kamata, K., Ohuchi, K., Kirisawa, H. (2000) Altered endothelium-dependent and independent hyperpolarization and endothelium-dependent relaxation in carotid artery from streptozotocin-induced diabetic rats. *Naunyn Schmiedebergs Arch. Phar-macol.* **362**: 52–59.

143. Keegan, A., Jack, A.M., Cotter, M.A., Cameron, N.E. (2000) Effects of aldose reduc-tase inhibition on responses of the corpus cavernosum and mesenteric vascular bed of diabetic rats. *J. Cardiovasc. Pharmacol.* **35**: 606–613.

144. Makino, A., Ohuchi, K., Kamata, K. (2000) Mechanisms underlying the attenuation of endothelium-dependent vasodilatation in the mesenteric arterial bed of the strep-tozotocin-induced rat. *Br. J. Pharmacol.* **130**: 549–556.

145. Wigg, S.J., Tare, M., Tonta, M.A., O'Brien, R.C., Meredith, I.T., Parkington, H.C. (2001) Comparison of effects of diabetes mellitus on an EDHF-dependent and an EDHF-independent artery. *Am. J. Physiol.* **281**: H232–H240.

146. Matsumoto, T., Kobayashi, T., Kamata, K. (2003) Alterations in EDHF-type relaxation and phosphodiesterase activity in mesenteric arteries from diabetic rats. *Am. J. Physiol.* **285**: H283–H291.

147. Matsumoto, T., Wakabayashi, K., Kobayashi, T., Kamata, K. (2004) Diabetes-related changes in cAMP-dependent protein kinase activity and decrease in relaxation response in rat mesenteric artery. *Am. J. Physiol.* **287**: H64–H71.

148. Griffith, T.M., Chaytor, A.T., Taylor, H.J., Giddings, B.D., Edwards, D.H. (2002) cAMP facilitates EDHF-type relaxations in conduit arteries by enhancing electrotonic conduction via gap junctions. *Proc. Natl. Acad. Sci. USA* **99**: 6392–6397.

149. Yousif, M.H., Oriowo, M.A., Cherian, A., Adeagbo, A.S. (2002) Histamine-induced vasodilatation in the perfused arterial bed of diabetic rats. *Vascul. Pharmacol.* **39**: 287–292.

150. Ang, C., Hillier, C., Johnston, F., Cameron, A., Greer, I., Lumsden, M.A. (2002) Endothelial function is preserved in pregnant women with well controlled type I diabetes. *BJOG* **109**: 699–707.

151. Reaven, G.M. (2002) Multiple CHD risk factors in type 2 diabetes: beyond hyperglycaemia. *Diabetes Obes. Metab.* **4**: S13–S18.

152. Natali, A., Ferrannini, E. (2004) Hypertension, insulin resistance, and the metabolic syndrome. *Endocrinol. Metab. Clin. North. Am.* **33**: 417–429.

153. Miller, A.W., Hoenig, M.E., Ujhelyi, M.R. (1998) Mechanisms of impaired endothelial function associated with insulin resistance. *J. Cardiovasc. Pharmacol. Ther.* **3**: 125–134.

154. Miller, A.W., Katakam, P.V., Ujhelyi, M.R. (1999) Impaired endothelium-mediated relaxation in coronary arteries from insulin-resistant rats. *J. Vasc. Res.* **36**: 385–392.

155. Katakam, P.V., Ujhelyi, M.R., Miller, A.W. (1999) EDHF-mediated responses is impaired in fructose-fed rats. *J. Cardiovasc. Pharmacol.* **34**: 461–467.

156. Katakam, P.V., Hoenig, M., Ujhelyi, M.R., Miller, A.W. (2000) Cytochrome P450 activity and endothelial dysfunction in insulin resistance. *J. Vasc. Res.* **37**: 426–434.

157. Arvola, P., Wu, X., Kahonen, M., Makynen, H., Riutta, A., Mucha, I., Solakivi, T., Kainulainen, H., Porsti, I. (1999) Exercise enhances vasorelaxation in experimental obesity associated hypertension. *Cardiovasc. Res.* **43**: 992–1002.

158. Kagota, S., Yamaguchi, Y., Nakamura, K., Kunitomo, M. (2000) Altered endothelium-dependent responsiveness in the aortas and renal arteries of Otsuka Long-Evans Tokushima fatty rats, a model of non-insulin-dependent diabetes mellitus. *Gen. Pharmacol.* **34**: 201–209.

159. Minami, A., Ishimura, N., Harada, N., Sakamoto, S., Niwa, Y., Nakaya, Y. (2002) Exercise training improves acetylcholine-induced endothelium-dependent hyperpolarization in type 2 diabetic rats, Otsuka Long Evans Tokushima fatty rats. *Atherosclerosis* **162**: 85–92.

160. Matsumoto, T., Wakabayashi, K., Kobayashi, T., Kamata, K. (2004) Alterations in vascular function in the aorta and mesenteric artery in type II diabetic rats. *Can. J. Physiol. Pharmacol.* **82**: 175–182.

161. Angulo, J., Cuevas, P., Fernandez, A., Gabancho, S., Allona, A., Martin-Morales, A., Moncada, I., Videla, S., Saenz de Tejada, I. (2003) Diabetes impairs endothelium-dependent relaxation of human penile vascular tissues mediated by NO and EDHF. *Biochem. Biophys. Res. Comm.* **312**: 1202–1208.

162. Martin, S., Tesse, A., Hugel, B., Martinez, M.C., Morel, O., Freyssinet, J.M., Andriantsitohaina, R. (2004) Shed membrane particles from T lymphocytes impair endothelial function and regulate endothelial protein expression. *Circulation* **109**: 1653–1659.
163. Pannirselvam, M., Verma, S., Anderson, T.J., Triggle, C.R. (2002) Cellular basis of endothelial dysfunction in small mesenteric arteries from spontaneously diabetic (db/db–/–) mice: role of decreased tetrahydrobiopterin bioavailability. *Br. J. Pharmacol.* **136**: 255–263.
164. Pannirselvam, M., Anderson, T.J., Triggle, C.R. (2003) Characterization of endothelium-derived hyperpolarizing factor-mediated relaxation of small mesenteric arteries from diabetic (d/db–/–) mice. In *EDHF 2002*, P.M. Vanhoutte, ed., Taylor and Francis, London, pp. 124–131.
165. Zimmerman, P.A., Knot, H.J., Stevenson, A.S., Nelson, M.T. (1997) Increased myogenic tone and diminished responsiveness to ATP-sensitive K^+ channel openers in cerebral arteries from diabetic rats. *Circ. Res.* **81**: 996–1004.
166. Miura, H., Wachtel, R.E., Loberiza, F.R., Jr., Saito, T., Miura, M., Nicolosi, A.C., Gutterman, D.D. (2003) Diabetes mellitus impairs vasodilation to hypoxia in human coronary arterioles: reduced activity of ATP-sensitive potassium channels. *Circ. Res.* **92**: 151–158.
167. Tesfamarian, B., Brown, M.L., Deykin, D., Cohen, R.A. (1990) Elevated glucose generation of endothelium-derived vasoconstrictor prostanoids in rabbit aorta. *J. Clin. Invest.* **85**: 929–932.
168. Félétou, M., Rasetti, C., Duhault, J. (1994) Magnesium modulates endothelial dysfunction produced by elevated glucose incubation. *J. Cardiovasc. Pharmacol.* **24**: 470–478.
169. Liu, Y., Terata, K., Rusch, N.J., Gutterman, D.D. (2002) High glucose impairs voltage-gated K(+) channels current in rat small coronary arteries. *Circ. Res.* **89**: 146–152.
170. Kinoshita, H., Azma, T., Nakahata, K., Kimono, Y., Dojo, M., Yuge, O., Hatano, Y. (2004) Inhibitory effect of high concentration of glucose on relaxations to activation of ATP-sensitive K^+ channels in human omental artery. *Arterioscler. Thromb. Vasc. Biol.* **24**: 2290–2295.
171. Sobrevia, L., Cesare, P., Yudilevich, D.L., Mann, G.E. (1995) Diabetes-induced activation of system y^+ and nitric oxide synthase in human endothelial cells: association with membrane hyperpolarization. *J. Physiol.* **489**: 183–192.
172. Sobrevia, L., Yudilevich, D.L., Mann, G.E. (1998) Elevated D-glucose induces insulin insensitivity in human umbilical endothelial cells isolated from gestational diabetic pregnancies. *J. Physiol.* **506**: 219–230.
173. Flores, C., Rojas, S., Aguayo, C., Parodi, J., Mann, G., Pearson, J.D., Casanello, P., Sobrevia, L. (2003) Rapid stimulation of L-arginine transport by D-glucose involves p42/44(mapk) and nitric oxide in human umbilical vein endothelium. *Circ. Res.* **92**: 64–72.
174. Brownlee, M. (2001) Biochemistry and molecular cell biology of diabetic complications. *Nature* **414**: 813–820.
175. Imaeda, K., Okayama, N., Okouchi, M., Omi, H., Kato, T., Akao, M., Imai, S., Uranishi, H., Takeuchi, Y., Ohara, H., Fukutomi, T., Joh, T., Itoh, M. (2004) Effects of insulin on the acetylcholine-induced hyperpolarization in the guinea pig mesenteric arterioles. *J. Diabetes Complications* **18**: 356–362.
176. Kimura, M., Jefferis, A.M., Watanabe, H., Chin-Dusting, J. (2002) Insulin inhibits acetylcholine responses in rat isolated mesenteric arteries via a non-nitric oxide non-prostanoid pathway. *Hypertension* **39**: 35–40.

177. Farias, N.C., Borelli-Montigny, G.L., Fauaz, G., Feres, T., Borges, A.C., Paiva, T.B. (2002) Different mechanism of LPS-induced vasodilatation in resistance and conductance arteries from SHR and normotensive rats. *Br. J. Pharmacol.* **137**: 213–220.

178. Schildknecht, S., Bachschmid, M., Weber, K., Maass, D., Ulrich, V. (2005) Endotoxin elicits nitric oxide release in rat but prostacyclin synthesis in human and bovine vascular smooth muscle cells. *Biochem. Biophys. Res. Commun.* **327**: 43–48.

179. Fleming, I., Gray, G.A., Julou-Schaeffer, G., Parratt, J.R., Stoclet, J.C. (1990) Incubation with endotoxin activates the L-arginine pathway in vascular tissue. *Biochem. Biophys. Res. Commun.* **171**: 562–568.

180. Kristof, A.S., Noorhosseini, H., Hussain, S.N. (1997) Attenuation of endothelium-dependent hyperpolarizing factor by bacterial lipopolysaccharides. *Eur. J. Pharmacol.* **328**: 69–73.

181. Mitsumizo, S., Nakashima, M., Hamada, T., Totoki, T. (2004) NOS II inhibition restores attenuation of endothelium-dependent hyperpolarization in rat mesenteric artery exposed to lipopolysaccharide. *J. Cardiovasc. Pharmacol.* **43**: 589–594.

182. Fukao, M., Hattori, Y., Kanno, M., Sakuma, I., Kitabatake, A. (1997) Evidence against a role of cytochrome P450-derived arachidonic acid metabolites in endothelium-dependent hyperpolarisation by acetylcholine in rat isolated mesenteric artery. *Br. J. Pharmacol.* **120**: 439–446.

183. Wimalasundera, R., Fexby, S., Regan, L., Thom, S.A., Hughes, A.D. (2003) Effect of tumour necrosis factor-alpha and interleukin 1beta on endothelium-dependent relaxation in rat mesenteric resistance arteries in vitro. *Br. J. Pharmacol.* **138**: 1285–1294.

184. Kameritsch, P., Khandoga, N., Nagel, W., Hundhausen, C., Lidington, D., Pohl, U. (2005) Nitric oxide specifically reduces the permeability of Cx37-containing gap junctions to small molecules. *J. Cell Physiol.* **203**: 233–242.

185. Fernandez-Cobo, M., Gingalewski, C., Drujan, D., De Maio, A. (1999) Downregulation of connexin 43 gene expression in rat heart during inflammation. The role of tumour necrosis factor. *Cytokines* **11**: 216–224.

186. Chen, S.J., Wu, C.C., Yang, S.N., Lin, C.I., Yen, M.H. (2000) Abnormal activation of K$^+$ channels in aortic smooth muscle of rats with endotoxic shock: electrophysiological and functional evidence. *Br. J. Pharmacol.* **131**: 213–222.

187. Chen, S.J., Wu, C.C., Yang, S.N., Lin, C.I., Yen, M.H. (2000) Hyperpolarization contributes to vascular hyporeactivity in rats with lipopolysaccharide-induced endotoxic shock. *Life Sci.* **68**: 659–668.

188. Chataigneau, T., Félétou, M., Duhault, J., Vanhoutte, P.M. (1998) Epoxyeicosatrienoic acids, potassium channel blockers and endothelium-dependent hyperpolarisation in the guinea-pig carotid artery. *Br. J. Pharmacol.* **123**: 574–580.

189. Köhler, R., Degenhardt, C., Kühn, M., Runkel, N., Paul, M., Hoyer, J. (2000) Expression and function of endothelial Ca^{2+}-activated K$^+$ channels in human mesenteric artery—a single cell reverse transcriptase-polymerase chain reaction and electrophysiological study in situ. *Circ. Res.* **87**: 496–503.

190. Sugihara, T., Hattori, Y., Yamamoto, Y., Qi, F., Ichikawa, R., Sato, A., Liu, M.Y., Abe, K., Kanno, M. (1999) Preferential impairment of nitric oxide-mediated endothelium-dependent relaxation in human cervical arteries after irradiation. *Circulation* **100**: 635–641.

191. Soloviev, A.I., Tishkin, S.M., Parshikov, A.V., Ivanova, I.V., Goncharov, E.V., Gurney, A.M. (2003) Mechanisms of endothelial dysfunction after ionized radiation: selective impairment of the nitric oxide component of endothelium-dependent vasodilatation. *Br. J. Pharmacol.* **138**: 837–844.

192. Oishi, H., Nakashima, M., Totoki, T., Tomokuni, K. (1999) Chronic exposure to lead inhibits endothelium-dependent hyperpolarization in the mesenteric artery of the rat. In *Endothelium-Dependent Hyperpolarizations*, P.M. Vanhoutte, ed., Harwood Academic Publishers, Amsterdam, The Netherlands, pp. 313–322.

193. Onaka, U., Fujii, K., Fujishima, M. (1998) Antihypertensive treatment improves endothelium-dependent hyperpolarization in the mesenteric artery of spontaneously hypertensive rats. *Circulation* **98**: 175–182.

CHAPTER 6

1. Taddei, S., Virdis, A., Ghiadoni, L., Sudano, I., Salvetti, A. (2002) Effects of antihypertensive drugs on endothelial dysfunction: clinical implications. *Drugs* **62**: 265–284.

2. Félétou, M., Teisseire, B. (1990) Converting enzyme inhibition in isolated porcine resistance artery potentiates bradykinin relaxation. *Eur. J. Pharmacol.* **190**: 159–166.

3. Félétou, M., Germain, M., Teisseire, B. (1992) Converting enzyme inhibitors potentiate bradykinin induced relaxation in vitro. *Am. J. Physiol.* **262**: H839–H845.

4. Mombouli, J.V., Illiano, S., Nagao, T., Vanhoutte, P.M. (1992) The potentiation of bradykinin-induced relaxations by perindoprilat in canine coronary arteries involves both nitric oxide and endothelium-derived hyperpolarizing factor. *Circ. Res.* **71**: 137–144.

5. Nakashima, M., Mombouli, J-V., Taylor, A.A., Vanhoutte, P.M. (1993) Endothelium-dependent hyperpolarisation caused by bradykinin in human coronary arteries. *J. Clin. Invest.* **92**: 2867–2871.

6. Enseleit, F., Hurlimann, D., Luscher, T.F. (2001) Vascular protective effects of angiotensin converting enzyme inhibitors and their relation to clinical events. *J. Cardiovasc. Pharmacol.* **37**: S21–S30.

7. Erdös, E.G. (1979) Kininase. In *Handbook of Experimental Pharmacology*, E.G. Erdös, ed., Springer-Verlag, Berlin, pp. 427–487.

8. Benzing, T., Fleming, I., Blaukat, A., Muller-Esterl, W., Busse, R. (1999) Angiotensin-converting enzyme inhibitor ramiprilat interferes with the sequestration of the B_2 kinin receptor within the plasma membrane of native endothelial cells. *Circulation* **99**: 2034–3040.

9. Mombouli, J.V., Ballard, K.D., Vanhoutte, P.M. (2002) Kininase-independent potentiation of endothelium-dependent relaxations to kinins by converting enzyme inhibitor perindoprilat. *Acta Pharmacol. Sin.* **23**: 203–207.

10. Matsuda, H., Hayasaki, K., Wakino, S., Kubota, E., Honda, M., Tokuyuma, H., Takamatsu, I., Tatematsu, S., Saruta, T. (2004) Role of endothelium-derived hyperpolarizing factor in ACE inhibitor-induced renal vasodilatation in vivo. *Hypertension* **43**: 603–609.

11. Fujiki, T., Shimokawa, H., Morikawa, K., Kubota, H., Hatanaka, M., Talukder, M.A., Matoba, T., Takeshita, A., Sunagawa, K. (2005) Endothelium-derived hydrogen peroxide accounts for the enhancing effect of an angiotensin-converting enzyme inhibitor on endothelium-derived hyperpolarizing factor-mediated responses in mice. *Arterioscler. Thromb. Vasc. Biol.* **25**: 766–771.

12. Hutri-Kahonen, N., Kahonen, M., Tolvanen, J.P., Wu, X., Sallinen, K., Porsti, I. (1997) Ramipril therapy improves arterial dilation in experimental hypertension. *Cardiovasc. Res.* **33**: 188–195.

13. Onaka, U., Fujii, K., Fujishima, M. (1998) Antihypertensive treatment improves endothelium-dependent hyperpolarization in the mesenteric artery of spontaneously hypertensive rats. *Circulation* **98**: 175–182.

14. Kahonen, M., Tolvanen, J.P., Kalliovalkama, J., Wu, X., Karjala, K., Makynen, H., Porsti, I. (1999) Losartan and enalapril therapies enhance vasodilatation in the mesenteric artery of spontaneously hypertensive rats. *Eur. J. Pharmacol.* **368**: 213–222.
15. Goto, K., Fujii, K., Onaka, U., Abe, I., Fujishima, M. (2000) Renin-angiotensin system blockade improves endothelial dysfunction in hypertension. *Hypertension* **36**: 575–580.
16. Liu, Y.H., Yang, X.P., Sharov, V.G., Nass, O., Sabbah, H.N., Peterson, E., Carretero, O.A. (1997) Effects of angiotensin-converting enzyme inhibitors and angiotensin II type I receptor antagonists in rats with heart failure. Role of kinins and angiotensin II type 2 receptors. *J. Clin. Invest.* **99**: 1926–1935.
17. Kansui, Y., Fujii, K., Nakamura, K., Goto, K., Oniki, H., Abe, I., Shibata, Y., Iida, M. (2004) Angiotensin II receptor blockade corrects altered expression of gap junctions in vascular endothelial cells from hypertensive rats. *Am. J. Physiol.* **287**: H216–H224.
18. Goto, K., Fujii, K., Onaka, U., Abe, I., Fujishima, M. (2000) Angiotensin-converting enzyme inhibitor prevents age-related endothelial dysfunction. *Hypertension* **36**: 581–587.
19. Kansui, Y., Fujii, K., Goto, K., Abe, I., Iida, M. (2002) Angiotensin II receptor antagonist improves age-related endothelial dysfunction. *J. Hypertens.* **20**: 439–446.
20. Maeso, R., Navarro-Cid, J., Rodrigo, E., Ruilope, L.M., Cachofeiro, V., Lahera, V. (1999) Effects of antihypertensive therapy on factors mediating endothelium-dependent relaxations in rats treated chronically with L-NAME. *J. Hypertens.* **17**: 221–227.
21. Kalliovalkama, J., Jolma, P., Tolvanen, J.P., Kahonen, M., Hutri-Kahonen, N., Wu, X., Holm, P., Porsti, I. (1999) Arterial function in nitric oxide-deficient hypertension: influence of long-term angiotensin II receptor antagonism. *Cardiovasc. Res.* **42**: 772–782.
22. Makynen, H., Kahonen, M., Wu, X., Hutri-Kahonen, N., Tolvanen, J.P., Porsti, I. (1997) Arterial function in deoxycorticosterone-NaCl hypertension: influence of angiotensin-converting enzyme inhibition. *Pharmacol. Toxicol.* **81**: 180–189.
23. Pickkers, P., Hughes, A.D., Russel, F.G., Thien, T., Smits, P. (1998) Thiazide-induced vasodilatation in humans is mediated by potassium channel activation. *Hypertension* **32**: 1071–1076.
24. Pickkers, P., Garcha, R.S., Schachter, M., Smits, P., Hughes, A.D. (1999) Inhibition of carbonic hydrolase accounts for the direct vascular effects of hydrochlorothiazide. *Hypertension* **33**: 1043–1048.
25. Pickkers, P., Russel, F.G., Thien, T., Hughes, A.D., Smits, P. (2003) Only weak vasorelaxant properties of loop diuretics in isolated resistance arteries from man, rat and guinea pig. *Eur. J. Pharmacol.* **466**: 281–287.
26. Liguori, A., Casini, A., Di Loreto, M., Andreini, I., Napoli, C. (1999) Loop diuretics enhance the secretion of prostacyclin in vitro, in healthy persons and in patients with chronic heart failure. *Eur. J. Pharmacol.* **55**: 117–124.
27. Pourageaud, F., Bappel-Gozalbes, C., Marthan, R., Freslon, J-L. (2000) Role of EDHF in the vasodilatory effect of loop diuretics in guinea-pig mesenteric arteries. *Br. J. Pharmacol.* **131**: 1211–1219.
28. Farquhason, C.A., Struthers, A.D. (2000) Spironolactone increases nitric oxide bioactivity, improves endothelial vasodilator function, and suppresses vascular angiotensinI/angiotensinII conversion in patients with heart failure. *Circulation* **101**: 594–597.
29. Lakkis, J., Lu, W.X., Weir, M.R. (2003) RAAS escape: a real clinical entity that may be important in the progression of cardiovascular and renal disease. *Curr. Hypertens. Rep.* **5**: 408–417.

30. Muhlen, B.V., Millgard, J., Lind, L. (2001) Effects of digoxin, furosemide, enalaprilat and metoprolol on endothelial function in young normotensive subjects. *Clin. Exp. Pharmacol. Physiol.* **28**: 381–385.

31. Muiesan, M.L., Salvetti, M., Monteduro, C., Rizzoni, D., Zulli, R., Corbellini, C., Brun, C., Agabiti-Rosei, E. (1999) Effect of treatment on flow-dependent vasodilatation of the brachial artery in essential hypertension. *Hypertension* **33**: 575–580.

32. Higashi, Y., Sasaki, S., Nakagawa, K., Ueda, T., Yoshimizu, A., Kurisu, S., Matsuura, H., Kajiyama, G., Oshima, T. (2000) A comparison of angiotensin-converting enzyme inhibitors, calcium antagonists, beta-blockers and diuretic agents on reactive hyperemia in patients with essential hypertension: a multicenter study. *J. Am. Coll. Cardiol.* **35**: 284–291.

33. Giannattasio, C., Achilli, F., Grappiolo, A., Failla, M., Meles, E., Gentile, G., Calchera, I., Capra, A., Baglivo, J., Vincenzi, A., Sala, L., Mancia, G. (2001) Radial artery flow-mediated dilation in heart failure patients: effects of pharmacological and non pharmacological treatments. *Hypertension.* **38**: 1451–1455.

34. Farquhason, C.A., Struthers, A.D. (2002) Increasing plasma potassium with amiloride shortens the QT interval and reduces ventricular extrasystoles but does not change endothelial function or heart rate variability in chronic heart failure. *Heart.* **88**: 475–480.

35. Klingbeil, A.U., John, S., Schneider, M.P., Jacobi, J., Handrock, R., Schmieder, R.E. (2003) Effect of AT1 receptor blockade on endothelial function in essential hypertension. *Am. J. Hypertens.* **16**: 123–128.

36. Luckhoff, A., Pohl, U., Mulsch, A., Busse, R. (1988) Differential role of extra- and intracellular calcium in the release of EDRF and prostacyclin from cultured endothelial cells. *Br. J. Pharmacol.* **95**: 189–196.

37. Johns, A., Freay, A.D., Adams, D.J., Lategan, T.W., Ryan, U.S., Van Breemen, C. (1988) Role of calcium in the activation of endothelial cells. *J. Cardiovasc. Pharmacol.* **12**: S119–S123.

38. Illiano, S.C., Nagao, T., Vanhoutte, P.M. (1992) Calmidazolium, a calmodulin inhibitor, inhibits endothelium-dependent relaxations resistant to nitro-L-arginine in the canine coronary artery. *Br. J. Pharmacol.* **107**: 387–392.

39. Fukao, M., Hattori, Y., Kanno, M., Sakuma, I., Kitabatake, A. (1995) Thapsigargin- and cyclopiazonic acid-induced endothelium-dependent hyperpolarization in rat mesenteric artery. *Br. J. Pharmacol.* **115**: 987–992.

40. Chen, G., Suzuki, H. (1990) Calcium dependency of the endothelium-dependent hyperpolarisation in smooth muscle cells of the rabbit carotid artery. *J. Physiol.* **421**: 521–534.

41. Gluais, P., Edwards, G., Weston, A.H., Falck, J.R., Vanhoutte, P.M., Félétou, M. (2005) SK_{Ca} and IK_{Ca} in the endothelium-dependent hyperpolarization of the guinea-pig isolated carotid artery. *Br. J. Pharmacol.* **144**: 477–485.

42. Colden-Stanfield, M., Schilling, W.P., Ritchie, A.K., Eskin, S.G., Navarro, L.T., Kunze, D.L. (1987) Bradykinin-induced increases in cytosolic calcium and ionic currents in bovine aortic endothelial cells. *Circ. Res.* **61**: 632–640.

43. Uchida, H., Tanaka, Y., Ishii, K., Nakayama, K. (1999) L-type Ca^{2+} channels are not involved in coronary endothelial Ca^{2+} influx mechanism responsible for endothelium-dependent relaxations. *Res. Commun. Mol. Pathol. Pharmacol.* **104**: 127–144.

44. Fukao, M., Hattori, Y., Kanno, M., Sakuma, I., Kitabatake, A. (1997) Sources of Ca^{2+} in relation to generation of acetylcholine-induced endothelium-dependent hyperpolarization in rat mesenteric artery. *Br. J. Pharmacol.* **120**: 439–446.

45. Vilaine, J-P., Biondi, M.L., Villeneuve, N., Félétou, M., Peglion, J-L., Vanhoutte, P.M. (1991) The calcium channel antagonist S 11568 causes endothelium-dependent relaxation in canine arteries. *Eur. J. Pharmacol.* **197**: 41–48.

46. Boulanger, C.M., Nakashima, M., Olmos, L., Joly, G., Vanhoutte, P.M. (1994) Effects of Ca^{2+} antagonist RO 40-5967 on endothelium-dependent responses of isolated arteries. *J. Cardiovasc. Pharmacol.* **23**: 869–876.
47. Koshita, M., Takano, H., Nakahira, Y., Suzuki, H. (1999) Pranidipine enhances relaxation produced by endothelium-derived relaxing factor in carotid artery. *Eur. J. Pharmacol.* **385**: 191–197.
48. Dohi, Y., Kojima, M., Sato, K. (1996) Benidipine improves endothelial function in renal resistance arteries of hypertensive rats. *Hypertension* **28**: 58–63.
49. Fisslthaler, B., Hinsch, N., Chataigneau, T., Popp, R., Kiss, L., Busse, R., Fleming, I. (2000) Nifedipine increases cytochrome P4502C expression and endothelium-derived hyperpolarizing factor-mediated responses in coronary arteries. *Hypertension* **36**: 270–275.
50. Dhein, S., Salameh, A., Berkels, R., Klaus, W. (1999) Dual action of dihydropyridine calcium antagonists: a role for nitric oxide. *Drugs* **58**: 397–404.
51. Taddei, S., Virdis, A., Ghiadoni, L., Versari, D., Salvetti, G., Magagna, A., Salvetti, A. (2003) Calcium antagonist treatment by lercanidipine prevents hyperpolarization in essential hypertension. *Hypertension* **41**: 950–955.
52. Taddei, S., Ghiadoni, L., Virdis, A., Versari, D., Salvetti, A. (2003) Mechanism of endothelial dysfunction: clinical significance and preventive non-pharmacological therapeutic strategies. *Curr. Pharm. Des.* **9**: 2385–2402.
53. Shimamura, K., Sekiguchi, F., Matsuda, K., Yamamoto, K., Tanaka, S., Sunano, S., Shibutani, T., Hashimoto, H., Tanaka, M. (1998) Membrane potential of mesenteric artery from carvedilol-treated spontaneously hypertensive rats. *Eur. J. Pharmacol.* **344**: 161–168.
54. Tolvanen, J.P., Wu, X., Kahonen, M., Sallinen, K., Makynen, H., Pekki, A., Porsti, I. (1996) Effects of celiprolol therapy on arterial dilation in experimental hypertension. *Br. J. Pharmacol.* **119**: 1137–1144.
55. Schiffrin, E.L., Park, J.B., Pu, Q. (2002) Effect of crossing over hypertensive patients from a beta-blocker to an angiotensin receptor antagonist on resistance artery structure and on endothelial function. *J. Hypertens.* **20**: 71–78.
56. Dessy, C., Moniotte, S., Ghisdal, P., Havaux, X., Noirhomme, P., Balligand, J.L. (2004) Endothelial beta3-adrenoceptors mediate vasorelaxation of human coronary arteries through nitric oxide and endothelium-dependent hyperpolarization. *Circulation* **110**: 948–954.
57. Gao, Y.S., Nagao, T., Bond, R.A., Janssens, W.J., Vanhoutte, P.M. (1991) Nebivolol induces endothelium-dependent relaxations of canine coronary arteries. *J. Cardiovasc. Pharmacol.* **17**: 964–969.
58. Cockcroft, J. (2004) Nebivolol, a review. *Expert Opin. Pharmacother.* **5**: 893–899.
59. Wolfrum, S., Jensen, K.S., Liao, J.K. (2003) Endothelium-dependent effects of statins. *Arterioscler. Throm. Vasc. Biol.* **23**: 729–736.
60. Kuhlmann, C.R., Gast, C., Li, F., Schafer, M., Tillmanns, H., Waldecker, B., Wiecha, J. (2004) Cerivastatin activates endothelial calcium-activated potassium channels and thereby modulates endothelial nitric oxide production and cell proliferation. *J. Am. Soc. Nephrol.* **15**: 868–875.
61. Mukai, Y., Shimokawa, H., Matoba, T., Kunihiro, I., Fujiki, T., Takeshita, A. (2003) Acute vascular effects of HMG-CoA reductase inhibitors: involvement of PI3-kinase/Akt pathway and K_V channels. *J. Cardiovasc. Pharmacol.* **42**: 118–124.
62. Brandes, R.P., Behra, A., Lebherz, C., Boger, R.H., Bode-Boger, S.M., Mugge, A. (1999) Lovastatin maintains nitric oxide—but not EDHF-mediated endothelium-dependent relaxation in the hypercholesterolemic rabbit carotid artery. *Atherosclerosis* **142**: 97–104.

63. Cohen, R.A., Plane, F., Najibi, S., Huk, I., Malinski, T., Garland, C.J. (1997) Nitric oxide is the mediator of both endothelium-dependent relaxation and hyperpolarisation of the rabbit carotid artery. *Proc. Natl. Acad. Sci. USA* **94**: 4193–4198.

64. Kansui, Y., Fujii, K., Goto, K., Abe, I., Iida, M. (2004) Effects of fluvastatin on endothelium-derived hyperpolarizing factor- and nitric oxide-mediated relaxations in arteries of hypertensive rats. *Clin. Exp. Pharmacol. Physiol.* **31**: 354–359.

65. Cameron, N.E., Cotter, M., Inkster, M., Nangle, M. (2003) Looking to the future: diabetic neuropathy and effects of rosuvastatin on neurovascular function in diabetes models. *Diabetes Res. Clin. Pract.* **61**: S35–S39.

66. Nakashima, M., Vanhoutte, P.M. (1993) Age-dependent decrease in endothelium-dependent hyperpolarizations to endothelin-3 in the rat mesenteric artery. *J. Cardiovasc. Pharmacol.* **22**: S352–S354.

67. Nakashima, M., Vanhoutte, P.M. (1993) Endothelin-1 and -3 cause endothelium-dependent hyperpolarization in the rat mesenteric artery. *Am. J. Physiol.* **265**: H2137–H2141.

68. Iglarz, M., Levy, B.I., Henrion, D. (1999) Prolonged blockade of endothelin ET(A) receptors decreases vascular reactivity in the aorta of spontaneously hypertensive rats in vitro. *J. Cardiovasc. Pharmacol.* **34**: 354–358.

69. Guerci, B., Kearney-Schwartz, A., Bohme, P., Zannad, F., Drouin, P. (2001) Endothelial dysfunction and type 2 diabetes. Part 1: physiology and methods for exploring the endothelial function. *Diabetes Metab.* **27**: 425–434.

70. Guerci, B., Bohme, P., Kearney-Schwartz, A., Zannad, F., Drouin, P. (2001) Endothelial dysfunction and type 2 diabetes. Part 2: altered endothelial function and the effects of treatment in type 2 diabetes mellitus. *Diabetes Metab.* **27**: 436–447.

71. Vehkavaara, S., Makimattila, S., Schlenzka, A., Vakkilainen, J., Westerbacka, J., Yki-Jarvinen, H. (2000) Insulin therapy improves endothelial function in type 2 diabetes. *Arterioscler. Thromb. Vasc. Biol.* **20**: 545–550.

72. Gaenzer, H., Neumayr, G., Marschang, P., Sturm, W., Lechleitner, M., Foger, B., Kirchmair, R., Patsch, J. (2002) Effect of insulin therapy on endothelium-dependent dilation in type 2 diabetes mellitus. *Am. J. Cardiol.* **89**: 431–434.

73. Bagg, W., Whalley, G.A., Gamble, G., Drury, P.L., Sharpe, N., Braatverdt, G.D. (2001) Effects of improved glycaemic control on endothelial function in patients with type 2 diabetes. *Intern. Med. J.* **31**: 322–328.

74. Sharma, A.M. (2004) Is there a rationale for angiotensin blockade in the management of obesity hypertension. *Hypertension* **44**: 12–19.

75. Danesh, F.R., Kanwar, Y.S. (2004) Modulatory effects of HMG-CoA reductase inhibitors in diabetic angiopathy. *FASEB J.* 18: 805–815.

76. Mather, K.J., Verma, S., Anderson, T.J. (2001) Improved endothelial function with metformin in type 2 diabetes mellitus. *Am. J. Coll. Cardiol.* **37**: 1344–1350.

77. Schafers, R.F. (2003) Do effects on blood pressure contribute to improve clinical outcomes with metformin? *Diabetes Metab.* **29**: S62–S70.

78. Dhindsa, P., Davis, K.R., Donelly, R. (2003) Comparison of the micro- and macro-vascular effects of glimepiride and gliclazide in metformin-treated patients with type 2 diabetes: a double-blind crossover study. *Br. J. Clin. Pharmacol.* **55**: 616–619.

79. Renier, G., Mamputu, J.C., Serri, O. (2003) Benefits of gliclazide in the atherosclerotic process: decrease in monocyte adhesion to endothelial cells. *Metabolism* **52**: 13–18.

80. Sidhu, J.S., Cowan, D., Kashi, J.C. (2004) Effects of rosiglitazone on endothelial function in men with coronary artery disease without diabetes mellitus. *Am. J. Cardiol.* **94**: 151–156.

81. Wang, T.D., Chen, W.J., Lin, J.W., Chen, M.F., Lee, Y.T. (2004) Effects of rosiglitazone on endothelial function, C-reactive protein and components of the metabolic syndrome in non-diabetic patients with the metabolic syndrome. *Am. J. Cardiol.* **93**: 362–365.

82. Pistrosch, F., Passauer, J., Fischer, S., Fuecker, K., Hanefeld, M., Gross, P. (2004) In type 2 diabetes, rosiglitazone therapy for insulin resistance ameliorates endothelial dysfunction independent of glucose control. *Diabetes Care* **27**: 484–490.

83. Vinik, A.I., Stransberry, K.B., Barlow, P.M. (2003) Rosiglitazone treatment increases nitric oxide production in human peripheral skin: a controlled clinical trial in patients with type 2 diabetes mellitus. *J. Diabetes Complications* **17**: 279–285.

84. Stuhlinger, M.C., Abbasi, F., Chu, J.W., Lamendola, C., McLaughlin, T.L., Cooke, J.P., Reaven, G.M., Tsao, P.S. (2002) Relationship between insulin resistance and an endogenous nitric oxide synthase inhibitor. *JAMA* **287**: 1420–1426.

85. Fisman, E.Z., Tenebaum, A., Motro, M., Adler, Y. (2004) Oral antidiabetic therapy in patients with heart disease. A cardiologic standpoint. *Herz* **29**: 290–298.

86. Evans, M., Anderson, R.A., Graham, J., Ellis, G.R., Morris, K., Davies, S., Jackson, S.K., Lewis, M.J., Frenneaux, M.P., Rees, A. (2000) Ciprofibrate therapy improves endothelial function and reduces postprandial lipemia and oxidative stress in type 2 diabetes mellitus. *Circulation* **101**: 1773–1779.

87. Katakam, P.V., Ujhelyi, M.R., Hoenig, M., Miller, A.W. (2000) Metformin improves vascular function in insulin-resistant rats. *Hypertension* **35**: 108–112.

88. Walker, A.B., Chattington, P.D., Buckingham, R.E., Williams, G. (1999) The thiazolidinedione rosiglitazone (BRL-49653) lowers blood pressure and protects against impairment of endothelial function in Zucker fatty rats. *Diabetes* **48**: 1448–1453.

89. Ryan, M.J., Didion, S.P., Mathur, S., Faraci, F.M., Sigmund, C.D. (2004) PPAR(gamma) agonist rosiglitazone improves vascular function and lowers blood pressure in hypertensive transgenic mice. *Hypertension* **43**: 661–666.

90. Brownlee, M. (2001) Biochemistry and molecular cell biology of diabetic complications. *Nature* **414**: 813–820.

91. Keegan, A., Jack, A.M., Cotter, M.A., Cameron, N.E. (2000) Effects of aldose reductase inhibition on responses of the corpus cavernosum and mesenteric vascular bed of diabetic rats. *J. Cardiovasc. Pharmacol.* **35**: 606–613.

92. Cotter, M.A., Jack, A.M., Cameron, N.E. (2002) Effects of the protein kinase C inhibitor LY333531 on neural and vascular function in rats with streptozotocin-induced diabetes. *Clin. Sci. (Lond.)* **103**: 311–321.

93. Soriano, F.G., Virag, L., Jagtap, P., Szabo, E., Mabley, J.G., Liaudet, L., Marton, A., Hoyt, D.G., Murthy, K.G.K., Salzman, A.L., Southan, G.J., Szabo, C. (2001) Diabetic endothelial dysfunction: the role of poly(ADP-ribose) polymerase activation. *Nat. Med.* **7**: 108–113.

94. Ruiz, E., Lorente, R., Tejerina, T. (1997) Effects of calcium dobesilate on the synthesis of endothelium-dependent relaxing factors in rabbit isolated aorta. *Br. J. Pharmacol.* **121**: 711–716.

95. Angulo, J., Cuevas, P., Fernandez, A., Gabancho, S., Allona, A., Martin-Morales, A., Moncada, I., Videla, S., Saenz de Tejada, I. (2003) Diabetes impairs endothelium-dependent relaxation of human penile vascular tissues mediated by NO and EDHF. *Biochem. Biophys. Res. Comm.* **312**: 1202–1208.

96. Antkowiak, B. (2001) How do general anaesthetics work? *Naturwissenschaften* **88**: 201–213.

97. Blaise, G., Sill, J.C., Nugent, M., Van Dyke, R.A., Vanhoutte, P.M. (1987) Isoflurane causes endothelium-dependent inhibition of contractile responses of canine coronary arteries. *Anesthesiology* **67**: 513–517.

98. Witzeling, T.M., Sill, J.C., Hughes, J.M., Blaise, G.A., Nugent, M., Rorie, D.K. (1990) Isoflurane and halothane attenuate coronary artery constriction evoked by serotonin in isolated porcine vessels and intact pigs. *Anesthesioogy* **73**: 100–108.

99. Zhou, X., Abboud, W., Manabat, N.C., Salem, M.R., Crystal, G.J. (1998) Isoflurane-induced dilation of porcine coronary arterioles is mediated by ATP-sensitive potassium channels. *Anesthesiology* **89**: 182–189.

100. Gamperl, A.K., Hein, T.W., Kuo, L., Cason, B.A. (2002) Isoflurane-induced dilation of porcine coronary microvessels is endothelium dependent and inhibited by glibenclamide. *Anesthesiology* **96**: 1465–1471.

101. Akata, T., Nakashima, M., Kodama, K., Boyle, W.A., 3rd, Takahashi, S. (1995) Effects of volatile anesthetics on acetylcholine-induced relaxation in the rabbit mesenteric artery. *Anesthesiology* **82**: 188–204.

102. Iranami, H., Hatano, Y., Tsukiyama, Y., Yamamoto, M., Maeda, H., Mizumoto, K. (1997) Halothane inhibition of acetylcholine-induced relaxation in rat mesenteric artery and aorta. *Can. J. Anaesth.* **44**: 1196–1203.

103. Lischke, V., Busse, R., Hecker, M. (1995) Inhalation anesthetics inhibit the release of endothelium-derived hyperpolarizing factor in the rabbit carotid artery. *Anesthesiology* **83**: 574–582.

104. Patel, A.J., Honore, E., Lesage, F., Fink, M., Romey, G., Lazdunski, M. (1999) Inhalational anaesthetics activate two-pore domain background K^+ channels. *Nat. Neurosci.* **2**: 422–426.

105. Terrenoire, C., Lauritzen, I., Lesage, F., Romey, G., Lazdunski, M.A. (2001) TREK-1-like potassium channel in atrial cells inhibited by beta-adrenergic stimulation and activated by volatile anesthetics. *Circ. Res.* **89**: 336–342.

106. Iida, H., Ohata, H., Iida, M., Watanabe, Y., Dohi, S. (1998) Isoflurane and sevoflurane induce vasodilation of cerebral vessels via ATP-sensitive K^+ channel activation. *Anesthesiology* **89**: 954–960.

107. Kokita, N., Stekiel, T.A., Yamazaki, M., Bosnjak, Z.J., Kampine, J.P., Stekiel, W.J. (1999) Potassium channel-mediated hyperpolarization of mesenteric vascular smooth muscle by isoflurane. *Anesthesiology* **90**: 779–788.

108. Park, K.W., Dai, H.B., Comunale, M.E., Gopal, A., Sellke, F.W. (2000) Dilation by isoflurane of preconstricted, very small arterioles from human right atrium is mediated in part by K(+)-ATP channel opening. *Anesth. Analg.* **91**: 76–81.

109. Kehl, F., Krolikowski, J.G., Tessmer, J.P., Pagel, P.S., Warltier, D.C., Kersten, J.R. (2002) Increases in coronary collateral blood flow produced by sevoflurane are mediated by calcium-activated potassium (BKCa) channels in vivo. *Anesthesiology* **97**: 725–731.

110. Yamakura, T., Lewohl, J.M., Harris, R.A. (2001) Differential effects of general anesthetics on G protein-coupled inwardly rectifying and other potassium channels. *Anesthesiology* **95**:144–153.

111. Friederich, P., Benzenberg, D., Trellakis, S., Urban, B.W. (2001) Interaction of volatile anesthetics with human Kv channels in relation to clinical concentrations. *Anesthesiology* **95**: 954–958.

112. Huneke, R., Jungling, E., Skasa, M., Rossaint, R., Luckhoff, A. (2001) Effects of the anesthetic gases xenon, halothane, and isoflurane on calcium and potassium currents in human atrial cardiomyocytes. *Anesthesiology* **95**: 999–1006.

113. Chen, X., Yamakage, M., Namiki, A. (2002) Inhibitory effects of volatile anesthetics on K^+ and Cl^- channel currents in porcine tracheal and bronchial smooth muscle. *Anesthesiology* **96**: 458–466.

114. Chen, X., Yamakage, M., Yamada, Y., Tohse, N., Namiki, A. (2002) Inhibitory effects of volatile anesthetics on currents produced on heterologous expression of KvLQT1 and minK in Xenopus oocytes. *Vascul. Pharmacol.* **39**: 33–38.

115. Namba, T., Ishii, T.M., Ikeda, M., Hisano, T., Itoh, T., Hirota, K., Adelman, J.P., Fukuda, K. (2000) Inhibition of the human intermediate conductance Ca(2+)-activated K(+) channel, hIK1, by volatile anesthetics. *Eur. J. Pharmacol.* **395**: 95–101.

116. Dreixler, J.C., Jenkins, A., Cao, Y.J., Roizen, J.D., Houamed, K.M. (2000) Patch-clamp analysis of anesthetic interactions with recombinant SK2 subtype neuronal calcium-activated potassium channels. *Anesth. Analg.* **90**: 727–732.

117. Hashiguchi-Ikeda, M., Namba, T., Ishii, T.M., Hisano, T., Fukuda, K. (2003) Halothane inhibits an intermediate conductance Ca^{2+}-activated K^+ channel by acting at the extracellular side of the ionic pore. *Anesthesiology* **99**: 1340–1345.

118. Beny, J-L., Pacicca, C. (1994) Bidirectional electrical communication between smooth muscle and endothelial cells in the pig coronary artery. *Am. J. Physiol.* **266**: H1465–H1472.

119. Lischke, V., Busse, R., Hecker, M. (1995) Volatile and intravenous anesthetics selectively attenuate the release of endothelium-derived hyperpolarizing factor elicited by bradykinin in the coronary microcirculation. *Naunyn Schmiedebergs Arch. Pharmacol.* **352**: 346–349.

120. Kessler, P., Lischke, V., Hecker, M. (1996) Etomidate and thiopental inhibit the release of endothelium-derived-hyperpolarizing factor in the human renal artery. *Anesthesiology* **84**: 1485–1488.

121. Sohn, J.T., Murray, P.A. (2003) Inhibitory effects of etomidate and ketamine on adenosine triphosphate-sensitive potassium channel relaxation in canine pulmonary artery. *Anesthesiology* **98**:104–113.

122. Nagakawa, T., Yamazaki, M., Hatakeyama, N., Stekiel, T.A. (2003) The mechanisms of propofol-mediated hyperpolarization of in situ rat mesenteric vascular smooth muscle. *Anesth. Analg.* **97**: 1639–1645.

123. Klockgether-Radke, A.P., Schulze, H., Neumann, P., Hellige, G. (2004) Activation of the K^+ channel BK(Ca) is involved in the relaxing effect of propofol on coronary arteries. *Eur. J. Anaesthesiol.* **21**: 226–230.

124. Bodelsson, G., Sandstrom, K., Wallerstedt, S.M., Hidestal, J., Tornebrandt, K., Bodelsson, M. (2000) Effects of propofol on substance P-induced relaxation in isolated human omental arteries and veins. *Eur. J. Anesthesiol.* **17**: 720–728.

125. Horibe, M., Ogawa, K., Sohn, J.T., Murray, P.A. (2000) Propofol attenuates acetylcholine-induced pulmonary vasorelaxation: role of nitric oxide and endothelium-derived hyperpolarizing factors. *Anesthesiology* **93**: 447–455.

126. Simoneau, C., Thuringer, D., Cai, S., Garneau, L., Blaise, G., Sauve, R. (1996) Effect of halothane and isoflurane on bradykinin-evoked Ca^{2+} influx in bovine aortic endothelial cells. *Anesthesiology* **85**: 366–379.

127. Ogawa, K., Tanaka, S., Murray, P.A. (2001) Inhibitory effects of etomidate and ketamine on endothelium-dependent relaxation in canine pulmonary artery. *Anesthesiology* **94**: 668–677.

128. Kanna, T., Akata, T., Izumi, K., Nakashima, M., Yonemitsu, Y., Hashizume, M., Takahashi, S. (2002) Sevoflurane and bradykinin-induced calcium mobilization in pulmonary arterial valvular endothelial cells in situ: sevoflurane stimulates plasmalemmal calcium influx into endothelial cells. *J. Cardiovasc. Pharmacol.* **40**: 714–724.

129. Hoebel, B.G., Kostner, G.M., Graier, W.F. (1997) Activation of microsomal P450 mono-oxygenase by Ca^{2+} store depletion and its contribution to Ca^{2+} entry in porcine aortic endothelial cells. *Br. J. Pharmacol.* **121**: 1579–1588.

130. Johns, R.A. (1989) Local anesthetics inhibit endothelium-dependent vasodilation. *Anesthesiology* **70**: 805–811.

131. Meyer, P., Flammer, J., Luscher, T.F. (1993) Local anesthetic drugs reduce endothelium-dependent relaxations of porcine ciliary arteries. *Invest. Ophthalmol. Vis. Sci.* **34**: 2730–2736.

132. Minamoto, Y., Nakamura, K., Toda, H., Miyawaki, I., Kitamura, R., Vinh, V.H., Hatano, Y., Mori, K. (1997) Suppression of acetylcholine-induced relaxation by local anesthetics and vascular NO-cyclic GMP system. *Acta Anaesthesiol. Scand.* **41**: 1054–1060.

133. Kindler, C.H., Paul, M., Zou, H., Liu, C., Winegar, B.D., Gray, A.T., Yost, C.S. (2003) Amide local anesthetics potently inhibit the human tandem pore domain background K^+ channel TASK-2 (KCNK5). *J. Pharmacol. Exp. Ther.* **306**: 84–92.

134. Nilsson, J., Madeja, M., Arhem, P. (2003) Local anesthetic block of K_V channels: role of the S6 helix and the S5-S6 linker for bupivacaine action. *Mol. Pharmacol.* **63**: 1417–1429.

135. Punke, M.A., Licher, T., Pongs, O., Friederich, P. (2003) Inhibition of human TREK-1 channels by bupivacaine. *Anesth. Analg.* **96**: 1665–1673.

136. Friederich, P., Solth, A., Schillemeit, S., Isbrandt, D. (2004) Local anaesthetic sensitivities of cloned HERG channels from human heart: comparison with HERG/MiRP1 and HERG/MiRP1 T8A. *Br. J. Anaesth.* **92**: 93–101.

137. Kawano, T., Oshita, S., Takahashi, A., Tsutsumi, Y., Tomiyama, Y., Kitahata, H., Kuroda, Y., Nakaya, Y. (2004) Molecular mechanisms of the inhibitory effects of bupivacaine, levobupivacaine, and ropivacaine on sarcolemmal adenosine triphosphate-sensitive potassium channels in the cardiovascular system. *Anesthesiology* **101**: 390–398.

138. Kinoshita, H., Nakahata, K., Dojo, M., Kimoto, Y., Hatano, Y. (2004) Lidocaine impairs vasodilation mediated by adenosine triphosphate-sensitive K^+ channels but not by inward rectifier K^+ channels in rat cerebral microvessels. *Anesth. Analg.* **99**: 904–909.

139. Huang, Y., Lau, C.W., Chan, F.L., Yao, X.Q. (1999) Contribution of nitric oxide and K^+ channel activation to vasorelaxation of isolated rat aorta induced by procaine. *Eur. J. Pharmacol.* **367**: 231–237.

140. Yang, Q., Yim, A.P., Arifi, A.A., He, G.W. (2002) Procaine in cardioplegia: the effect on EDHF-mediated function in porcine coronary arteries. *J. Card. Surg.* **17**: 470–475.

141. Loeb, A.L., Godeny, I., Longnecker, D.E. (1997) Anesthetics alter relative contributions of NO and EDHF in rat cremaster muscle microcirculation. *Am. J. Physiol.* **273**: H618–H627.

142. De Wit, C., Esser, N., Lehr, H.A., Bolz, S.S., Pohl, U. (1999) Pentobarbital-sensitive EDHF comediates ACh-induced arteriolar dilation in the hamster microcirculation. *Am. J. Physiol.* **276**: H1527–H1534.

143. Shimokawa, H., Lam, J.Y., Chesebro, J.H., Bowie, E.J., Vanhoutte, P.M. (1987) Effects of dietary supplementation with cod-liver oil on endothelium-dependent responses in porcine coronary arteries. *Circulation* **76**: 898–905.

144. Shimokawa, H., Vanhoutte, P.M. (1988) Dietary cod liver oil improves endothelium-dependent responses in hypercholesterolemic and atherosclerotic porcine coronary arteries. *Circulation* **78**: 1421–1430.

145. Nagao, T., Nakashima, M., Smart, F.W., Bond, R.A., Morrison, K.J., Vanhoutte, P.M. (1995) Potentiation of endothelium-dependent hyperpolarization to serotonin by dietary intake of NC 020, a defined fish oil, in the porcine coronary artery. *J. Cardiovasc. Pharmacol.* **26**: 679–681.

146. Jack, A.M., Kegan, A., Cotter, M.A., Cameron, N.E. (2002) Effects of diabetes and evening primrose oil treatment on responses of aorta, corpus cavernosum and mesenteric vasculature in rats. *Life Sci.* **71**: 1863–1877.

147. De Vriese, A.S., Van de Voorde, J., Blom, H.J., Vanhoutte, P.M., Verbeke, M., Lameire, N.H. (2000) The impaired renal vasodilatator response attributed to endothelium-derived hyperpolarizing factor in streptozotocin-induced diabetic rats is restored by 5-methyltetrahydrofolate. *Diabetes* **43**: 1116–1125.

148. Cameron, N.E., Jack, A.M., Cotter, M.A. (2001) Effect of alpha-lipoic acid on vascular responses and nociception in diabetic rats. *Free Radic. Biol. Med.* **31**: 125–135.

149. Inkster, M.E., Cotter, M.A., Cameron, N.E. (2002) Effects of trientine, a metal chelator, on defective endothelium-dependent relaxation in the mesenteric vasculature of diabetic rats. *Free Radic. Res.* **36**: 1091–1099.

150. Adeagbo, A.S., Joshua, I.G., Falkner, C., Matheson, P.J. (2003) Tempol, an antioxidant, restores endothelium-derived hyperpolarizing factor-mediated vasodilation during hypertension. *Eur. J. Pharmacol.* **481**: 91–100.

151. Fitzpatrick, D.F., Hirschfield, S.L., Riici, T., Jantzen, P., Coffey, R.G. (1995) Endothelium-dependent vasorelaxation caused by various plant extracts. *J. Cardiovasc. Pharmacol.* **26**: 90–95.

152. Kwan, C.Y., Zhang, W.B., Deyama, T., Nishibe, S. (2004) Endothelium-dependent vascular relaxation induced by Eucommia ulmoides Oliv. bark extract is mediated by NO and EDHF in small vessels. *Naunyn Schmiedebergs Arch. Pharmacol.* **369**: 206–211.

153. Kwan, C.Y., Zhang, W.B., Sim, S.M., Deyama, T., Nishibe, S. (2004) Vascular effects of Siberian ginseng (Eleutherococcus senticosus): endothelium-dependent NO- and EDHF-mediated relaxation depending on vessel size. *Naunyn Schmiedebergs Arch. Pharmacol.* **369**: 473–480.

154. Runnie, I., Salleh, M.N., Mohamed, S., Head, R.J., Abeywardena, M.Y. (2004) Vasorelaxation induced by common edible plant extracts in isolated aorta and mesenteric vascular bed. *J. Ethnopharmacol.* **92**: 311–316.

155. Stoclet, J.-C., Chataigneau, T., Ndiaye, M., Oak, M-H., El Bedoui, J., Chataigneau, M., Schini-Kerth, V. (2004) Vascular protection by dietary polyphenols. *Eur. J. Pharmacol.* **500**: 299–313.

156. Andriambelosan, E., Magnier, C., Haan-Archipoff, G., Lobstein, A., Anton, R., Beretz, A., Stoclet, J-C., Andriantsitohaina, R. (1998) Natural dietary polyphenolic compounds cause endothelium-dependent vasorelaxation in rat thoracic aorta. *J. Nutr.* **128**: 2324–2333.

157. Chataigneau, T., Ndiaye, M., Stoclet, J-C., Schini-Kerth, V.B. (2003) Red wine polyphenolic compounds induce EDHF-mediated relaxation and hyperpolarization in the porcine coronary artery: involvement of REDOX-sensitive mechanisms. In *EDHF 2002*, P.M. Vanhoutte, ed., Taylor and Francis, London, pp. 165–173.

158. Ndiaye, M., Chataigneau, T., Chataigneau, M., Schini-Kerth, V.B. (2004) Red wine polyphenols induce EDHF-mediated relaxations through the redox-sensitive activation of the PI3-kinase/Akt pathway. *Br. J. Pharmacol.* **142**: 1131–1136.

159. Ndiaye, M., Chataigneau, M., Lobysheva, I., Chataigneau, T., Schini-Kerth, V.B. (2005) Red wine polyphenols-induced endothelium-dependent NO-mediated relaxation is due to the redox-sensitive PI3-kinase/Akt-dependent phosphorylation of endothelial NO synthase in the isolated porcine coronary arteries. *FASEB J.* **19**: 455–457.

160. de Moura, R.S., Miranda, D.Z., Pinto, A.C., Sicca, R.F., Souza, M.A., Rubenich, L.M., Carvalho, L.C., Rangel, B.M., Tano, T., Madeira, S.V., Resende, A.C. (2004) Mechanism of the endothelium-dependent vasodilation and the antihypertensive effect of Brazilian red wine. *J. Cardiovasc. Pharmacol.* **44**: 302–309.

161. Fitzpatrick, D.F., Hirschfield, S.L., Coffey, R.G. (1993) Endothelium-dependent vasorelaxing activity of wine and other grape products. *Am. J. Physiol.* **265**: H774–H778.

162. Andriambelosan, E., Kleyschyov, A.L., Muller, B., Beretz, A., Stoclet, J-C., Andriantsitohaina, R. (1997) Nitric oxide production and endothelium-dependent relaxation induced by wine polyphenols in rat aorta. *Br. J. Pharmacol.* **120**: 1053–1058.

163. Ndiaye, M., Chataigneau, T., Andriantsitohaina, R., Stoclet, J-C., Schini-Kerth, V.B. (2003) Red wine polyphenols cause endothelium-dependent EDHF-mediated relaxation in porcine coronary arteries via a redox-sensitive mechanism. *Biochem. Biophys. Res. Commun.* **17**: 371–377.

164. Tachibana, H., Koga, K., Fujimura, Y., Yamada, K. (2004) A receptor for green tea polyphenol EGCG. *Nat. Struct. Biol.* **11**: 380–381.

165. Walker, H.A., Dean, T.S., Sanders, T.A., Jackson, G., Ritter, J.M., Chowienczyk, P.J. (2001) The phytoestrogen genistein produces acute nitric oxide-dependent dilation of human forearm vasculature with similar potency to 17beta-estradiol. *Circulation* **103**: 258–262.

166. Squadrito, F., Altavilla, D., Crisafulli, A., Saitta, A., Cucinotta, D., Morabito, N., D'Anna, R., Corrado, F., Ruggeri, P., Frisina, N., Squadrito, G. (2003) Effect of genistein on endothelial function in post-menopausal women: a randomized, double-blind study. *Am. J. Med.* **114**: 470–476.

167. Wallerath, T., Deckert, G., Ternes, T., Anderson, H., Li, H., Witte, K., Forstermann, U. (2002) Resveratrol, a polyphenolic phytoalexin present in red wine, enhances expression and activity of endothelial nitric oxide synthase. *Circulation* **106**: 1652–1658.

168. Lee, M.Y., Man, R.Y. (2003) The phytoestrogen genistein enhances endothelium-independent relaxation in the porcine coronary artery. *Eur. J. Pharmacol.* **481**: 227–232.

169. Au, A.L., Kwok, C.C., Kwan, Y.W., Lee, M.M., Zhang, R.Z., Ngai, S.M., Lee, S.M., He, G.W., Fung, K.P. (2004) Activation of iberiotoxin-sensitive, Ca^{2+}-activated K^+ channels of porcine isolated left anterior descending coronary artery by diosgenin. *Eur. J. Pharmacol.* **502**: 123–133.

170. Woodman, O.L., Boujaoude, M. (2004) Chronic treatment of male rats with daidzein and 17beta-oestradiol induces the contribution of EDHF to endothelium-dependent relaxation. *Br. J. Pharmacol.* **141**: 322–328.

171. Wu, X., Makynen, H., Korpela, R., Porsti, I. (1996) Whey mineral supplementation and arterial tone in mineralocorticoid-NaCl hypertension. *Cardiovasc. Res.* **32**: 1115–1122.

172. Wu, X., Tolvanen, J.P., Hutri-Kahonen, N., Kahonen, M., Makynen, H., Korpela, R., Ruskoaho, H., Karjala, K., Porsti, I. (1998) Comparison of the effects of supplementation with whey mineral and potassium on arterial tone in experimental hypertension. *Cardiovasc. Res.* **40**: 364–374.

173. Raij, L., Luscher, T.F., Vanhoutte, P.M. (1988) High potassium diet augments endothelium-dependent relaxation in the Dahl rat. *Hypertension* **12**: 562–567.

174. Makynen, H., Kahonen, M., Wu, X., Arvola, P., Porsti, I. (1996) Endothelial function in deoxycorticosterone-NaCl hypertension: effect of calcium supplementation. *Circulation* **93**: 1000–1008.

175. Makynen, H., Kahonen, M., Wu, X., Wuorela, H., Porsti, I. (1996) Reversal of hypertension and endothelial dysfunction in deoxycorticoterone-NaCl-treated rats by high-Ca^{2+} diet. *Am. J. Physiol.* **270**: H1250–H1257.

176. Tolvanen, J.P., Makynen, H., Wu, X., Hutri-Kahonen, N., Ruskoaho, H., Karjala, K., Porsti, I. (1998) Effects of calcium and potassium supplements on arterial tone in vitro in spontaneously hypertensive rats. *Br. J. Pharmacol.* **124**: 119–128.

177. Jolma, P., Kalliovalkama, J., Tolvanen, J.P., Koobi, P., Kahonen, M., Hutri-Kahonen, N., Wu, X., Porsti, I. (2000) High-calcium diet enhances vasorelaxation in nitric oxide-deficient hypertension. *Am. J. Physiol.* **279**: H1036–H1043.

178. Kahonen, M., Nappi, S., Jolma, P., Hutri-Kahonen, N., Tolvanen, J.P., Saha, H., Koivisto, P., Krogerus, L., Kalliovalkama, J., Porsti, I. (2003) Vascular influences of calcium supplementation and vitamin D3-induced hypercalcemia in NaCl-hypertensive rats. *J. Cardiovasc. Pharmacol.* **42**: 319–328.

179. Borges, A.C.R., Feres, T., Vianna, L.M., Paiva, T.B. (1999) Recovery of impaired K+ channels in mesenteric arteries from spontaneously hypertensive rats by prolonged treatment with cholecalciferol. *Br. J. Pharmacol.* **127**: 772–778.

180. Miller, V.M., Vanhoutte, P.M. (1988) Enhanced release of endothelium-derived factor(s) by chronic increases in blood flow. *Am. J. Physiol.* **255**: H446–H451.

181. Sessa, W.C., Pritchard, K., Seyedi, N., Wang, J., Hintze, T.H. (1994) Chronic exercise in dogs increases coronary vascular nitric oxide production and endothelial cell nitric oxide gene expression. *Circ. Res.* **74**: 349–353.

182. Kingwell, B.A., Sherrard, B., Jennings, G.L., Dart, A.M. (1997) Four weeks of cycle training increases basal production of nitric oxide from the forearm. *Am. J. Physiol.* **272**: H1070–H1077.

183. Mombouli, J.V., Nakashima, M., Hamra, M., Vanhoutte, P.M. (1996) Endothelium-dependent relaxation and hyperpolarization evoked by bradykinin in canine coronary arteries: enhancement by exercise-training. *Br. J. Pharmacol.* **117**: 413–418.

184. Yen, M.H., Yang, J.H., Sheu, J.R., Lee, Y.M., Ding, Y.A. (1995) Chronic exercise enhances endothelium-mediated dilation in spontaneously hypertensive rats. *Life Sci.* **57**: 2205–2213.

185. Griffin, K.L., Laughlin, M.H., Parker, J.L. (1999) Exercise training improves endothelium-mediated vasorelaxation after chronic coronary occlusion. *J. Appl. Physiol.* **87**: 1948–1956.

186. Minami, A., Ishimura, N., Harada, N., Sakamoto, S., Niwa, Y., Nakaya, Y. (2002) Exercise training improves acetylcholine-induced endothelium-dependent hyperpolarization in type 2 diabetic rats, Otsuka Long Evans Tokushima fatty rats. *Atherosclerosis* **162**: 85–92.

187. Sherman, D.L. (2000) Exercise and endothelial function. *Coron. Artery Dis.* **11**: 117–122.

188. Moyna, N.M., Thompson, P.D. (2004) The effect of physical activity on endothelial function in man. *Acta Physiol. Scand.* **180**: 113–123.

Index

C